GENE CLONING

AND DNA

ANALYSIS

GENE CLONING
AND DNA
ANALYSIS

An Introduction

T.A. BROWN

Faculty of Life Sciences
University of Manchester
Manchester

Sixth Edition

WILEY-BLACKWELL

A John Wiley & Sons, Ltd., Publication

This edition first published 2010, © 2010, 2006 by T.A. Brown
First, second and third editions published by Chapman & Hall 1986, 1990, 1995
Fourth and fifth editions published by Blackwell Publishing Ltd 2001, 2006

Blackwell Publishing was acquired by John Wiley & Sons in February 2007. Blackwell's
publishing program has been merged with Wiley's global Scientific, Technical and
Medical business to form Wiley-Blackwell.

Registered office: John Wiley & Sons Ltd, The Atrium, Southern Gate, Chichester,
West Sussex, PO19 8SQ, UK

Editorial offices: 9600 Garsington Road, Oxford, OX4 2DQ, UK
The Atrium, Southern Gate, Chichester, West Sussex, PO19 8SQ, UK
111 River Street, Hoboken, NJ 07030-5774, USA

For details of our global editorial offices, for customer services and for information about
how to apply for permission to reuse the copyright material in this book please see our website
at www.wiley.com/wiley-blackwell

Library of Congress Cataloguing-in-Publication Data
Brown, T.A. (Terence A.)
  Gene cloning and DNA analysis : an introduction / T.A. Brown.—6th ed.
    p. cm.
  ISBN 978-1-4051-8173-0 (pbk. : alk. paper) – ISBN 978-1-4443-3407-4 (hbk. : alk. paper)
  1. Molecular cloning.   2. Nucleotide sequence.   3. DNA—Analysis.   I. Title.
  QH442.2.B76 2010
  572.8′633—dc22

                                                                    2009038739

ISBN: 9781405181730 (paperback) and 9781444334074 (hardback)

A catalog record for this book is available from the British Library.

Set in 10/12pt Classical Garamond
by Graphicraft Limited, Hong Kong
Printed and bound in Malaysia by Vivar Printing Sdn Bhd

1   2010

Brief Contents

BRIEF CONTENTS

Companion website available at www.wiley.com/go/brown/cloning

Contents

CONTENTS

Part II The Applications of Gene Cloning and DNA Analysis in Research 163

Companion website available at www.wiley.com/go/brown/cloning

Preface to the Sixth Edition

PREFACE TO THE SIXTH EDITION

During the four years since publication of the Fifth Edition of *Gene Cloning and DNA Analysis: An Introduction* there have been important advances in DNA sequencing technology, in particular the widespread adoption of high throughput approaches based on pyrosequencing. Inclusion of these new techniques in the Sixth Edition has prompted me to completely rewrite the material on DNA sequencing and to place all the relevant information—both on the methodology itself and its application to genome sequencing —into a single chapter. This has enabled me to devote another entire chapter to the post-sequencing methods used to study genomes. The result is, I hope, a more balanced treatment of the various aspects of genomics and post-genomics than I had managed in previous editions.

A second important development of the last few years has been the introduction of real-time PCR as a means of quantifying the amount of a particular DNA sequence present in a preparation. This technique is now described as part of Chapter 9. Elsewhere, I have made various additions, such as inclusion of topoisomerase-based methods for blunt end ligation in Chapter 4, and generally tidied up parts of chapters that had become slightly unwieldy due to the cumulative effects of modifications made over the 25 years since the First Edition of this book. The Sixth Edition is almost twice as long as the First, but retains the philosophy of that original edition. It is still an introductory text that begins at the beginning and does not assume that the reader has any prior knowledge of the techniques used to study genes and genomes.

I would like to thank Nigel Balmforth and Andy Slade at Wiley-Blackwell for helping me to make this new edition a reality. As always I must also thank my wife Keri for the unending support that she has given to me in my decision to use up evenings and weekends writing this and other books.

T.A. Brown
Faculty of Life Sciences
University of Manchester

PART I

The Basic Principles of Gene Cloning and DNA Analysis

Chapter 1

Why Gene Cloning and DNA Analysis are Important

Chapter contents

In the middle of the 19th century, Gregor Mendel formulated a set of rules to explain the inheritance of biological characteristics. The basic assumption of these rules is that each heritable property of an organism is controlled by a factor, called a gene, that is a physical particle present somewhere in the cell. The rediscovery of Mendel's laws in 1900 marks the birth of genetics, the science aimed at understanding what these genes are and exactly how they work.

1.1 The early development of genetics

For the first 30 years of its life this new science grew at an astonishing rate. The idea that genes reside on chromosomes was proposed by W. Sutton in 1903, and received experimental backing from T.H. Morgan in 1910. Morgan and his colleagues then developed the techniques for gene mapping, and by 1922 had produced a comprehensive analysis of the relative positions of over 2000 genes on the 4 chromosomes of the fruit fly, *Drosophila melanogaster*.

Despite the brilliance of these classical genetic studies, there was no real understanding of the molecular nature of the gene until the 1940s. Indeed, it was not until

Gene Cloning and DNA Analysis: An Introduction. 6th edition. By T.A. Brown. Published 2010 by Blackwell Publishing.

the experiments of Avery, MacLeod, and McCarty in 1944, and of Hershey and Chase in 1952, that anyone believed that deoxyribonucleic acid (DNA) is the genetic material: up until then it was widely thought that genes were made of protein. The discovery of the role of DNA was a tremendous stimulus to genetic research, and many famous biologists (Delbrück, Chargaff, Crick, and Monod were among the most influential) contributed to the second great age of genetics. In the 14 years between 1952 and 1966, the structure of DNA was elucidated, the genetic code cracked, and the processes of transcription and translation described.

1.2 The advent of gene cloning and the polymerase chain reaction

These years of activity and discovery were followed by a lull, a period of anticlimax when it seemed to some molecular biologists (as the new generation of geneticists styled themselves) that there was little of fundamental importance that was not understood. In truth there was a frustration that the experimental techniques of the late 1960s were not sophisticated enough to allow the gene to be studied in any greater detail.

Then in the years 1971–1973 genetic research was thrown back into gear by what at the time was described as a revolution in experimental biology. A whole new methodology was developed, enabling previously impossible experiments to be planned and carried out, if not with ease, then at least with success. These methods, referred to as recombinant DNA technology or genetic engineering, and having at their core the process of gene cloning, sparked another great age of genetics. They led to rapid and efficient DNA sequencing techniques that enabled the structures of individual genes to be determined, reaching a culmination at the turn of the century with the massive genome sequencing projects, including the human project which was completed in 2000. They led to procedures for studying the regulation of individual genes, which have allowed molecular biologists to understand how aberrations in gene activity can result in human diseases such as cancer. The techniques spawned modern biotechnology, which puts genes to work in production of proteins and other compounds needed in medicine and industrial processes.

During the 1980s, when the excitement engendered by the gene cloning revolution was at its height, it hardly seemed possible that another, equally novel and equally revolutionary process was just around the corner. According to DNA folklore, Kary Mullis invented the polymerase chain reaction (PCR) during a drive along the coast of California one evening in 1985. His brainwave was an exquisitely simple technique that acts as a perfect complement to gene cloning. PCR has made easier many of the techniques that were possible but difficult to carry out when gene cloning was used on its own. It has extended the range of DNA analysis and enabled molecular biology to find new applications in areas of endeavor outside of its traditional range of medicine, agriculture, and biotechnology. Archaeogenetics, molecular ecology, and DNA forensics are just three of the new disciplines that have become possible as a direct consequence of the invention of PCR, enabling molecular biologists to ask questions about human evolution and the impact of environmental change on the biosphere, and to bring their powerful tools to bear in the fight against crime. Forty years have passed since the dawning of the age of gene cloning, but we are still riding the rollercoaster and there is no end to the excitement in sight.

1 Construction of a recombinant DNA molecule

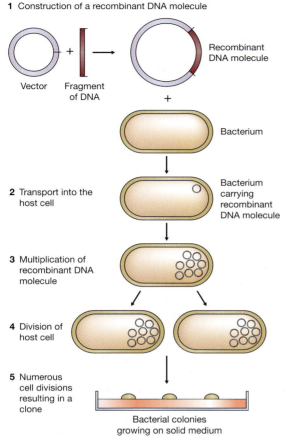

Recombinant DNA molecule

Vector   Fragment
          of DNA

+

Bacterium

2 Transport into the
host cell

Bacterium
carrying
recombinant
DNA molecule

3 Multiplication of
recombinant DNA
molecule

4 Division of
host cell

5 Numerous
cell divisions
resulting in a
clone

Bacterial colonies
growing on solid medium

Figure 1.1

The basic steps in gene cloning.

1.3 What is gene cloning?

What exactly is gene cloning? The easiest way to answer this question is to follow through the steps in a gene cloning experiment (Figure 1.1):

1 A fragment of DNA, containing the gene to be cloned, is inserted into a circular DNA molecule called a vector, to produce a recombinant DNA molecule.

2 The vector transports the gene into a host cell, which is usually a bacterium, although other types of living cell can be used.

3 Within the host cell the vector multiplies, producing numerous identical copies, not only of itself but also of the gene that it carries.

4 When the host cell divides, copies of the recombinant DNA molecule are passed to the progeny and further vector replication takes place.

5 After a large number of cell divisions, a colony, or clone, of identical host cells is produced. Each cell in the clone contains one or more copies of the recombinant DNA molecule; the gene carried by the recombinant molecule is now said to be cloned.

1.4 What is PCR?

The polymerase chain reaction is very different from gene cloning. Rather than a series of manipulations involving living cells, PCR is carried out in a single test tube simply by mixing DNA with a set of reagents and placing the tube in a thermal cycler, a piece of equipment that enables the mixture to be incubated at a series of temperatures that are varied in a preprogrammed manner. The basic steps in a PCR experiment are as follows (Figure 1.2):

Figure 1.2

The basic steps in the polymerase chain reaction.

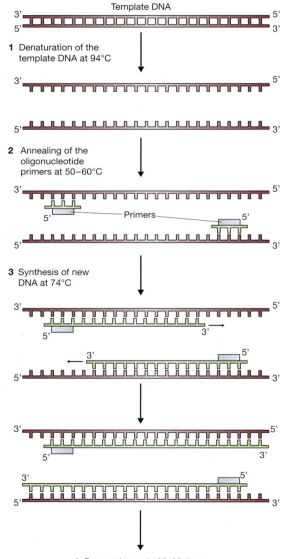

1   The mixture is heated to 94°C, at which temperature the hydrogen bonds that hold together the two strands of the double-stranded DNA molecule are broken, causing the molecule to **denature**.

2   The mixture is cooled down to 50–60°C. The two strands of each molecule could join back together at this temperature, but most do not because the mixture contains a large excess of short DNA molecules, called **oligonucleotides** or **primers**, which **anneal** to the DNA molecules at specific positions.

3   The temperature is raised to 74°C. This is a good working temperature for the *Taq* **DNA polymerase** that is present in the mixture. We will learn more about **DNA polymerases** on p. 48. All we need to understand at this stage is that the *Taq* DNA polymerase attaches to one end of each primer and synthesizes new strands of DNA, complementary to the **template** DNA molecules, during this step of the PCR. Now we have four stands of DNA instead of the two that there were to start with.

4   The temperature is increased back to 94°C. The double-stranded DNA molecules, each of which consists of one strand of the original molecule and one new strand of DNA, denature into single strands. This begins a second cycle of denaturation–annealing–synthesis, at the end of which there are eight DNA strands. By repeating the cycle 30 times the double-stranded molecule that we began with is converted into over 130 million new double-stranded molecules, each one a copy of the region of the starting molecule delineated by the annealing sites of the two primers.

1.5   Why gene cloning and PCR are so important

As you can see from Figures 1.1 and 1.2, gene cloning and PCR are relatively straightforward procedures. Why, then, have they assumed such importance in biology? The answer is largely because both techniques can provide a pure sample of an individual gene, separated from all the other genes in the cell.

1.5.1   Obtaining a pure sample of a gene by cloning

To understand exactly how cloning can provide a pure sample of a gene, consider the basic experiment from Figure 1.1, but drawn in a slightly different way (Figure 1.3). In this example the DNA fragment to be cloned is one member of a mixture of many different fragments, each carrying a different gene or part of a gene. This mixture could indeed be the entire genetic complement of an organism—a human, for instance. Each of these fragments becomes inserted into a different vector molecule to produce a family of recombinant DNA molecules, one of which carries the gene of interest. Usually only one recombinant DNA molecule is transported into any single host cell, so that although the final set of clones may contain many different recombinant DNA molecules, each individual clone contains multiple copies of just one molecule. The gene is now separated away from all the other genes in the original mixture, and its specific features can be studied in detail.

In practice, the key to the success or failure of a gene cloning experiment is the ability to identify the particular clone of interest from the many different ones that are obtained. If we consider the **genome** of the bacterium *Escherichia coli*, which contains

Figure 1.3
Cloning allows individual fragments of DNA to be purified.

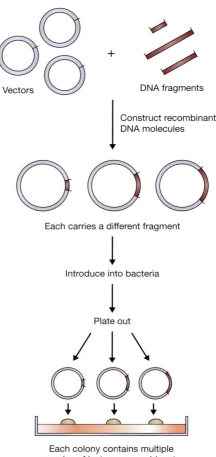

Vectors   +   DNA fragments

Construct recombinant DNA molecules

Each carries a different fragment

Introduce into bacteria

Plate out

Each colony contains multiple copies of just one recombinant DNA molecule

just over 4000 different genes, we might at first despair of being able to find just one gene among all the possible clones (Figure 1.4). The problem becomes even more overwhelming when we remember that bacteria are relatively simple organisms and that the human genome contains about five times as many genes. However, as explained in Chapter 8, a variety of different strategies can be used to ensure that the correct gene can be obtained at the end of the cloning experiment. Some of these strategies involve modifications to the basic cloning procedure, so that only cells containing the desired recombinant DNA molecule can divide and the clone of interest is automatically selected. Other methods involve techniques that enable the desired clone to be identified from a mixture of lots of different clones.

Once a gene has been cloned there is almost no limit to the information that can be obtained about its structure and expression. The availability of cloned material has stimulated the development of analytical methods for studying genes, with new techniques being introduced all the time. Methods for studying the structure and expression of a cloned gene are described in Chapters 10 and 11, respectively.

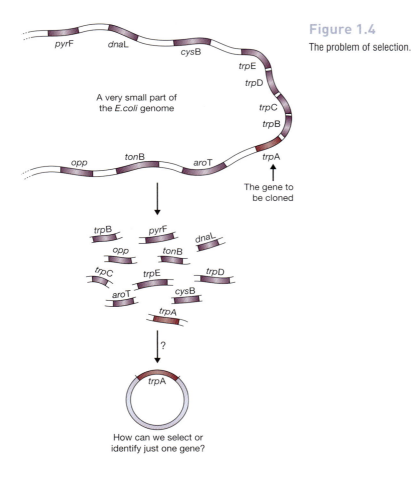

Figure 1.4

The problem of selection.

1.5.2 PCR can also be used to purify a gene

The polymerase chain reaction can also be used to obtain a pure sample of a gene. This is because the region of the starting DNA molecule that is copied during PCR is the segment whose boundaries are marked by the annealing positions of the two oligonucleotide primers. If the primers anneal either side of the gene of interest, many copies of that gene will be synthesized (Figure 1.5). The outcome is the same as with a gene cloning experiment, although the problem of selection does not arise because the desired gene is automatically "selected" as a result of the positions at which the primers anneal.

A PCR experiment can be completed in a few hours, whereas it takes weeks if not months to obtain a gene by cloning. Why then is gene cloning still used? This is because PCR has two limitations:

- In order for the primers to anneal to the correct positions, either side of the gene of interest, the sequences of these annealing sites must be known. It is easy to synthesize a primer with a predetermined sequence (see p. 139), but if the sequences of the annealing sites are unknown then the appropriate primers cannot be made. This means that PCR cannot be used to isolate genes that have not been studied before—that has to be done by cloning.

Figure 1.5

Gene isolation by PCR.

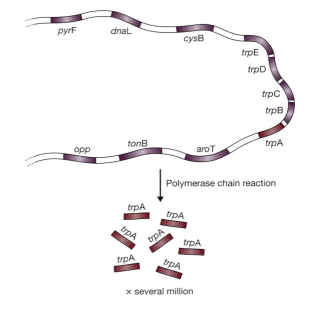

- There is a limit to the length of DNA sequence that can be copied by PCR. Five kilobases (kb) can be copied fairly easily, and segments up to forty kb can be dealt with by using specialized techniques, but this is shorter than the lengths of many genes, especially those of humans and other vertebrates. Cloning must be used if an intact version of a long gene is required.

Gene cloning is therefore the only way of isolating long genes or those that have never been studied before. But PCR still has many important applications. For example, even if the sequence of a gene is not known, it may still be possible to determine the appropriate sequences for a pair of primers, based on what is known about the sequence of the equivalent gene in a different organism. A gene that has been isolated and sequenced from, say, mouse could therefore be used to design a pair of primers for isolation of the equivalent gene from humans.

In addition, there are many applications where it is necessary to isolate or detect genes whose sequences are already known. A PCR of human globin genes, for example, is used to test for the presence of mutations that might cause the blood disease called thalassaemia. Design of appropriate primers for this PCR is easy because the sequences of the human globin genes are known. After the PCR, the gene copies are sequenced or studied in some other way to determine if any of the thalassaemia mutations are present.

Another clinical application of PCR involves the use of primers specific for the DNA of a disease-causing virus. A positive result indicates that a sample contains the virus and that the person who provided the sample should undergo treatment to prevent onset of the disease. The polymerase chain reaction is tremendously sensitive: a carefully set up reaction yields detectable amounts of DNA, even if there is just one DNA molecule in the starting mixture. This means that the technique can detect viruses at the earliest stages of an infection, increasing the chances of treatment being successful. This great sensitivity means that PCR can also be used with DNA from forensic material such as hairs and dried bloodstains or even from the bones of long-dead humans (Chapter 16).

1.6 How to find your way through this book

This book explains how gene cloning, PCR and other DNA analysis techniques are carried out and describes the applications of these techniques in modern biology. The applications are covered in the second and third parts of the book. Part II describes how genes and genomes are studied and Part III gives accounts of the broader applications of gene cloning and PCR in biotechnology, medicine, agriculture, and forensic science.

In Part I we deal with the basic principles. Most of the nine chapters are devoted to gene cloning because this technique is more complicated than PCR. When you have understood how cloning is carried out you will have understood many of the basic principles of how DNA is analyzed. In Chapter 2 we look at the central component of a gene cloning experiment—the vector—which transports the gene into the host cell and is responsible for its replication. To act as a cloning vector a DNA molecule must be capable of entering a host cell and, once inside, replicating to produce multiple copies of itself. Two naturally occurring types of DNA molecule satisfy these requirements:

- **Plasmids**, which are small circles of DNA found in bacteria and some other organisms. Plasmids can replicate independently of the host cell chromosome.

- **Virus chromosomes**, in particular the chromosomes of **bacteriophages**, which are viruses that specifically infect bacteria. During infection the bacteriophage DNA molecule is injected into the host cell where it undergoes replication.

Chapter 3 describes how DNA is purified from living cells—both the DNA that will be cloned and the vector DNA—and Chapter 4 covers the various techniques for handling purified DNA molecules in the laboratory. There are many such techniques, but two are particularly important in gene cloning. These are the ability to cut the vector at a specific point and then to repair it in such a way that the gene is inserted (see Figure 1.1). These and other DNA manipulations were developed as an offshoot of basic research into DNA synthesis and modification in living cells, and most of the manipulations make use of purified enzymes. The properties of these enzymes, and the way they are used in DNA studies, are described in Chapter 4.

Once a recombinant DNA molecule has been constructed, it must be introduced into the host cell so that replication can take place. Transport into the host cell makes use of natural processes for uptake of plasmid and viral DNA molecules. These processes and the ways they are utilized in gene cloning are described in Chapter 5, and the most important types of cloning vector are introduced, and their uses examined, in Chapters 6 and 7. To conclude the coverage of gene cloning, in Chapter 8 we investigate the problem of selection (see Figure 1.4), before returning in Chapter 9 to a more detailed description of PCR and its related techniques.

Further reading

FURTHER READING

Blackman, K. (2001) The advent of genetic engineering. *Trends in Biochemical Science*, 26, 268–270. [An account of the early days of gene cloning.]

Brock, T.D. (1990) *The Emergence of Bacterial Genetics*. Cold Spring Harbor Laboratory Press, New York. [Details the discovery of plasmids and bacteriophages.]

Brown, T.A. (2006) *Genomes*, 3rd edn. Garland Science, Oxford. [An introduction to modern genetics and molecular biology.]

Cherfas, J. (1982) *Man Made Life*. Blackwell, Oxford. [A history of the early years of genetic engineering.]

Judson, H.F. (1979) *The Eighth Day of Creation*. Penguin Science, London. [A very readable account of the development of molecular biology in the years before the gene cloning revolution.]

Mullis, K.B. (1990) The unusual origins of the polymerase chain reaction. *Scientific American*, 262(4), 56–65. [An entertaining account of how PCR was invented.]

Chapter 2

Vectors for Gene Cloning: Plasmids and Bacteriophages

A DNA molecule needs to display several features to be able to act as a vector for gene cloning. Most importantly it must be able to replicate within the host cell, so that numerous copies of the recombinant DNA molecule can be produced and passed to the daughter cells. A cloning vector also needs to be relatively small, ideally less than 10 kb in size, as large molecules tend to break down during purification, and are also more difficult to manipulate. Two kinds of DNA molecule that satisfy these criteria can be found in bacterial cells: plasmids and bacteriophage chromosomes.

2.1 Plasmids

Plasmids are circular molecules of DNA that lead an independent existence in the bacterial cell (Figure 2.1). Plasmids almost always carry one or more genes, and often these genes are responsible for a useful characteristic displayed by the host bacterium. For example, the ability to survive in normally toxic concentrations of antibiotics such as chloramphenicol or ampicillin is often due to the presence in the bacterium of a plasmid carrying antibiotic resistance genes. In the laboratory, antibiotic resistance is often used as a selectable marker to ensure that bacteria in a culture contain a particular plasmid (Figure 2.2).

Most plasmids possess at least one DNA sequence that can act as an origin of replication, so they are able to multiply within the cell independently of the main bacterial

Gene Cloning and DNA Analysis: An Introduction. 6th edition. By T.A. Brown. Published 2010 by Blackwell Publishing.

Figure 2.1

Plasmids: independent genetic elements found in bacterial cells.

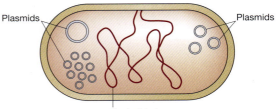

Figure 2.2

The use of antibiotic resistance as a selectable marker for a plasmid. RP4 (top) carries genes for resistance to ampicillin, tetracycline and kanamycin. Only those *E. coli* cells that contain RP4 (or a related plasmid) are able to survive and grow in a medium that contains toxic amounts of one or more of these antibiotics.

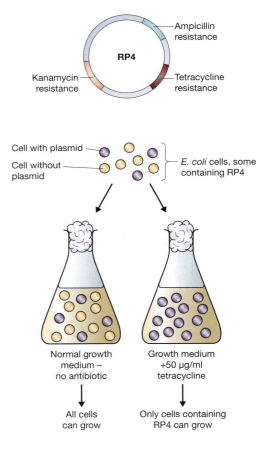

chromosome (Figure 2.3a). The smaller plasmids make use of the host cell's own DNA replicative enzymes in order to make copies of themselves, whereas some of the larger ones carry genes that code for special enzymes that are specific for plasmid replication. A few types of plasmid are also able to replicate by inserting themselves into the bacterial chromosome (Figure 2.3b). These integrative plasmids or **episomes** may be stably maintained in this form through numerous cell divisions, but always at some stage exist as independent elements.

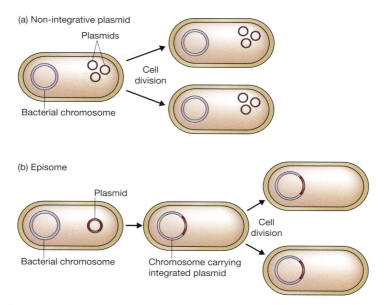

Figure 2.3

Replication strategies for (a) a non-integrative plasmid, and (b) an episome.

Table 2.1

Sizes of representative plasmids.

	SIZE		
PLASMID	**NUCLEOTIDE LENGTH (kb)**	**MOLECULAR MASS (MDa)**	**ORGANISM**
pUC8	2.1	1.8	*E. coli*
ColE1	6.4	4.2	*E. coli*
RP4	54.0	36.0	*Pseudomonas* and others
F	95.0	63.0	*E. coli*
TOL	117.0	78.0	*Pseudomonas putida*
pTiAch5	213.0	142.0	*Agrobacterium tumefaciens*

2.1.1 Size and copy number

The size and copy number of a plasmid are particularly important as far as cloning is concerned. We have already mentioned the relevance of plasmid size and stated that less than 10 kb is desirable for a cloning vector. Plasmids range from about 1.0 kb for the smallest to over 250 kb for the largest plasmids (Table 2.1), so only a few are useful for cloning purposes. However, as we will see in Chapter 7, larger plasmids can be adapted for cloning under some circumstances.

The copy number refers to the number of molecules of an individual plasmid that are normally found in a single bacterial cell. The factors that control copy number are not well understood. Some plasmids, especially the larger ones, are stringent and have a

Figure 2.4

Plasmid transfer by conjugation between bacterial cells. The donor and recipient cells attach to each other by a pilus, a hollow appendage present on the surface of the donor cell. A copy of the plasmid is then passed to the recipient cell. Transfer is thought to occur through the pilus, but this has not been proven and transfer by some other means (e.g. directly across the bacterial cell walls) remains a possibility.

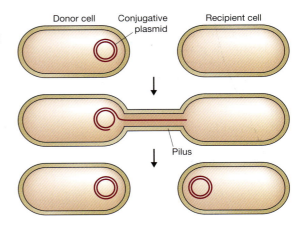

low copy number of perhaps just one or two per cell; others, called relaxed plasmids, are present in multiple copies of 50 or more per cell. Generally speaking, a useful cloning vector needs to be present in the cell in multiple copies so that large quantities of the recombinant DNA molecule can be obtained.

2.1.2 Conjugation and compatibility

Plasmids fall into two groups: conjugative and non-conjugative. Conjugative plasmids are characterized by the ability to promote sexual conjugation between bacterial cells (Figure 2.4), a process that can result in a conjugative plasmid spreading from one cell to all the other cells in a bacterial culture. Conjugation and plasmid transfer are controlled by a set of transfer or *tra* genes, which are present on conjugative plasmids but absent from the non-conjugative type. However, a non-conjugative plasmid may, under some circumstances, be cotransferred along with a conjugative plasmid when both are present in the same cell.

Several different kinds of plasmid may be found in a single cell, including more than one different conjugative plasmid at any one time. In fact, cells of *E. coli* have been known to contain up to seven different plasmids at once. To be able to coexist in the same cell, different plasmids must be compatible. If two plasmids are incompatible then one or the other will be rapidly lost from the cell. Different types of plasmid can therefore be assigned to different incompatibility groups on the basis of whether or not they can coexist, and plasmids from a single incompatibility group are often related to each other in various ways. The basis of incompatibility is not well understood, but events during plasmid replication are thought to underlie the phenomenon.

2.1.3 Plasmid classification

The most useful classification of naturally occurring plasmids is based on the main characteristic coded by the plasmid genes. The five major types of plasmid according to this classification are as follows:

- Fertility or F plasmids carry only *tra* genes and have no characteristic beyond the ability to promote conjugal transfer of plasmids. A well-known example is the F plasmid of *E. coli*.

- **Resistance** or **R plasmids** carry genes conferring on the host bacterium resistance to one or more antibacterial agents, such as chloramphenicol, ampicillin, and mercury. R plasmids are very important in clinical microbiology as their spread through natural populations can have profound consequences in the treatment of bacterial infections. An example is RP4, which is commonly found in *Pseudomonas*, but also occurs in many other bacteria.
- **Col plasmids** code for colicins, proteins that kill other bacteria. An example is ColE1 of *E. coli*.
- **Degradative plasmids** allow the host bacterium to metabolize unusual molecules such as toluene and salicylic acid, an example being TOL of *Pseudomonas putida*.
- **Virulence plasmids** confer pathogenicity on the host bacterium; these include the **Ti plasmids** of ***Agrobacterium tumefaciens***, which induce crown gall disease on dicotyledonous plants.

2.1.4 Plasmids in organisms other than bacteria

Although plasmids are widespread in bacteria they are by no means as common in other organisms. The best characterized eukaryotic plasmid is the 2 μm circle that occurs in many strains of the yeast *Saccharomyces cerevisiae*. The discovery of the 2 μm plasmid was very fortuitous as it allowed the construction of cloning vectors for this very important industrial organism (p. 105). However, the search for plasmids in other eukaryotes (such as filamentous fungi, plants and animals) has proved disappointing, and it is suspected that many higher organisms simply do not harbor plasmids within their cells.

2.2 Bacteriophages

Bacteriophages, or phages as they are commonly known, are viruses that specifically infect bacteria. Like all viruses, phages are very simple in structure, consisting merely of a DNA (or occasionally ribonucleic acid (RNA)) molecule carrying a number of genes, including several for replication of the phage, surrounded by a protective coat or **capsid** made up of protein molecules (Figure 2.5).

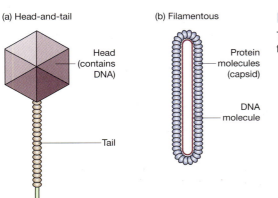

(a) Head-and-tail

Head (contains DNA)

Tail

(b) Filamentous

Protein molecules (capsid)

DNA molecule

Figure 2.5

The two main types of phage structure: (a) head-and-tail (e.g. λ); (b) filamentous (e.g. M13).

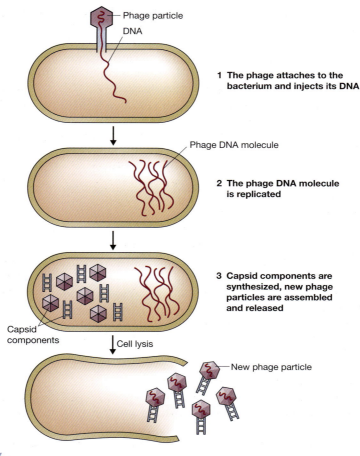

Figure 2.6
The general pattern of infection of a bacterial cell by a bacteriophage.

2.2.1 The phage infection cycle

The general pattern of infection, which is the same for all types of phage, is a three-step process (Figure 2.6):

1 The phage particle attaches to the outside of the bacterium and injects its DNA chromosome into the cell.
2 The phage DNA molecule is replicated, usually by specific phage enzymes coded by genes in the phage chromosome.
3 Other phage genes direct synthesis of the protein components of the capsid, and new phage particles are assembled and released from the bacterium.

With some phage types the entire infection cycle is completed very quickly, possibly in less than 20 minutes. This type of rapid infection is called a lytic cycle, as release of the new phage particles is associated with lysis of the bacterial cell. The characteristic feature of a lytic infection cycle is that phage DNA replication is immediately followed by synthesis of capsid proteins, and the phage DNA molecule is never maintained in a stable condition in the host cell.

2.2.2 Lysogenic phages

In contrast to a lytic cycle, lysogenic infection is characterized by retention of the phage DNA molecule in the host bacterium, possibly for many thousands of cell divisions. With many lysogenic phages the phage DNA is inserted into the bacterial genome, in a manner similar to episomal insertion (see Figure 2.3b). The integrated form of the phage DNA (called the prophage) is quiescent, and a bacterium (referred to as a lysogen) that carries a prophage is usually physiologically indistinguishable from an uninfected cell. However, the prophage is eventually released from the host genome and the phage reverts to the lytic mode and lyses the cell. The infection cycle of lambda (λ), a typical lysogenic phage of this type, is shown in Figure 2.7.

A limited number of lysogenic phages follow a rather different infection cycle. When M13 or a related phage infects *E. coli*, new phage particles are continuously assembled and released from the cell. The M13 DNA is not integrated into the bacterial genome and does not become quiescent. With these phages, cell lysis never occurs, and the infected bacterium can continue to grow and divide, albeit at a slower rate than uninfected cells. Figure 2.8 shows the M13 infection cycle.

Although there are many different varieties of bacteriophage, only λ and M13 have found a major role as cloning vectors. We will now consider the properties of these two phages in more detail.

Gene organization in the λ DNA molecule

λ is a typical example of a head-and-tail phage (see Figure 2.5a). The DNA is contained in the polyhedral head structure and the tail serves to attach the phage to the bacterial surface and to inject the DNA into the cell (see Figure 2.7).

The λ DNA molecule is 49 kb in size and has been intensively studied by the techniques of gene mapping and DNA sequencing. As a result the positions and identities of all of the genes in the λ DNA molecule are known (Figure 2.9). A feature of the λ genetic map is that genes related in terms of function are clustered together in the genome. For example, all of the genes coding for components of the capsid are grouped together in the left-hand third of the molecule, and genes controlling integration of the prophage into the host genome are clustered in the middle of the molecule. Clustering of related genes is profoundly important for controlling expression of the λ genome, as it allows genes to be switched on and off as a group rather than individually. Clustering is also important in the construction of λ-based cloning vectors, as we will discover when we return to this topic in Chapter 6.

The linear and circular forms of λ DNA

A second feature of λ that turns out to be of importance in the construction of cloning vectors is the conformation of the DNA molecule. The molecule shown in Figure 2.9 is linear, with two free ends, and represents the DNA present in the phage head structure. This linear molecule consists of two complementary strands of DNA, base-paired according to the Watson–Crick rules (that is, double-stranded DNA). However, at either end of the molecule is a short 12-nucleotide stretch in which the DNA is single-stranded (Figure 2.10a). The two single strands are complementary, and so can base pair with one another to form a circular, completely double-stranded molecule (Figure 2.10b).

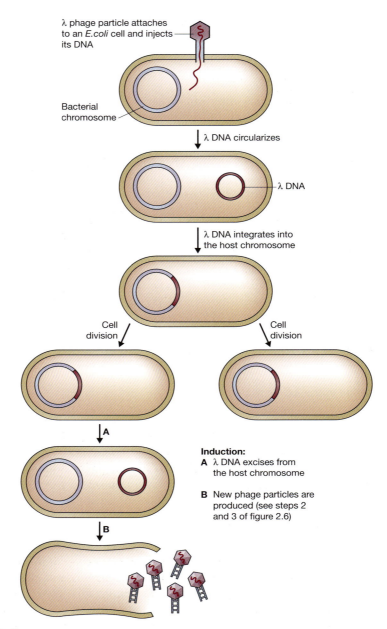

Figure 2.7

The lysogenic infection cycle of bacteriophage λ.

Complementary single strands are often referred to as "sticky" ends or cohesive ends, because base pairing between them can "stick" together the two ends of a DNA molecule (or the ends of two different DNA molecules). The λ cohesive ends are called the *cos sites* and they play two distinct roles during the λ infection cycle. First, they allow the linear DNA molecule that is injected into the cell to be circularized, which is a necessary prerequisite for insertion into the bacterial genome (see Figure 2.7).

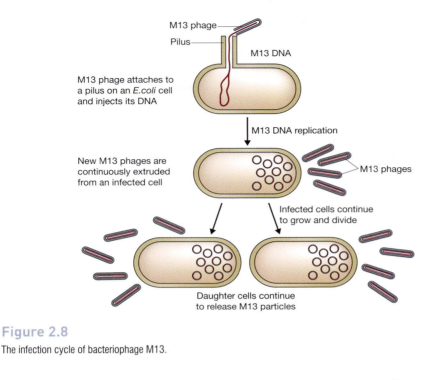

Figure 2.8

The infection cycle of bacteriophage M13.

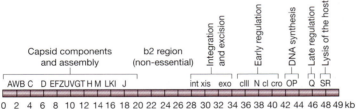

Figure 2.9

The λ genetic map, showing the positions of the important genes and the functions of the gene clusters.

The second role of the *cos* sites is rather different, and comes into play after the prophage has excised from the host genome. At this stage a large number of new λ DNA molecules are produced by the rolling circle mechanism of replication (Figure 2.10c), in which a continuous DNA strand is "rolled off" the template molecule. The result is a catenane consisting of a series of linear λ genomes joined together at the *cos* sites. The role of the *cos* sites is now to act as recognition sequences for an endonuclease that cleaves the catenane at the *cos* sites, producing individual λ genomes. This endonuclease, which is the product of gene *A* on the λ DNA molecule, creates the single-stranded sticky ends, and also acts in conjunction with other proteins to package each λ genome into a phage head structure. The cleavage and packaging processes recognize just the *cos* sites and the DNA sequences to either side of them, so changing the structure of the internal regions of the λ genome, for example by inserting new genes, has no effect on these events so long as the overall length of the λ genome is not altered too greatly.

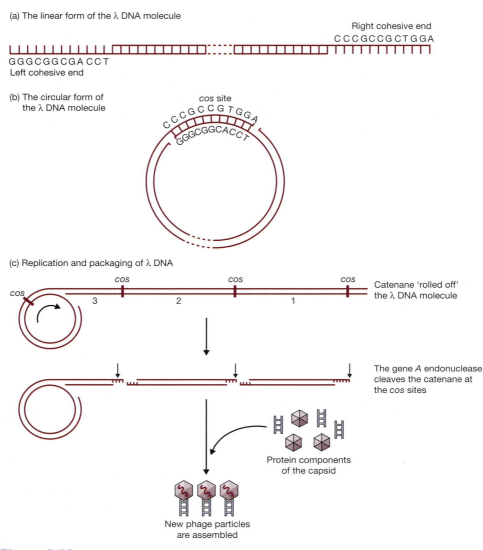

(a) The linear form of the λ DNA molecule

Right cohesive end
C C C G C C G C T G G A

G G G C G G C G A C C T
Left cohesive end

(b) The circular form of the λ DNA molecule

cos site

C C C G C C G T G G A

G G G C G G C A C C T

(c) Replication and packaging of λ DNA

cos   cos   cos   cos

Catenane 'rolled off' the λ DNA molecule

3   2   1

The gene A endonuclease cleaves the catenane at the cos sites

Protein components of the capsid

New phage particles are assembled

Figure 2.10

The linear and circular forms of λ DNA. (a) The linear form, showing the left and right cohesive ends. (b) Base pairing between the cohesive ends results in the circular form of the molecule. (c) Rolling circle replication produces a catenane of new linear λ DNA molecules, which are individually packaged into phage heads as new λ particles are assembled.

M13—a filamentous phage

M13 is an example of a filamentous phage (see Figure 2.5b) and is completely different in structure from λ. Furthermore, the M13 DNA molecule is much smaller than the λ genome, being only 6407 nucleotides in length. It is circular and is unusual in that it consists entirely of single-stranded DNA.

The smaller size of the M13 DNA molecule means that it has room for fewer genes than the λ genome. This is possible because the M13 capsid is constructed from multiple copies of just three proteins (requiring only three genes), whereas synthesis of the λ

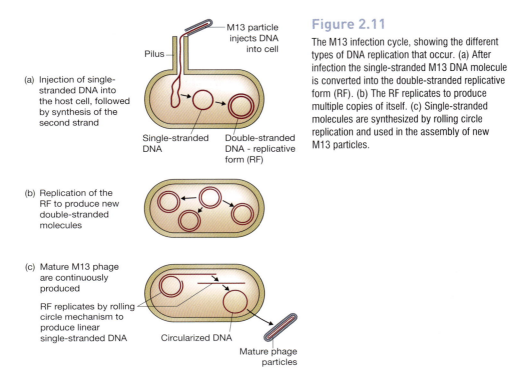

(a) Injection of single-stranded DNA into the host cell, followed by synthesis of the second strand

(b) Replication of the RF to produce new double-stranded molecules

(c) Mature M13 phage are continuously produced

RF replicates by rolling circle mechanism to produce linear single-stranded DNA

Figure 2.11

The M13 infection cycle, showing the different types of DNA replication that occur. (a) After infection the single-stranded M13 DNA molecule is converted into the double-stranded replicative form (RF). (b) The RF replicates to produce multiple copies of itself. (c) Single-stranded molecules are synthesized by rolling circle replication and used in the assembly of new M13 particles.

head-and-tail structure involves over 15 different proteins. In addition, M13 follows a simpler infection cycle than λ, and does not need genes for insertion into the host genome.

Injection of an M13 DNA molecule into an *E. coli* cell occurs via the **pilus**, the structure that connects two cells during sexual conjugation (see Figure 2.4). Once inside the cell the single-stranded molecule acts as the template for synthesis of a complementary strand, resulting in normal double-stranded DNA (Figure 2.11a). This molecule is not inserted into the bacterial genome, but instead replicates until over 100 copies are present in the cell (Figure 2.11b). When the bacterium divides, each daughter cell receives copies of the phage genome, which continues to replicate, thereby maintaining its overall numbers per cell. As shown in Figure 2.11c, new phage particles are continuously assembled and released, about 1000 new phages being produced during each generation of an infected cell.

Several features of M13 make this phage attractive as a cloning vector. The genome is less than 10 kb in size, well within the range desirable for a potential vector. In addition, the double-stranded **replicative form (RF)** of the M13 genome behaves very much like a plasmid, and can be treated as such for experimental purposes. It is easily prepared from a culture of infected *E. coli* cells (p. 43) and can be reintroduced by **transfection** (p. 81). Most importantly, genes cloned with an M13-based vector can be obtained in the form of single-stranded DNA. Single-stranded versions of cloned genes are useful for several techniques, notably DNA sequencing and *in vitro* mutagenesis (pp. 169 and 203). Cloning in an M13 vector is an easy and reliable way of obtaining single-stranded DNA for this type of work. M13 vectors are also used in **phage display**, a technique for identifying pairs of genes whose protein products interact with one another (p. 220).

2.2.3 Viruses as cloning vectors for other organisms

Most living organisms are infected by viruses and it is not surprising that there has been great interest in the possibility that viruses might be used as cloning vectors for higher organisms. This is especially important when it is remembered that plasmids are not commonly found in organisms other than bacteria and yeast. Several eukaryotic viruses have been employed as cloning vectors for specialized applications: for example, human adenoviruses are used in gene therapy (p. 259), baculoviruses are used to synthesize important pharmaceutical proteins in insect cells (p. 240), and caulimoviruses and geminiviruses have been used for cloning in plants (p. 120). These vectors are discussed more fully in Chapter 7.

Further reading

Dale, J.W. & Park, S.T. (2004) *Molecular Genetics of Bacteria*, 4th edn. Wiley Blackwell, Chichester. [Provides a detailed description of plasmids and bacteriophages.]

Willey, J., Sherwood, L. & Woolverton, C. (2007) *Prescott's Microbiology*, 7th edn. McGraw Hill Higher Education, Maidenhead. [A good introduction to microbiology, including plasmids and phages.]

Chapter 3

Purification of DNA from Living Cells

Chapter contents

The genetic engineer will, at different times, need to prepare at least three distinct kinds of DNA. First, total cell DNA will often be required as a source of material from which to obtain genes to be cloned. Total cell DNA may be DNA from a culture of bacteria, from a plant, from animal cells, or from any other type of organism that is being studied. It consists of the genomic DNA of the organism along with any additional DNA molecules, such as plasmids, that are present.

The second type of DNA that will be required is pure plasmid DNA. Preparation of plasmid DNA from a culture of bacteria follows the same basic steps as purification of total cell DNA, with the crucial difference that at some stage the plasmid DNA must be separated from the main bulk of chromosomal DNA also present in the cell.

Finally, phage DNA will be needed if a phage cloning vector is to be used. Phage DNA is generally prepared from bacteriophage particles rather than from infected cells, so there is no problem with contaminating bacterial DNA. However, special techniques are needed to remove the phage capsid. An exception is the double-stranded replicative form of M13, which is prepared from *E. coli* cells in the same way as a bacterial plasmid.

3.1 Preparation of total cell DNA

The fundamentals of DNA preparation are most easily understood by first considering the simplest type of DNA purification procedure, that where the entire DNA complement

Gene Cloning and DNA Analysis: An Introduction. 6th edition. By T.A. Brown. Published 2010 by Blackwell Publishing.

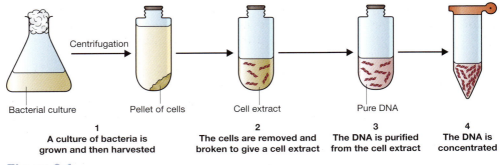

Bacterial culture	Pellet of cells	Cell extract	Pure DNA
1	**2**	**3**	**4**
A culture of bacteria is grown and then harvested	**The cells are removed and broken to give a cell extract**	**The DNA is purified from the cell extract**	**The DNA is concentrated**

Figure 3.1

The basic steps in preparation of total cell DNA from a culture of bacteria.

of a bacterial cell is required. The modifications needed for plasmid and phage DNA preparation can then be described later.

The procedure for total DNA preparation from a culture of bacterial cells can be divided into four stages (Figure 3.1):

1 A culture of bacteria is grown and then **harvested**.
2 The cells are broken open to release their contents.
3 This **cell extract** is treated to remove all components except the DNA.
4 The resulting DNA solution is concentrated.

3.1.1 Growing and harvesting a bacterial culture

Most bacteria can be grown without too much difficulty in a liquid medium (**broth culture**). The culture medium must provide a balanced mixture of the essential nutrients at concentrations that will allow the bacteria to grow and divide efficiently. Two typical growth media are detailed in Table 3.1.

M9 is an example of a **defined medium** in which all the components are known. This medium contains a mixture of inorganic nutrients to provide essential elements such as

Table 3.1

The composition of two typical media for the growth of bacterial cultures.

MEDIUM	COMPONENT	g/l OF MEDIUM
M9 medium	Na_2HPO_4	6.0
	KH_2PO_4	3.0
	NaCl	0.5
	NH_4Cl	1.0
	$MgSO_4$	0.5
	Glucose	2.0
	$CaCl_2$	0.015
LB (Luria-Bertani medium)	Tryptone	10.0
	Yeast extract	5.0
	NaCl	10.0

nitrogen, magnesium, and calcium, as well as glucose to supply carbon and energy. In practice, additional growth factors such as trace elements and vitamins must be added to M9 before it will support bacterial growth. Precisely which supplements are needed depends on the species concerned.

The second medium described in Table 3.1 is rather different. Luria-Bertani (LB) is a complex or undefined medium, meaning that the precise identity and quantity of its components are not known. This is because two of the ingredients, tryptone and yeast extract, are complicated mixtures of unknown chemical compounds. Tryptone in fact supplies amino acids and small peptides, while yeast extract (a dried preparation of partially digested yeast cells) provides the nitrogen requirements, along with sugars and inorganic and organic nutrients. Complex media such as LB need no further supplementation and support the growth of a wide range of bacterial species.

Defined media must be used when the bacterial culture has to be grown under precisely controlled conditions. However, this is not necessary when the culture is being grown simply as a source of DNA, and under these circumstances a complex medium is appropriate. In LB medium at 37°C, aerated by shaking at 150–250 rpm on a rotary platform, *E. coli* cells divide once every 20 min or so until the culture reaches a maximum density of about $2-3 \times 10^9$ cells/ml. The growth of the culture can be monitored by reading the optical density (OD) at 600 nm (Figure 3.2), at which wavelength 1 OD unit corresponds to about 0.8×10^9 cells/ml.

In order to prepare a cell extract, the bacteria must be obtained in as small a volume as possible. Harvesting is therefore performed by spinning the culture in a centrifuge (Figure 3.3). Fairly low centrifugation speeds will pellet the bacteria at the bottom of the centrifuge tube, allowing the culture medium to be poured off. Bacteria from a

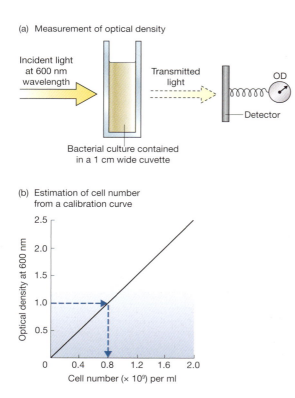

(a) Measurement of optical density

Incident light at 600 nm wavelength

Transmitted light

OD

Detector

Bacterial culture contained in a 1 cm wide cuvette

(b) Estimation of cell number from a calibration curve

Optical density at 600 nm

Cell number (× 10⁹) per ml

Figure 3.2

Estimation of bacterial cell number by measurement of optical density (OD). (a) A sample of the culture is placed in a glass cuvette and light with a wavelength of 600 nm shone through. The amount of light that passes through the culture is measured and the OD (also called the absorbance) calculated as:

$$1 \text{ OD unit} = \log_{10} \frac{\text{intensity of transmitted light}}{\text{intensity of incident light}}$$

The operation is performed with a spectrophotometer. (b) The cell number corresponding to the OD reading is calculated from a calibration curve. This curve is plotted from the OD values of a series of cultures of known cell density. For *E. coli*, 1 OD unit = 0.8×10^9 cells/ml.

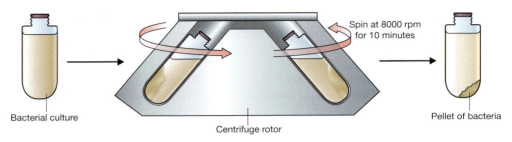

Figure 3.3
Harvesting bacteria by centrifugation.

1000 ml culture at maximum cell density can then be resuspended into a volume of 10 ml or less.

3.1.2 Preparation of a cell extract

The bacterial cell is enclosed in a cytoplasmic membrane and surrounded by a rigid cell wall. With some species, including *E. coli*, the cell wall may itself be enveloped by a second, outer membrane. All of these barriers have to be disrupted to release the cell components.

Techniques for breaking open bacterial cells can be divided into physical methods, in which the cells are disrupted by mechanical forces, and chemical methods, where cell lysis is brought about by exposure to chemical agents that affect the integrity of the cell barriers. Chemical methods are most commonly used with bacterial cells when the object is DNA preparation.

Chemical lysis generally involves one agent attacking the cell wall and another disrupting the cell membrane (Figure 3.4a). The chemicals that are used depend on the species of bacterium involved, but with *E. coli* and related organisms, weakening of the cell wall is usually brought about by lysozyme, ethylenediamine tetraacetate (EDTA), or a combination of both. Lysozyme is an enzyme that is present in egg white and in secretions such as tears and saliva, and which digests the polymeric compounds that give the cell wall its rigidity. EDTA removes magnesium ions that are essential for preserving the overall structure of the cell envelope, and also inhibits cellular enzymes that could degrade DNA. Under some conditions, weakening the cell wall with lysozyme or

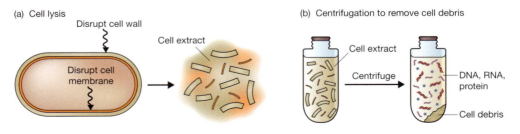

Figure 3.4
Preparation of a cell extract. (a) Cell lysis. (b) Centrifugation of the cell extract to remove insoluble debris.

EDTA is sufficient to cause bacterial cells to burst, but usually a detergent such as sodium dodecyl sulphate (SDS) is also added. Detergents aid the process of lysis by removing lipid molecules and thereby cause disruption of the cell membranes.

Having lysed the cells, the final step in preparation of a cell extract is removal of insoluble cell debris. Components such as partially digested cell wall fractions can be pelleted by centrifugation (Figure 3.4b), leaving the cell extract as a reasonably clear supernatant.

3.1.3 Purification of DNA from a cell extract

In addition to DNA, a bacterial cell extract contains significant quantities of protein and RNA. A variety of methods can be used to purify the DNA from this mixture. One approach is to treat the mixture with reagents which degrade the contaminants, leaving a pure solution of DNA (Figure 3.5a). Other methods use ion-exchange chromatography to separate the mixture into its various components, so the DNA is removed from the proteins and RNA in the extract (Figure 3.5b).

Removing contaminants by organic extraction and enzyme digestion

The standard way to deproteinize a cell extract is to add phenol or a 1 : 1 mixture of phenol and chloroform. These organic solvents precipitate proteins but leave the nucleic acids (DNA and RNA) in aqueous solution. The result is that if the cell extract is mixed gently with the solvent, and the layers then separated by centrifugation, precipitated protein molecules are left as a white coagulated mass at the interface between the aqueous and organic layers (Figure 3.6). The aqueous solution of nucleic acids can then be removed with a pipette.

With some cell extracts the protein content is so great that a single phenol extraction is not sufficient to completely purify the nucleic acids. This problem could be solved by carrying out several phenol extractions one after the other, but this is undesirable as each mixing and centrifugation step results in a certain amount of breakage of the DNA molecules. The answer is to treat the cell extract with a protease such as pronase or

(a) Degradation of contaminants

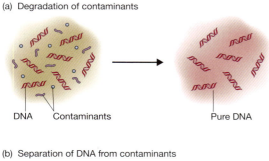

DNA   Contaminants   Pure DNA

(b) Separation of DNA from contaminants

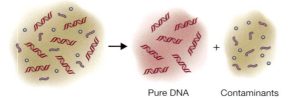

Pure DNA   Contaminants

Figure 3.5

Two approaches to DNA purification. (a) Treating the mixture with reagents which degrade the contaminants, leaving a pure solution of DNA. (b) Separating the mixture into different fractions, one of which is pure DNA.

Figure 3.6

Removal of protein contaminants by phenol extraction.

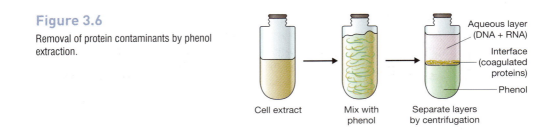

Cell extract   Mix with phenol   Separate layers by centrifugation

Aqueous layer (DNA + RNA)
Interface (coagulated proteins)
Phenol

proteinase K before phenol extraction. These enzymes break polypeptides down into smaller units, which are more easily removed by phenol.

Some RNA molecules, especially **messenger RNA (mRNA)**, are removed by phenol treatment, but most remain with the DNA in the aqueous layer. The only effective way to remove the RNA is with the enzyme **ribonuclease**, which rapidly degrades these molecules into ribonucleotide subunits.

Using ion-exchange chromatography to purify DNA from a cell extract

Biochemists have devised various methods for using differences in electrical charge to separate mixtures of chemicals into their individual components. One of these methods is ion-exchange chromatography, which separates molecules according to how tightly they bind to electrically charged particles present in a chromatographic matrix or **resin**. DNA and RNA are both negatively charged, as are some proteins, and so bind to a positively charged resin. The electrical attachment is disrupted by salt (Figure 3.7a), removal of the more tightly bound molecules requiring higher concentrations of salt. By gradually increasing the salt concentration, different types of molecule can be detached from the resin one after another.

The simplest way to carry out ion-exchange chromatography is to place the resin in a glass or plastic column and then add the cell extract to the top (Figure 3.7b). The extract passes through the column, and because this extract contains very little salt all the negatively charged molecules bind to the resin and are retained in the column. If a salt solution of gradually increasing concentration is now passed through the column, the different types of molecule will **elute** (i.e., become unbound) in the sequence protein, RNA, and finally DNA. However, such careful separation is usually not needed so just two salt solutions are used, one whose concentration is sufficient to elute the protein and RNA, leaving just the DNA bound, followed by a second of a higher concentration which elutes the DNA, now free from protein and RNA contaminants.

3.1.4 Concentration of DNA samples

Organic extraction often results in a very thick solution of DNA that does not need to be concentrated any further. Other purification methods give more dilute solutions and it is therefore important to consider methods for increasing the DNA concentration.

The most frequently used method of concentration is **ethanol precipitation**. In the presence of salt (strictly speaking, monovalent cations such as sodium ions (Na^+)), and at a temperature of $-20°C$ or less, absolute ethanol efficiently precipitates polymeric nucleic acids. With a thick solution of DNA the ethanol can be layered on top of the sample, causing molecules to precipitate at the interface. A spectacular trick is to push a glass rod through the ethanol into the DNA solution. When the rod is removed,

(a) Attachment of DNA to ion-exchange particles

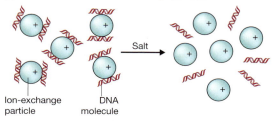

Ion-exchange particle

DNA molecule

Figure 3.7

DNA purification by ion-exchange chromatography. (a) Attachment of DNA to ion-exchange particles. (b) DNA is purified by column chromatography. The solutions passing through the column can be collected by gravity flow or by the **spin column** method, in which the column is placed in a low-speed centrifuge.

(b) DNA purification by ion-exchange chromatography

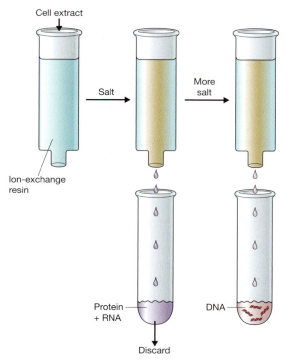

Cell extract

Salt

More salt

Ion-exchange resin

Protein + RNA

DNA

Discard

DNA molecules adhere and can be pulled out of the solution in the form of a long fiber (Figure 3.8a). Alternatively, if ethanol is mixed with a dilute DNA solution, the precipitate can be collected by centrifugation (Figure 3.8b), and then redissolved in an appropriate volume of water. Ethanol precipitation has the added advantage of leaving short-chain and monomeric nucleic acid components in solution. Ribonucleotides produced by ribonuclease treatment are therefore lost at this stage.

3.1.5 Measurement of DNA concentration

It is crucial to know exactly how much DNA is present in a solution when carrying out a gene cloning experiment. Fortunately DNA concentrations can be accurately measured by **ultraviolet (UV) absorbance spectrophotometry**. The amount of UV radiation absorbed by a solution of DNA is directly proportional to the amount of DNA in the

Figure 3.8

Collecting DNA by ethanol precipitation.
(a) Absolute ethanol is layered on top of a
concentrated solution of DNA. Fibers of DNA
can be withdrawn with a glass rod. (b) For less
concentrated solutions ethanol is added (at a ratio
of 2.5 volumes of absolute ethanol to 1 volume of
DNA solution) and precipitated DNA collected by
centrifugation.

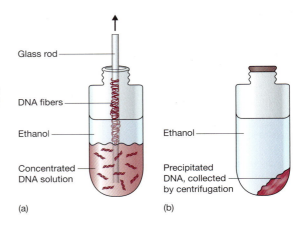

sample. Usually absorbance is measured at 260 nm, at which wavelength an absorbance (A_{260}) of 1.0 corresponds to 50 mg of double-stranded DNA per ml. Measurements of as little as 1 μl of a DNA solution can be carried out in spectrophotometers designed especially for this purpose.

Ultraviolet absorbance can also be used to check the purity of a DNA preparation. With a pure sample of DNA, the ratio of the absorbances at 260 and 280 nm (A_{260}/A_{280}) is 1.8. Ratios of less than 1.8 indicate that the preparation is contaminated, either with protein or with phenol.

3.1.6 Other methods for the preparation of total cell DNA

Bacteria are not the only organisms from which DNA may be required. Total cell DNA from, for example, plants or animals will be needed if the aim of the genetic engineering project is to clone genes from these organisms. Although the basic steps in DNA purification are the same whatever the organism, some modifications may have to be introduced to take account of the special features of the cells being used.

Obviously growth of cells in liquid medium is appropriate only for bacteria, other microorganisms, and plant and animal cell cultures. The major modifications, however, are likely to be needed at the cell breakage stage. The chemicals used for disrupting bacterial cells do not usually work with other organisms: lysozyme, for example, has no effect on plant cells. Specific degradative enzymes are available for most cell wall types, but often physical techniques, such as grinding frozen material with a mortar and pestle, are more efficient. On the other hand, most animal cells have no cell wall at all, and can be lysed simply by treating with detergent.

Another important consideration is the biochemical content of the cells from which DNA is being extracted. With most bacteria the main biochemicals present in a cell extract are protein, DNA and RNA, so phenol extraction and/or protease treatment, followed by removal of RNA with ribonuclease, leaves a pure DNA sample. These treatments may not, however, be sufficient to give pure DNA if the cells also contain significant quantities of other biochemicals. Plant tissues are particularly difficult in this respect as they often contain large amounts of carbohydrates that are not removed by phenol extraction. Instead a different approach must be used. One method makes use of a detergent called cetyltrimethylammonium bromide (CTAB), which forms an insoluble complex with nucleic acids. When CTAB is added to a plant cell extract the

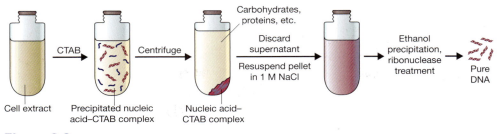

Figure 3.9

The CTAB method for purification of plant DNA.

nucleic acid–CTAB complex precipitates, leaving carbohydrate, protein, and other contaminants in the supernatant (Figure 3.9). The precipitate is then collected by centrifugation and resuspended in 1 M sodium chloride, which causes the complex to break down. The nucleic acids can now be concentrated by ethanol precipitation and the RNA removed by ribonuclease treatment.

The need to adapt organic extraction methods to take account of the biochemical contents of different types of starting material has stimulated the search for DNA purification methods that can be used with any species. This is one of the reasons why ion-exchange chromatography has become so popular. A similar method involves a compound called guanidinium thiocyanate, which has two properties that make it useful for DNA purification. First, it denatures and dissolves all biochemicals other than nucleic acids and can therefore be used to release DNA from virtually any type of cell or tissue. Second, in the presence of guanidinium thiocyanate, DNA binds tightly to silica particles (Figure 3.10a). This provides an easy way of recovering the DNA from the denatured mix of biochemicals. One possibility is to add the silica directly to the cell extract but, as with the ion-exchange methods, it is more convenient to use a chromatography column. The silica is placed in the column and the cell extract added (Figure 3.10b). DNA binds to the silica and is retained in the column, whereas the denatured biochemicals pass straight through. After washing away the last contaminants with guanidinium thiocyanate solution, the DNA is recovered by adding water, which destabilizes the interactions between the DNA molecules and the silica.

3.2 Preparation of plasmid DNA

Purification of plasmids from a culture of bacteria involves the same general strategy as preparation of total cell DNA. A culture of cells, containing plasmids, is grown in liquid medium, harvested, and a cell extract prepared. The protein and RNA are removed, and the DNA probably concentrated by ethanol precipitation. However, there is an important distinction between plasmid purification and preparation of total cell DNA. In a plasmid preparation it is always necessary to separate the plasmid DNA from the large amount of bacterial chromosomal DNA that is also present in the cells.

Separating the two types of DNA can be very difficult, but is nonetheless essential if the plasmids are to be used as cloning vectors. The presence of the smallest amount of contaminating bacterial DNA in a gene cloning experiment can easily lead to undesirable results. Fortunately several methods are available for removal of bacterial DNA during

Figure 3.10

DNA purification by the guanidinium thiocyanate and silica method. (a) In the presence of guanidinium thiocyanate, DNA binds to silica particles. (b) DNA is purified by column chromatography.

(a) Attachment of DNA to silica particles

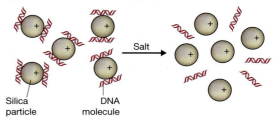

Silica particle   DNA molecule

(b) DNA purification by column chromatography

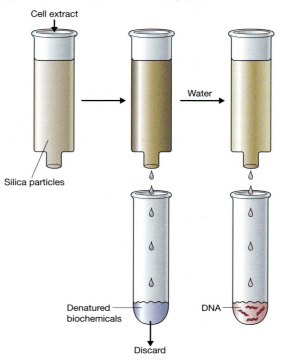

plasmid purification, and the use of these methods, individually or in combination, can result in isolation of very pure plasmid DNA.

The methods are based on the several physical differences between plasmid DNA and bacterial DNA, the most obvious of which is size. The largest plasmids are only 8% of the size of the *E. coli* chromosome, and most are much smaller than this. Techniques that can separate small DNA molecules from large ones should therefore effectively purify plasmid DNA.

In addition to size, plasmids and bacterial DNA differ in conformation. When applied to a polymer such as DNA, the term conformation refers to the overall spatial configuration of the molecule, with the two simplest conformations being linear and circular. Plasmids and the bacterial chromosome are circular, but during preparation of the cell extract the chromosome is always broken to give linear fragments. A method for separating circular from linear molecules will therefore result in pure plasmids.

3.2.1 Separation on the basis of size

The usual stage at which size fractionation is performed is during preparation of the cell extract. If the cells are lysed under very carefully controlled conditions, only a minimal amount of chromosomal DNA breakage occurs. The resulting DNA fragments are still very large—much larger than the plasmids—and can be removed with the cell debris by centrifugation. This process is aided by the fact that the bacterial chromosome is physically attached to the cell envelope, so fragments of the chromosome sediment with the cell debris if these attachments are not broken.

Cell disruption must therefore be carried out very gently to prevent wholesale break-age of the bacterial DNA. For *E. coli* and related species, controlled lysis is performed as shown in Figure 3.11. Treatment with EDTA and lysozyme is carried out in the presence of sucrose, which prevents the cells from bursting immediately. Instead, sphaeroplasts are formed, cells with partially degraded cell walls that retain an intact cytoplasmic membrane. Cell lysis is now induced by adding a non-ionic detergent such as Triton X-100 (ionic detergents, such as SDS, cause chromosomal breakage). This method causes very little breakage of the bacterial DNA, so centrifugation leaves a cleared lysate, consisting almost entirely of plasmid DNA.

A cleared lysate will, however, invariably retain some chromosomal DNA. Further-more, if the plasmids themselves are large molecules, they may also sediment with the

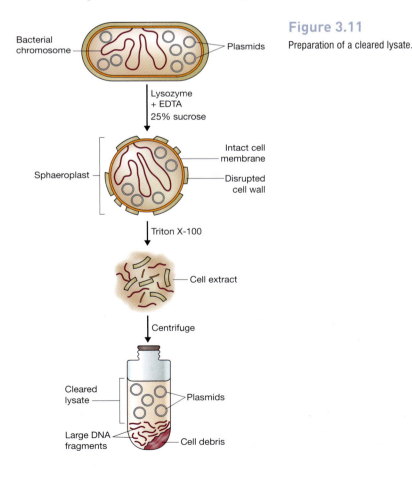

Figure 3.11

Preparation of a cleared lysate.

Figure 3.12

Two conformations of circular double-stranded DNA:
(a) supercoiled—both strands are intact; (b) open-circular
—one or both strands are nicked.

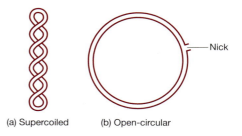

(a) Supercoiled   (b) Open-circular

cell debris. Size fractionation is therefore rarely sufficient on its own, and we must consider alternative ways of removing the bacterial DNA contaminants.

3.2.2 Separation on the basis of conformation

Before considering the ways in which conformational differences between plasmids and bacterial DNA can be used to separate the two types of DNA, we must look more closely at the overall structure of plasmid DNA. It is not strictly correct to say that plasmids have a circular conformation, because double-stranded DNA circles can take up one of two quite distinct configurations. Most plasmids exist in the cell as supercoiled molecules (Figure 3.12a). Supercoiling occurs because the double helix of the plasmid DNA is partially unwound during the plasmid replication process by enzymes called topoisomerases (p. 69). The supercoiled conformation can be maintained only if both polynucleotide strands are intact, hence the more technical name of covalently closed-circular (ccc) DNA. If one of the polynucleotide strands is broken the double helix reverts to its normal relaxed state, and the plasmid takes on the alternative conformation, called open-circular (oc) (Figure 3.12b).

Supercoiling is important in plasmid preparation because supercoiled molecules can be fairly easily separated from non-supercoiled DNA. Two different methods are commonly used. Both can purify plasmid DNA from crude cell extracts, although in practice best results are obtained if a cleared lysate is first prepared.

Alkaline denaturation

The basis of this technique is that there is a narrow pH range at which non-supercoiled DNA is denatured, whereas supercoiled plasmids are not. If sodium hydroxide is added to a cell extract or cleared lysate, so that the pH is adjusted to 12.0–12.5, then the hydrogen bonding in non-supercoiled DNA molecules is broken, causing the double helix to unwind and the two polynucleotide chains to separate (Figure 3.13). If acid is now added, these denatured bacterial DNA strands reaggregate into a tangled mass. The insoluble network can be pelleted by centrifugation, leaving plasmid DNA in the supernatant. An additional advantage of this procedure is that, under some circumstances (specifically cell lysis by SDS and neutralization with sodium acetate), most of the protein and RNA also becomes insoluble and can be removed by the centrifugation step. Further purification by organic extraction or column chromatography may therefore not be needed if the alkaline denaturation method is used.

Ethidium bromide–caesium chloride density gradient centrifugation

This is a specialized version of the more general technique of equilibrium or density gradient centrifugation. A density gradient is produced by centrifuging a solution of

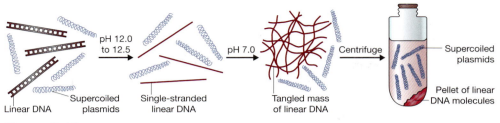

Figure 3.13

Plasmid purification by the alkaline denaturation method.

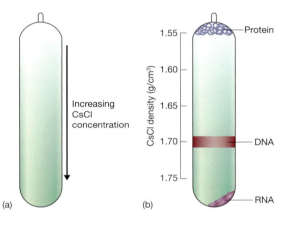

Figure 3.14

Caesium chloride density gradient centrifugation.
(a) A CsCl density gradient produced by high
speed centrifugation. (b) Separation of protein,
DNA, and RNA in a density gradient.

caesium chloride (CsCl) at a very high speed (Figure 3.14a). Macromolecules present in the CsCl solution when it is centrifuged form bands at distinct points in the gradient (Figure 3.14b). Exactly where a particular molecule bands depends on its buoyant density: DNA has a buoyant density of about 1.70 g/cm³, and therefore migrates to the point in the gradient where the CsCl density is also 1.70 g/cm³. In contrast, protein molecules have much lower buoyant densities, and so float at the top of the tube, whereas RNA forms a pellet at the bottom (Figure 3.14b). Density gradient centrifugation can therefore separate DNA, RNA, and protein and is an alternative to organic extraction or column chromatography for DNA purification.

More importantly, density gradient centrifugation in the presence of ethidium bromide (EtBr) can be used to separate supercoiled DNA from non-supercoiled molecules. Ethidium bromide binds to DNA molecules by intercalating between adjacent base pairs, causing partial unwinding of the double helix (Figure 3.15). This unwinding results in a decrease in the buoyant density, by as much as 0.125 g/cm³ for linear DNA. However, supercoiled DNA, with no free ends, has very little freedom to unwind, and can only bind a limited amount of EtBr. The decrease in buoyant density of a supercoiled molecule is therefore much less, only about 0.085 g/cm³. As a consequence, supercoiled molecules form a band in an EtBr–CsCl gradient at a different position to linear and open-circular DNA (Figure 3.16a).

Ethidium bromide–caesium chloride density gradient centrifugation is a very efficient method for obtaining pure plasmid DNA. When a cleared lysate is subjected to this procedure, plasmids band at a distinct point, separated from the linear bacterial DNA,

Figure 3.15

Partial unwinding of the DNA double helix by EtBr intercalation between adjacent base pairs. The normal DNA molecule shown on the left is partially unwound by taking up four EtBr molecules, resulting in the "stretched" structure on the right.

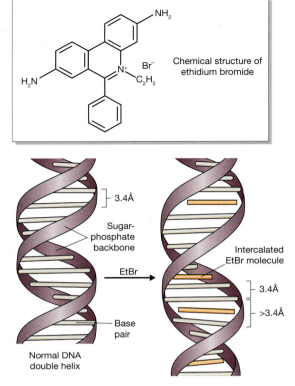

Figure 3.16

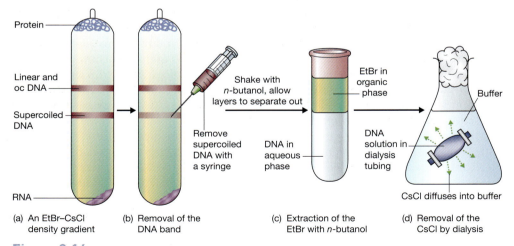

Purification of plasmid DNA by EtBr–CsCl density gradient centrifugation.

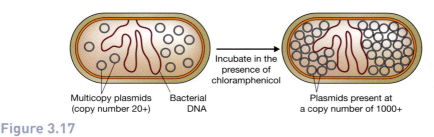

Figure 3.17
Plasmid amplification.

with the protein floating on the top of the gradient and RNA pelleted at the bottom. The position of the DNA bands can be seen by shining ultraviolet radiation on the tube, which causes the bound EtBr to fluoresce. The pure plasmid DNA is removed by puncturing the side of the tube and withdrawing a sample with a syringe (Figure 3.16b). The EtBr bound to the plasmid DNA is extracted with *n*-butanol (Figure 3.16c) and the CsCl removed by dialysis (Figure 3.16d). The resulting plasmid preparation is virtually 100% pure and ready for use as a cloning vector.

3.2.3 Plasmid amplification

Preparation of plasmid DNA can be hindered by the fact that plasmids make up only a small proportion of the total DNA in the bacterial cell. The yield of DNA from a bacterial culture may therefore be disappointingly low. Plasmid amplification offers a means of increasing this yield.

   The aim of amplification is to increase the copy number of a plasmid. Some multi-copy plasmids (those with copy numbers of 20 or more) have the useful property of being able to replicate in the absence of protein synthesis. This contrasts with the main bacterial chromosome, which cannot replicate under these conditions. This property can be utilized during the growth of a bacterial culture for plasmid DNA purification. After a satisfactory cell density has been reached, an inhibitor of protein synthesis (e.g., chloramphenicol) is added, and the culture incubated for a further 12 hours. During this time the plasmid molecules continue to replicate, even though chromosome replication and cell division are blocked (Figure 3.17). The result is that plasmid copy numbers of several thousand may be attained. Amplification is therefore a very efficient way of increasing the yield of multicopy plasmids.

3.3 Preparation of bacteriophage DNA

The key difference between phage DNA purification and the preparation of either total cell DNA or plasmid DNA is that for phages the starting material is not normally a cell extract. This is because bacteriophage particles can be obtained in large numbers from the extracellular medium of an infected bacterial culture. When such a culture is centrifuged, the bacteria are pelleted, leaving the phage particles in suspension (Figure 3.18). The phage particles are then collected from the suspension and their DNA extracted by a single deproteinization step to remove the phage capsid.

   This overall process is more straightforward than the procedure used to prepare total cell or plasmid DNA. Nevertheless, successful purification of significant quantities of

Figure 3.18

Preparation of a phage suspension from an infected culture of bacteria.

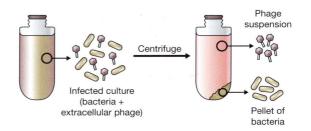

phage DNA is subject to several pitfalls. The main difficulty, especially with λ, is growing an infected culture in such a way that the extracellular phage titer (the number of phage particles per ml of culture) is sufficiently high. In practical terms, the maximum titer that can reasonably be expected for λ is 10^{10} per ml; yet 10^{10} λ particles will yield only 500 ng of DNA. Large culture volumes, in the range of 500–1000 ml, are therefore needed if substantial quantities of λ DNA are to be obtained.

3.3.1 Growth of cultures to obtain a high λ titer

Growing a large volume culture is no problem (bacterial cultures of 100 liters and over are common in biotechnology), but obtaining the maximum phage titer requires a certain amount of skill. The naturally occurring λ phage is lysogenic (p. 19), and an infected culture consists mainly of cells carrying the prophage integrated into the bacterial DNA (see Figure 2.7). The extracellular λ titer is extremely low under these circumstances.

To get a high yield of extracellular λ, the culture must be **induced**, so that all the bacteria enter the lytic phase of the infection cycle, resulting in cell death and release of λ particles into the medium. Induction is normally very difficult to control, but most laboratory strains of λ carry a **temperature-sensitive (ts) mutation** in the *c*I gene. This is one of the genes that are responsible for maintaining the phage in the integrated state. If inactivated by a mutation, the *c*I gene no longer functions correctly and the switch to lysis occurs. In the *c*I*ts* mutation, the *c*I gene is functional at 30°C, at which temperature normal lysogeny can occur. But at 42°C, the *c*I*ts* gene product does not work properly, and lysogeny cannot be maintained. A culture of *E. coli* infected with a λ phages carrying the *c*I*ts* mutation can therefore be induced to produce extracellular phages by transferring from 30°C to 42°C (Figure 3.19).

3.3.2 Preparation of non-lysogenic λ phages

Although most λ strains are lysogenic, many cloning vectors derived from λ are modified, by deletions of the *c*I and other genes, so that lysogeny never occurs. These phages cannot integrate into the bacterial genome and can infect cells only by a lytic cycle (p. 18).

With these phages the key to obtaining a high titer lies in the way in which the culture is grown, in particular the stage at which the cells are infected by adding phage particles. If phages are added before the cells are dividing at their maximal rate, then all the cells are lysed very quickly, resulting in a low titer (Figure 3.20a). On the other hand, if the cell density is too high when the phages are added, then the culture will never be completely lysed, and again the phage titer will be low (Figure 3.20b). The ideal situation is when the age of the culture, and the size of the phage inoculum, are balanced such that the culture continues to grow, but eventually all the cells are infected

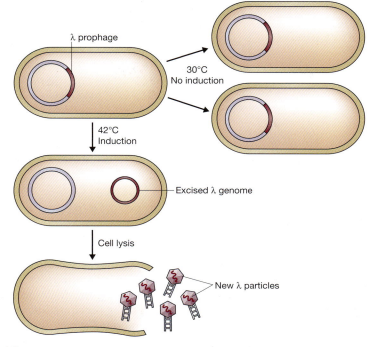

Figure 3.19
Induction of a λ *clts* lysogen by transferring from 30°C to 42°C.

Bacteria

All the cells are rapidly
lysed = low phage titer

(a) Culture density is too low

Culture is never completely
lysed = low phage titer

(b) Culture density is too high

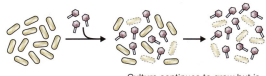

Culture continues to grow but is
eventually lysed = high phage titer

(c) Culture density is just right

Figure 3.20
Achieving the right balance between culture age
and inoculum size when preparing a sample of a
non-lysogenic phage.

Figure 3.21

Collection of phage particles by polyethylene glycol (PEG) precipitation.

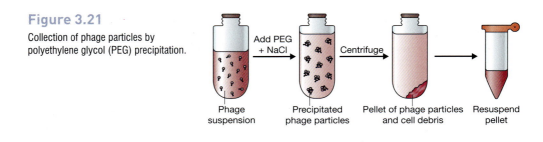

Phage suspension — Add PEG + NaCl → Precipitated phage particles — Centrifuge → Pellet of phage particles and cell debris — Resuspend pellet

and lysed (Figure 3.20c). As can be imagined, skill and experience are needed to judge the matter to perfection.

3.3.3 Collection of phages from an infected culture

The remains of lysed bacterial cells, along with any intact cells that are inadvertently left over, can be removed from an infected culture by centrifugation, leaving the phage particles in suspension (see Figure 3.18). The problem now is to reduce the size of the suspension to 5 ml or less, a manageable size for DNA extraction.

Phage particles are so small that they are pelleted only by very high speed centrifugation. Collection of phages is therefore usually achieved by precipitation with **polyethylene glycol (PEG)**. This is a long-chain polymeric compound which, in the presence of salt, absorbs water, thereby causing macromolecular assemblies such as phage particles to precipitate. The precipitate can then be collected by centrifugation, and redissolved in a suitably small volume (Figure 3.21).

3.3.4 Purification of DNA from λ phage particles

Deproteinization of the redissolved PEG precipitate is sometimes sufficient to extract pure phage DNA, but usually λ phages are subjected to an intermediate purification step. This is necessary because the PEG precipitate also contains a certain amount of bacterial debris, possibly including unwanted cellular DNA. These contaminants can be separated from the λ particles by CsCl density gradient centrifugation. The λ particles band in a CsCl gradient at 1.45–1.50 g/cm^3 (Figure 3.22), and can be withdrawn from

Figure 3.22

Purification of λ phage particles by CsCl density gradient centrifugation.

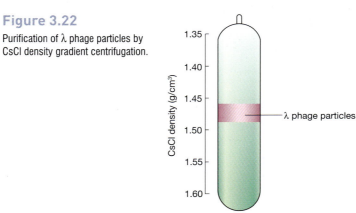

CsCl density (g/cm^3)

1.35
1.40
1.45
1.50
1.55
1.60

λ phage particles

the gradient as described previously for DNA bands (see Figure 3.16). Removal of CsCl by dialysis leaves a pure phage preparation from which the DNA can be extracted by either phenol or protease treatment to digest the phage protein coat.

3.3.5 Purification of M13 DNA causes few problems

Most of the differences between the M13 and λ infection cycles are to the advantage of the molecular biologist wishing to prepare M13 DNA. First, the double-stranded replicative form of M13 (p. 23), which behaves like a high copy number plasmid, is very easily purified by the standard procedures for plasmid preparation. A cell extract is prepared from cells infected with M13, and the replicative form separated from bacterial DNA by, for example, EtBr–CsCl density gradient centrifugation.

However, the single-stranded form of the M13 genome, contained in the extracellular phage particles, is frequently required. In this respect the big advantage compared with λ is that high titers of M13 are very easy to obtain. As infected cells continually secrete M13 particles into the medium (see Figure 2.8), with lysis never occurring, a high M13 titer is achieved simply by growing the infected culture to a high cell density. In fact, titers of 10^{12} per ml and above are quite easy to obtain without any special tricks being used. Such high titers mean that significant amounts of single-stranded M13 DNA can be prepared from cultures of small volume—5 ml or less. Furthermore, as the infected cells are not lysed, there is no problem with cell debris contaminating the phage suspension. Consequently the CsCl density gradient centrifugation step, needed for λ phage preparation, is rarely required with M13.

In summary, single-stranded M13 DNA preparation involves growth of a small volume of infected culture, centrifugation to pellet the bacteria, precipitation of the phage particles with PEG, phenol extraction to remove the phage protein coats, and ethanol precipitation to concentrate the resulting DNA (Figure 3.23).

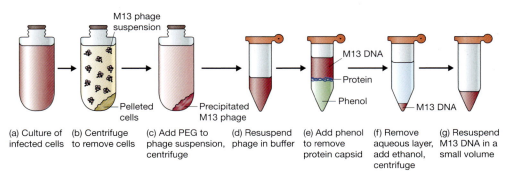

Figure 3.23

Preparation of single-stranded M13 DNA from an infected culture of bacteria.

Further reading

Birnboim, H.C. & Doly, J. (1979) A rapid alkaline extraction procedure for screening recombinant plasmid DNA. *Nucleic Acids Research*, 7, 1513–1523. [A method for preparing plasmid DNA.]

Boom, R., Sol, C.J.A., Salimans, M.M.M., Jansen, C.L., Wertheim van Dillen, P.M.E. & van der Noordaa, J. (1990) Rapid and simple method for purification of nucleic acids. *Journal of Clinical Microbiology*, 28, 495–503. [The guanidinium thiocyanate and silica method for DNA purification.]

Clewell, D.B. (1972) Nature of ColE1 plasmid replication in *Escherichia coli* in the presence of chloramphenicol. *Journal of Bacteriology*, 110, 667–676. [The biological basis of plasmid amplification.]

Marmur, J. (1961) A procedure for the isolation of deoxyribonucleic acid from microorganisms. *Journal of Molecular Biology*, 3, 208–218. [Genomic DNA preparation.]

Radloff, R., Bauer, W. & Vinograd, J. (1967) A dye-buoyant-density method for the detection and isolation of closed-circular duplex DNA. *Proceedings of the National Academy of Sciences of the USA*, 57, 1514–1521. [The original description of ethidium bromide density gradient centrifugation.]

Rogers, S.O. & Bendich, A.J. (1985) Extraction of DNA from milligram amounts of fresh, herbarium and mummified plant tissues. *Plant Molecular Biology*, 5, 69–76. [The CTAB method.]

Yamamoto, K.R., Alberts, B.M., Benzinger, R., Lawhorne, L. & Trieber, G. (1970) Rapid bacteriophage sedimentation in the presence of polyethylene glycol and its application to large scale virus preparation. *Virology*, 40, 734–744. [Preparation of λ DNA.]

Zinder, N.D. & Boeke, J.D. (1982) The filamentous phage (Ff) as vectors for recombinant DNA. *Gene*, 19, 1–10. [Methods for phage growth and DNA preparation.]

Chapter 4

Manipulation of Purified DNA

Chapter contents

Once pure samples of DNA have been prepared, the next step in a gene cloning experiment is construction of the recombinant DNA molecule (see Figure 1.1). To produce this recombinant molecule, the vector, as well as the DNA to be cloned, must be cut at specific points and then joined together in a controlled manner. Cutting and joining are two examples of DNA manipulative techniques, a wide variety of which have been developed over the past few years. As well as being cut and joined, DNA molecules can be shortened, lengthened, copied into RNA or into new DNA molecules, and modified by the addition or removal of specific chemical groups. These manipulations, all of which can be carried out in the test tube, provide the foundation not only for gene cloning, but also for studies of DNA biochemistry, gene structure, and the control of gene expression.

Almost all DNA manipulative techniques make use of purified enzymes. Within the cell these enzymes participate in essential processes such as DNA replication and transcription, breakdown of unwanted or foreign DNA (e.g., invading virus DNA), repair of mutated DNA, and recombination between different DNA molecules. After purification from cell extracts, many of these enzymes can be persuaded to carry out their natural reactions, or something closely related to them, under artificial conditions. Although these enzymatic reactions are often straightforward, most are absolutely impossible to perform by standard chemical methods. Purified enzymes are therefore crucial to genetic engineering and an important industry has sprung up around their preparation, characterization, and marketing. Commercial suppliers of high purity enzymes provide an essential service to the molecular biologist.

Gene Cloning and DNA Analysis: An Introduction. 6th edition. By T.A. Brown. Published 2010 by Blackwell Publishing.

The cutting and joining manipulations that underlie gene cloning are carried out by enzymes called restriction endonucleases (for cutting) and ligases (for joining). Most of this chapter will be concerned with the ways in which these two types of enzyme are used. First, however, we must consider the whole range of DNA manipulative enzymes, to see exactly what types of reaction can be performed. Many of these enzymes will be mentioned in later chapters when procedures that make use of them are described.

4.1 The range of DNA manipulative enzymes

DNA manipulative enzymes can be grouped into four broad classes, depending on the type of reaction that they catalyze:

- Nucleases are enzymes that cut, shorten, or degrade nucleic acid molecules.
- Ligases join nucleic acid molecules together.
- Polymerases make copies of molecules.
- Modifying enzymes remove or add chemical groups.

Before considering in detail each of these classes of enzyme, two points should be made. The first is that, although most enzymes can be assigned to a particular class, a few display multiple activities that span two or more classes. Most importantly, many polymerases combine their ability to make new DNA molecules with an associated DNA degradative (i.e., nuclease) activity.

Second, it should be appreciated that, as well as the DNA manipulative enzymes, many similar enzymes able to act on RNA are known. The ribonuclease used to remove contaminating RNA from DNA preparations (p. 30) is an example of such an enzyme. Although some RNA manipulative enzymes have applications in gene cloning and will be mentioned in later chapters, we will in general restrict our thoughts to those enzymes that act on DNA.

4.1.1 Nucleases

Nucleases degrade DNA molecules by breaking the phosphodiester bonds that link one nucleotide to the next in a DNA strand. There are two different kinds of nuclease (Figure 4.1):

- Exonucleases remove nucleotides one at a time from the end of a DNA molecule.
- Endonucleases are able to break internal phosphodiester bonds within a DNA molecule.

The main distinction between different exonucleases lies in the number of strands that are degraded when a double-stranded molecule is attacked. The enzyme called Bal31 (purified from the bacterium *Alteromonas espejiana*) is an example of an exonuclease that removes nucleotides from both strands of a double-stranded molecule (Figure 4.2a). The greater the length of time that Bal31 is allowed to act on a group of DNA molecules, the shorter the resulting DNA fragments will be. In contrast, enzymes such as *E. coli* exonuclease III degrade just one strand of a double-stranded molecule, leaving single-stranded DNA as the product (Figure 4.2b).

The same criterion can be used to classify endonucleases. S1 endonuclease (from the fungus *Aspergillus oryzae*) only cleaves single strands (Figure 4.3a), whereas deoxyribonuclease I (DNase I), which is prepared from cow pancreas, cuts both single- and double-stranded molecules (Figure 4.3b). DNase I is non-specific in that it attacks DNA at any internal phosphodiester bond, so the end result of prolonged DNase I

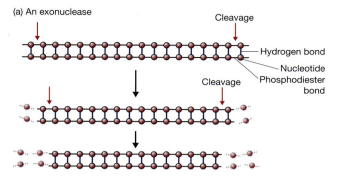

(a) An exonuclease

Figure 4.1

The reactions catalyzed by the two different kinds of nuclease. (a) An exonuclease, which removes nucleotides from the end of a DNA molecule. (b) An endonuclease, which breaks internal phosphodiester bonds.

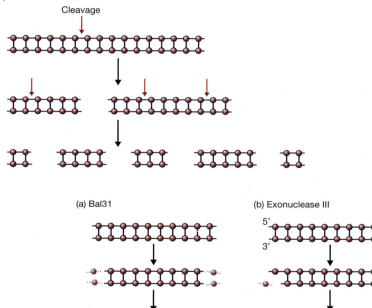

Figure 4.2

The reactions catalyzed by different types of exonuclease. (a) Bal31, which removes nucleotides from both strands of a double-stranded molecule. (b) Exonuclease III, which removes nucleotides only from the 3′ terminus.

action is a mixture of mononucleotides and very short oligonucleotides. On the other hand, the special group of enzymes called restriction endonucleases cleave double-stranded DNA only at a limited number of specific recognition sites (Figure 4.3c). These important enzymes are described in detail on p. 50.

4.1.2 Ligases

In the cell the function of DNA ligase is to repair single-stranded breaks ("discontinuities") that arise in double-stranded DNA molecules during, for example, DNA replication. DNA ligases from most organisms can also join together two individual fragments of

Figure 4.3

The reactions catalyzed by different types of endonuclease. (a) S1 nuclease, which cleaves only single-stranded DNA, including single-stranded nicks in mainly double-stranded molecules. (b) DNase I, which cleaves both single- and double-stranded DNA. (c) A restriction endonuclease, which cleaves double-stranded DNA, but only at a limited number of sites.

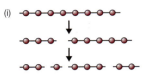

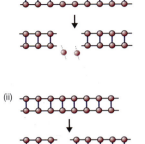

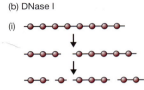

Figure 4.4

The two reactions catalyzed by DNA ligase. (a) Repair of a discontinuity—a missing phosphodiester bond in one strand of a double-stranded molecule. (b) Joining two molecules together.

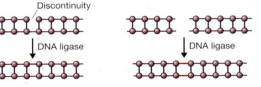

double-stranded DNA (Figure 4.4). The role of these enzymes in construction of recombinant DNA molecules is described on p. 63.

4.1.3 Polymerases

DNA polymerases are enzymes that synthesize a new strand of DNA complementary to an existing DNA or RNA template (Figure 4.5a). Most polymerases can function only if the template possesses a double-stranded region that acts as a primer for initiation of polymerization.

Four types of DNA polymerase are used routinely in genetic engineering. The first is DNA polymerase I, which is usually prepared from *E. coli*. This enzyme attaches to a short single-stranded region (or nick) in a mainly double-stranded DNA molecule, and then synthesizes a completely new strand, degrading the existing strand as it proceeds (Figure 4.5b). DNA polymerase I is therefore an example of an enzyme with a dual activity—DNA polymerization and DNA degradation.

The polymerase and nuclease activities of DNA polymerase I are controlled by different parts of the enzyme molecule. The nuclease activity is contained in the first 323 amino acids of the polypeptide, so removal of this segment leaves a modified enzyme that retains the polymerase function but is unable to degrade DNA. This modified

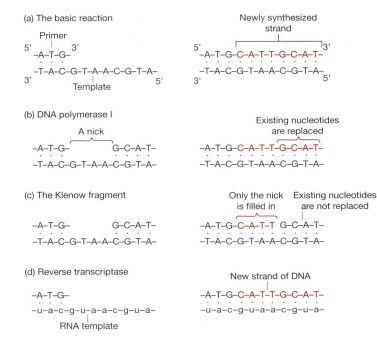

Figure 4.5

The reactions catalyzed by DNA polymerases. (a) The basic reaction: a new DNA strand is synthesized in the 5′ to 3′ direction. (b) DNA polymerase I, which initially fills in nicks but then continues to synthesize a new strand, degrading the existing one as it proceeds. (c) The Klenow fragment, which only fills in nicks. (d) Reverse transcriptase, which uses a template of RNA.

enzyme, called the Klenow fragment, can still synthesize a complementary DNA strand on a single-stranded template, but as it has no nuclease activity it cannot continue the synthesis once the nick is filled in (Figure 4.5c). Several other enzymes—natural polymerases and modified versions—have similar properties to the Klenow fragment. The major application of these polymerases is in DNA sequencing (p. 166).

The *Taq* DNA polymerase used in the polymerase chain reaction (PCR) (see Figure 1.2) is the DNA polymerase I enzyme of the bacterium *Thermus aquaticus*. This organism lives in hot springs, and many of its enzymes, including the *Taq* DNA polymerase, are thermostable, meaning that they are resistant to denaturation by heat treatment. This is the special feature of *Taq* DNA polymerase that makes it suitable for PCR, because if it was not thermostable it would be inactivated when the temperature of the reaction is raised to 94°C to denature the DNA.

The final type of DNA polymerase that is important in genetic engineering is reverse transcriptase, an enzyme involved in the replication of several kinds of virus. Reverse transcriptase is unique in that it uses as a template not DNA but RNA (Figure 4.5d). The ability of this enzyme to synthesize a DNA strand complementary to an RNA template is central to the technique called complementary DNA (cDNA) cloning (p. 133).

4.1.4 DNA modifying enzymes

There are numerous enzymes that modify DNA molecules by addition or removal of specific chemical groups. The most important are as follows:

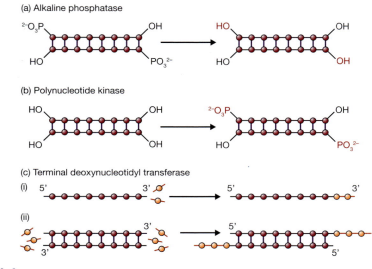

Figure 4.6

The reactions catalyzed by DNA modifying enzymes. (a) Alkaline phosphatase, which removes 5′-phosphate groups. (b) Polynucleotide kinase, which attaches 5′-phosphate groups. (c) Terminal deoxynucleotidyl transferase, which attaches deoxyribonucleotides to the 3′ termini of polynucleotides in either (i) single-stranded or (ii) double-stranded molecules.

- **Alkaline phosphatase** (from *E. coli*, calf intestinal tissue, or arctic shrimp), which removes the phosphate group present at the **5′ terminus** of a DNA molecule (Figure 4.6a).
- **Polynucleotide kinase** (from *E. coli* infected with T4 phage), which has the reverse effect to alkaline phosphatase, adding phosphate groups onto free 5′ termini (Figure 4.6b).
- **Terminal deoxynucleotidyl transferase** (from calf thymus tissue), which adds one or more deoxyribonucleotides onto the **3′ terminus** of a DNA molecule (Figure 4.6c).

4.2 Enzymes for cutting DNA—restriction endonucleases

Gene cloning requires that DNA molecules be cut in a very precise and reproducible fashion. This is illustrated by the way in which the vector is cut during construction of a recombinant DNA molecule (Figure 4.7a). Each vector molecule must be cleaved at a single position, to open up the circle so that new DNA can be inserted: a molecule that is cut more than once will be broken into two or more separate fragments and will be of no use as a cloning vector. Furthermore, each vector molecule must be cut at exactly the same position on the circle—as will become apparent in later chapters, random cleavage is not satisfactory. It should be clear that a very special type of nuclease is needed to carry out this manipulation.

Often it is also necessary to cleave the DNA that is to be cloned (Figure 4.7b). There are two reasons for this. First, if the aim is to clone a single gene, which may consist of only 2 or 3 kb of DNA, then that gene will have to be cut out of the large (often greater than 80 kb) DNA molecules produced by skilfull use of the preparative techniques

(a) Vector molecules

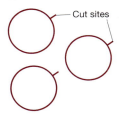
Cut sites

Each vector molecule must be cut
once, each at the same position

Figure 4.7

The need for very precise cutting manipulations in
a gene cloning experiment.

(b) The DNA molecule containing the gene to be cloned

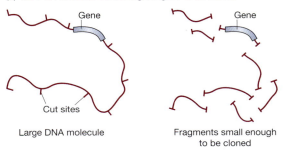

Gene   Gene

Cut sites

Large DNA molecule   Fragments small enough
                            to be cloned

described in Chapter 3. Second, large DNA molecules may have to be broken down
simply to produce fragments small enough to be carried by the vector. Most cloning
vectors exhibit a preference for DNA fragments that fall into a particular size range:
most plasmid-based vectors, for example, are very inefficient at cloning DNA molecules
more than 8 kb in length.

Purified restriction endonucleases allow the molecular biologist to cut DNA molecules in the precise, reproducible manner required for gene cloning. The discovery of
these enzymes, which led to Nobel Prizes for W. Arber, H. Smith, and D. Nathans in
1978, was one of the key breakthroughs in the development of genetic engineering.

4.2.1 The discovery and function of restriction endonucleases

The initial observation that led to the eventual discovery of restriction endonucleases was
made in the early 1950s, when it was shown that some strains of bacteria are immune
to bacteriophage infection, a phenomenon referred to as host-controlled restriction.

The mechanism of restriction is not very complicated, even though it took over
20 years to be fully understood. Restriction occurs because the bacterium produces
an enzyme that degrades the phage DNA before it has time to replicate and direct synthesis of new phage particles (Figure 4.8a). The bacterium's own DNA, the destruction
of which would of course be lethal, is protected from attack because it carries additional methyl groups that block the degradative enzyme action (Figure 4.8b).

These degradative enzymes are called restriction endonucleases and are synthesized
by many, perhaps all, species of bacteria: over 2500 different ones have been isolated
and more than 300 are available for use in the laboratory. Three different classes of
restriction endonuclease are recognized, each distinguished by a slightly different mode
of action. Types I and III are rather complex and have only a limited role in genetic

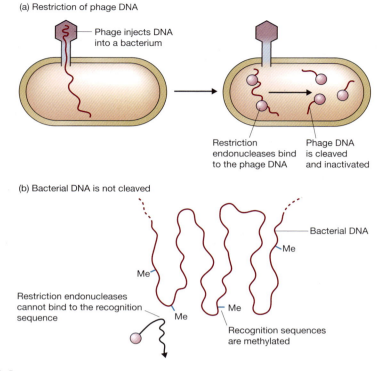

Figure 4.8

The function of a restriction endonuclease in a bacterial cell: (a) phage DNA is cleaved, but (b) bacterial DNA is not.

engineering. Type II restriction endonucleases, on the other hand, are the cutting enzymes that are so important in gene cloning.

4.2.2 Type II restriction endonucleases cut DNA at specific nucleotide sequences

The central feature of type II restriction endonucleases (which will be referred to simply as "restriction endonucleases" from now on) is that each enzyme has a specific recognition sequence at which it cuts a DNA molecule. A particular enzyme cleaves DNA at the recognition sequence and nowhere else. For example, the restriction endonuclease called *Pvu*I (isolated from *Proteus vulgaris*) cuts DNA only at the hexanucleotide CGATCG. In contrast, a second enzyme from the same bacterium, called *Pvu*II, cuts at a different hexanucleotide, in this case CAGCTG.

Many restriction endonucleases recognize hexanucleotide target sites, but others cut at four, five, eight, or even longer nucleotide sequences. *Sau*3A (from *Staphylococcus aureus* strain 3A) recognizes GATC, and *Alu*I (*Arthrobacter luteus*) cuts at AGCT. There are also examples of restriction endonucleases with degenerate recognition sequences, meaning that they cut DNA at any one of a family of related sites. *Hin*fI (*Haemophilus influenzae* strain R$_f$), for instance, recognizes GANTC, so cuts at GAATC, GATTC, GAGTC, and GACTC. The recognition sequences for some of the most frequently used restriction endonucleases are listed in Table 4.1.

Table 4.1

The recognition sequences for some of the most frequently used restriction endonucleases.

ENZYME	ORGANISM	RECOGNITION SEQUENCE*	BLUNT OR STICKY END
EcoRI	Escherichia coli	GAATTC	Sticky
BamHI	Bacillus amyloliquefaciens	GGATCC	Sticky
BglII	Bacillus globigii	AGATCT	Sticky
PvuI	Proteus vulgaris	CGATCG	Sticky
PvuII	Proteus vulgaris	CAGCTG	Blunt
HindIII	Haemophilus influenzae R_d	AAGCTT	Sticky
HinfI	Haemophilus influenzae R_f	GANTC	Sticky
Sau3A	Staphylococcus aureus	GATC	Sticky
AluI	Arthrobacter luteus	AGCT	Blunt
TaqI	Thermus aquaticus	TCGA	Sticky
HaeIII	Haemophilus aegyptius	GGCC	Blunt
NotI	Nocardia otitidis-caviarum	GCGGCCGC	Sticky
SfiI	Streptomyces fimbriatus	GGCCNNNNNGGCC	Sticky

*The sequence shown is that of one strand, given in the 5′ to 3′ direction. "N" indicates any nucleotide. Note that almost all recognition sequences are palindromes: when both strands are considered they read the same in each direction, for example:

```
        5′–GAATTC–3′
EcoRI   ||||||
        3′–CTTAAG–5′
```

4.2.3 Blunt ends and sticky ends

The exact nature of the cut produced by a restriction endonuclease is of considerable importance in the design of a gene cloning experiment. Many restriction endonucleases make a simple double-stranded cut in the middle of the recognition sequence (Figure 4.9a), resulting in a blunt end or flush end. PvuII and AluI are examples of blunt end cutters.

Other restriction endonucleases cut DNA in a slightly different way. With these enzymes the two DNA strands are not cut at exactly the same position. Instead the cleavage is staggered, usually by two or four nucleotides, so that the resulting DNA fragments have short single-stranded overhangs at each end (Figure 4.9b). These are called sticky or cohesive ends, as base pairing between them can stick the DNA molecule back together again (recall that sticky ends were encountered on p. 20 during the description of λ phage replication). One important feature of sticky end enzymes is that restriction endonucleases with different recognition sequences may produce the same sticky ends. BamHI (recognition sequence GGATCC) and BglII (AGATCT) are examples—both produce GATC sticky ends (Figure 4.9c). The same sticky end is also produced by Sau3A, which recognizes only the tetranucleotide GATC. Fragments of DNA produced by cleavage with either of these enzymes can be joined to each other, as each fragment carries a complementary sticky end.

4.2.4 The frequency of recognition sequences in a DNA molecule

The number of recognition sequences for a particular restriction endonuclease in a DNA molecule of known length can be calculated mathematically. A tetranucleotide sequence (e.g., GATC) should occur once every $4^4 = 256$ nucleotides, and a hexanucleotide

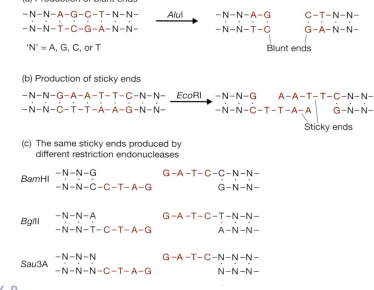

Figure 4.9

The ends produced by cleavage of DNA with different restriction endonucleases. (a) A blunt end produced by *Alu*I. (b) A sticky end produced by *Eco*RI. (c) The same sticky ends produced by *Bam*HI, *Bgl*II and *Sau*3A.

(e.g., GGATCC) once every $4^6 = 4096$ nucleotides. These calculations assume that the nucleotides are ordered in a random fashion and that the four different nucleotides are present in equal proportions (i.e., the GC content = 50%). In practice, neither of these assumptions is entirely valid. For example, the λ DNA molecule, at 49 kb, should contain about 12 sites for a restriction endonuclease with a hexanucleotide recognition sequence. In fact, many of these recognition sites occur less frequently (e.g., six for *Bgl*II, five for *Bam*HI, and only two for *Sal*I), a reflection of the fact that the GC content for λ is rather less than 50% (Figure 4.10a).

Furthermore, restriction sites are generally not evenly spaced out along a DNA molecule. If they were, then digestion with a particular restriction endonuclease would give fragments of roughly equal sizes. Figure 4.10b shows the fragments produced by cutting λ DNA with *Bgl*II, *Bam*HI, and *Sal*I. In each case there is a considerable spread of fragment sizes, indicating that in λ DNA the nucleotides are not randomly ordered.

The lesson to be learned from Figure 4.10 is that although mathematics may give an idea of how many restriction sites to expect in a given DNA molecule, only experimental analysis can provide the true picture. We must therefore move on to consider how restriction endonucleases are used in the laboratory.

4.2.5 Performing a restriction digest in the laboratory

As an example, we will consider how to digest a sample of λ DNA (concentration 125 mg/ml) with *Bgl*II.

First, the required amount of DNA must be pipetted into a test tube. The amount of DNA that will be restricted depends on the nature of the experiment. In this case we will digest 2 μg of λ DNA, which is contained in 16 μl of the sample (Figure 4.11a). Very accurate micropipettes will therefore be needed.

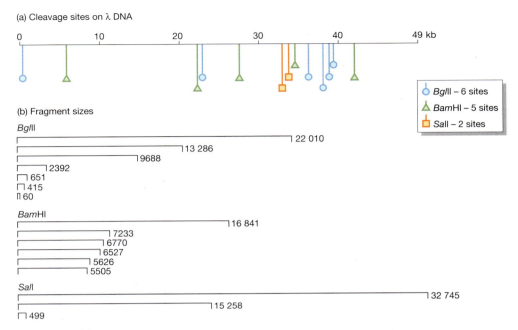

(a) Cleavage sites on λ DNA

Bg/II – 6 sites
BamHI – 5 sites
SalI – 2 sites

(b) Fragment sizes

Bg/II
22 010
13 286
9688
2392
651
415
60

BamHI
16 841
7233
6770
6527
5626
5505

SalI
32 745
15 258
499

Figure 4.10

Restriction of the λ DNA molecule. (a) The positions of the recognition sequences for *Bg/II*, *BamHI*, and *SalI*. (b) The fragments produced by cleavage with each of these restriction endonucleases. The numbers are the fragment sizes in base pairs.

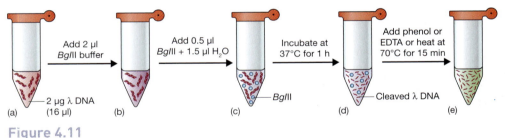

Add 2 µl
Bg/II buffer

Add 0.5 µl
Bg/II + 1.5 µl H₂O

Incubate at
37°C for 1 h

Add phenol or
EDTA or heat at
70°C for 15 min

2 µg λ DNA
(16 µl)

Bg/II

Cleaved λ DNA

(a)   (b)   (c)   (d)   (e)

Figure 4.11

Performing a restriction digest in the laboratory.

The other main component in the reaction will be the restriction endonuclease, obtained from a commercial supplier as a pure solution of known concentration. But before adding the enzyme, the solution containing the DNA must be adjusted to provide the correct conditions to ensure maximal activity of the enzyme. Most restriction endonucleases function adequately at pH 7.4, but different enzymes vary in their requirements for ionic strength (usually provided by sodium chloride (NaCl)) and magnesium (Mg^{2+}) concentration (all type II restriction endonucleases require Mg^{2+} in order to function). It is also advisable to add a reducing agent, such as dithiothreitol (DTT), which stabilizes the enzyme and prevents its inactivation. Providing the right conditions for the enzyme is very important—incorrect NaCl or Mg^{2+} concentrations not only decrease the activity of the restriction endonuclease, they might also cause

Table 4.2

A 10 × buffer suitable for restriction of DNA with *Bgl*II.

COMPONENT	CONCENTRATION (mM)
Tris–HCl, pH 7.4	500
MgCl$_2$	100
NaCl	500
Dithiothreitol	10

changes in the specificity of the enzyme, so that DNA cleavage occurs at additional, non-standard recognition sequences.

The composition of a suitable buffer for *Bgl*II is shown in Table 4.2. This buffer is ten times the working concentration, and is diluted by being added to the reaction mixture. In our example, a suitable final volume for the reaction mixture would be 20 μl, so we add 2 μl of 10 × *Bgl*II buffer to the 16 μl of DNA already present (Figure 4.11b).

The restriction endonuclease can now be added. By convention, 1 unit of enzyme is defined as the quantity needed to cut 1 μg of DNA in 1 hour, so we need 2 units of *Bgl*II to cut 2 μg of λ DNA. *Bgl*II is frequently obtained at a concentration of 4 units/μl, so 0.5 μl is sufficient to cleave the DNA. The final ingredients in the reaction mixture are therefore 0.5 μl *Bgl*II + 1.5 μl water, giving a final volume of 20 μl (Figure 4.11c).

The last factor to consider is incubation temperature. Most restriction endonucleases, including *Bgl*II, work best at 37°C, but a few have different requirements. *Taq*I, for example, is a restriction enzyme from *Thermus aquaticus* and, like *Taq* DNA polymerase, has a high working temperature. Restriction digests with *Taq*I must be incubated at 65°C to obtain maximum enzyme activity.

After 1 hour the restriction should be complete (Figure 4.11d). If the DNA fragments produced by restriction are to be used in cloning experiments, the enzyme must somehow be destroyed so that it does not accidentally digest other DNA molecules that may be added at a later stage. There are several ways of "killing" the enzyme. For many a short incubation at 70°C is sufficient, for others phenol extraction or the addition of ethylenediamine tetraacetate (EDTA), which binds Mg^{2+} ions preventing restriction endonuclease action, is used (Figure 4.11e).

4.2.6 Analyzing the result of restriction endonuclease cleavage

A restriction digest results in a number of DNA fragments, the sizes of which depend on the exact positions of the recognition sequences for the endonuclease in the original molecule (see Figure 4.10). A way of determining the number and sizes of the fragments is needed if restriction endonucleases are to be of use in gene cloning. Whether or not a DNA molecule is cut at all can be determined fairly easily by testing the viscosity of the solution. Larger DNA molecules result in a more viscous solution than smaller ones, so cleavage is associated with a decrease in viscosity. However, working out the number and sizes of the individual cleavage products is more difficult. In fact, for several years this was one of the most tedious aspects of experiments involving DNA. Eventually the problems were solved in the early 1970s when the technique of gel electrophoresis was developed.

(a) Standard electrophoresis

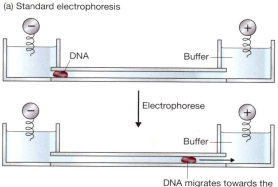

(b) Gel electrophoresis

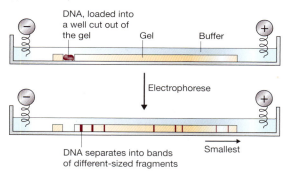

Figure 4.12
(a) Standard electrophoresis does not separate DNA fragments of different sizes, whereas (b) gel electrophoresis does.

Separation of molecules by gel electrophoresis

Electrophoresis, like ion-exchange chromatography (see p. 30), is a technique that uses differences in electrical charge to separate the molecules in a mixture. DNA molecules have negative charges, and so when placed in an electric field they migrate toward the positive pole (Figure 4.12a). The rate of migration of a molecule depends on two factors, its shape and its charge-to-mass ratio. Unfortunately, most DNA molecules are the same shape and all have very similar charge-to-mass ratios. Fragments of different sizes cannot therefore be separated by standard electrophoresis.

The size of the DNA molecule does, however, become a factor if the electrophoresis is performed in a gel. A gel, which is usually made of agarose, polyacrylamide, or a mixture of the two, comprises a complex network of pores, through which the DNA molecules must travel to reach the positive electrode. The smaller the DNA molecule, the faster it can migrate through the gel. Gel electrophoresis therefore separates DNA molecules according to their size (Figure 4.12b).

In practice the composition of the gel determines the sizes of the DNA molecules that can be separated. A 0.5 cm thick slab of 0.5% agarose, which has relatively large pores, would be used for molecules in the size range 1–30 kb, allowing, for example, molecules of 10 and 12 kb to be clearly distinguished. At the other end of the scale, a very thin (0.3 mm) 40% polyacrylamide gel, with extremely small pores, would be used to separate much smaller DNA molecules, in the range 1–300 bp, and could distinguish molecules differing in length by just a single nucleotide.

Figure 4.13

Visualizing DNA bands in an agarose gel by EtBr staining and ultraviolet (UV) irradiation.

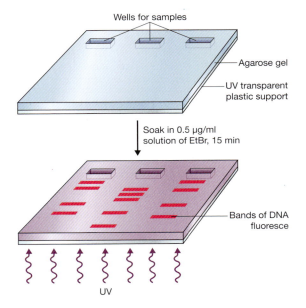

Wells for samples

Agarose gel

UV transparent plastic support

Soak in 0.5 µg/ml solution of EtBr, 15 min

Bands of DNA fluoresce

UV

Visualizing DNA molecules in an agarose gel

The easiest way to see the results of a gel electrophoresis experiment is to stain the gel with a compound that makes the DNA visible. Ethidium bromide (EtBr), already described on p. 37 as a means of visualizing DNA in caesium chloride gradients, is also routinely used to stain DNA in agarose and polyacrylamide gels (Figure 4.13). Bands showing the positions of the different size classes of DNA fragment are clearly visible under ultraviolet irradiation after EtBr staining, so long as sufficient DNA is present. Unfortunately, the procedure is very hazardous because ethidium bromide is a power-ful mutagen. EtBr staining also has limited sensitivity, and if a band contains less than about 10 ng of DNA then it might not be visible after staining.

For this reason, non-mutagenic dyes that stain DNA green, red, or blue are now used in many laboratories. Most of these dyes can be used either as a post-stain after electrophoresis, as illustrated in Figure 4.13 for EtBr, or alternatively, because they are non-hazardous, they can be included in the buffer solution in which the agarose or polyacrylamide is dissolved when the gel is prepared. Some of these dyes require ultraviolet irradiation in order to make the bands visible, but others are visualized by illumination at other wavelengths, for example under blue light, removing a second hazard as ultraviolet radiation can cause severe burns. The most sensitive dyes are able to detect bands that contain less than 1 ng DNA.

4.2.7 Estimation of the sizes of DNA molecules

Gel electrophoresis separates different sized DNA molecules, with the smallest molecules traveling the greatest distance toward the positive electrode. If several DNA fragments of varying sizes are present (the result of a successful restriction digest, for example), then a series of bands appears in the gel. How can the sizes of these fragments be determined?

The most accurate method is to make use of the mathematical relationship that links migration rate to molecular mass. The relevant formula is:

$$D = a - b(\log M)$$

(a) Rough estimation by eye

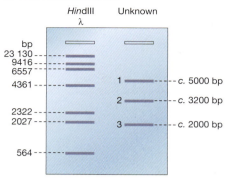

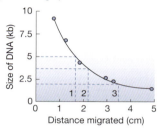

Figure 4.14

Estimation of the sizes of DNA fragments in an agarose gel.
(a) A rough estimate of fragment size can be obtained by eye.
(b) A more accurate measurement of fragment size is gained
by using the mobilities of the *Hind*III–λ fragments to construct
a calibration curve; the sizes of the unknown fragments can
then be determined from the distances they have migrated.

where D is the distance moved, M is the molecular mass, and a and b are constants that depend on the electrophoresis conditions.

Because extreme accuracy in estimating DNA fragment sizes is not always necessary, a much simpler though less precise method is more generally used. A standard restriction digest, comprising fragments of known size, is usually included in each electrophoresis gel that is run. Restriction digests of λ DNA are often used in this way as size markers. For example, *Hind*III cleaves λ DNA into eight fragments, ranging in size from 125 bp for the smallest to over 23 kb for the largest. As the sizes of the fragments in this digest are known, the fragment sizes in the experimental digest can be estimated by comparing the positions of the bands in the two tracks (Figure 4.14). Special mixtures of DNA fragments called **DNA ladders**, whose sizes are multiples of 100 bp or of 1 kb, can also be used as size markers. Although not precise, size estimation by comparison with DNA markers can be performed with as little as a 5% error, which is satisfactory for most purposes.

4.2.8 Mapping the positions of different restriction sites in a DNA molecule

So far we have considered how to determine the number and sizes of the DNA fragments produced by restriction endonuclease cleavage. The next step in **restriction analysis** is to construct a map showing the relative positions in the DNA molecule of the recognition sequences for a number of different enzymes. Only when a **restriction map** is available can the correct restriction endonucleases be selected for the particular cutting manipulation that is required (Figure 4.15).

Figure 4.15

Using a restriction map to work out which restriction endonucleases should be used to obtain DNA fragments containing individual genes.

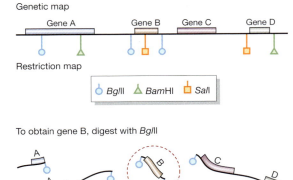

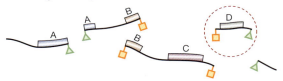

To construct a restriction map, a series of restriction digests must be performed. First, the number and sizes of the fragments produced by each restriction endonuclease must be determined by gel electrophoresis followed by comparison with size markers (Figure 4.16). This information must then be supplemented by a series of double digestions, in which the DNA is cut by two restriction endonucleases at once. It might be possible to perform a double digestion in one step if both enzymes have similar requirements for pH, Mg^{2+} concentration, etc. Alternatively, the two digestions may have to be carried out one after the other, adjusting the reaction mixture after the first digestion to provide a different set of conditions for the second enzyme.

Comparing the results of single and double digests will allow many, if not all, of the restriction sites to be mapped (Figure 4.16). Ambiguities can usually be resolved by partial digestion, carried out under conditions that result in cleavage of only a limited number of the restriction sites on any DNA molecule. Partial digestion is usually achieved by reducing the incubation period, so the enzyme does not have time to cut all the restriction sites, or by incubating at a low temperature (e.g., 4°C rather than 37°C), which limits the activity of the enzyme.

The result of a partial digestion is a complex pattern of bands in an electrophoresis gel. As well as the standard fragments, produced by total digestion, additional sizes are seen. These are molecules that comprise two adjacent restriction fragments, separated by a site that has not been cleaved. Their sizes indicate which restriction fragments in the complete digest are next to one another in the uncut molecule (Figure 4.16).

4.2.9 Special gel electrophoresis methods for separating larger molecules

During agarose gel electrophoresis, a DNA fragment migrates at a rate that is proportional to its size, but this relationship is not a direct one. The formula that links

Single and double digestions

Enzyme	Number of fragments	Sizes (kb)
*Xba*I	2	24.0, 24.5
*Xho*I	2	15.0, 33.5
*Kpn*I	3	1.5, 17.0, 30.0
*Xba*I + *Xho*I	3	9.0, 15.0, 24.5
*Xba*I + *Kpn*I	4	1.5, 6.0, 17.0, 24.0

Conclusions:

(1) As λ DNA is linear, the number of restriction sites for each enzyme is *Xba*I 1, *Xho*I 1, *Kpn*I 2.

(2) The *Xba*I and *Xho*I sites can be mapped:

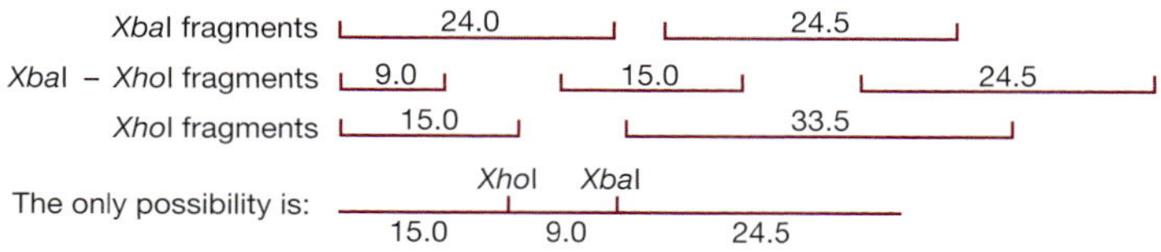

(3) All the *Kpn*I sites fall in the 24.5 kb *Xba*I fragment, as the 24.0 kb fragment is intact after *Xba*I–*Kpn*I double digestion. The order of the *Kpn*I fragments can be determined only by partial digestion.

Partial digestion

Enzyme	Fragment sizes (kb)
*Kpn*I – limiting conditions	1.5, 17.0, 18.5, 30.0, 31.5, 48.5

Conclusions:

(1) 48.5 kb fragment = uncut λ.
(2) 1.5, 17.0 and 30.0 kb fragments are products of complete digestion.
(3) 18.5 and 31.5 kb fragments are products of partial digestion.

The KpnI map must be:

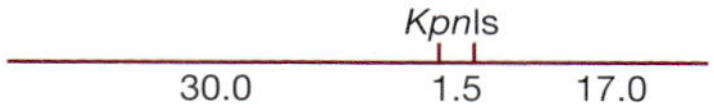

Therefore the complete map is:

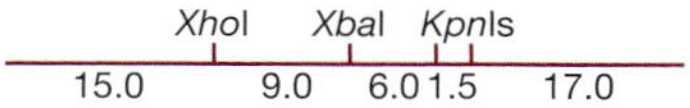

Figure 4.16

Restriction mapping. This example shows how the positions of the *Xba*I, *Xho*I and *Kpn*I sites on the λ DNA molecule can be determined.

migration rate to molecular mass has a logarithmic component (p. 59), which means that the difference in migration rates become increasingly small for larger molecules (Figure 4.17). In practice, molecules larger than about 50 kb cannot be resolved efficiently by standard gel electrophoresis.

This size limitation is not usually a problem when the restriction fragments being studied have been obtained by cutting the DNA with an enzyme with a tetranucleotide or hexanucleotide recognition sequence. Most, if not all, of the fragments produced in this way will be less than 30 kb in length and easily resolved by agarose gel electrophoresis. Difficulties might arise, however, if an enzyme with a longer recognition sequence is used, such as *Not*I, which cuts at an eight-nucleotide sequence (see Table 4.1). *Not*I would be expected, on average, to cut a DNA molecule once every

Figure 4.17

The influence of DNA size on migration rate during conventional gel electrophoresis.

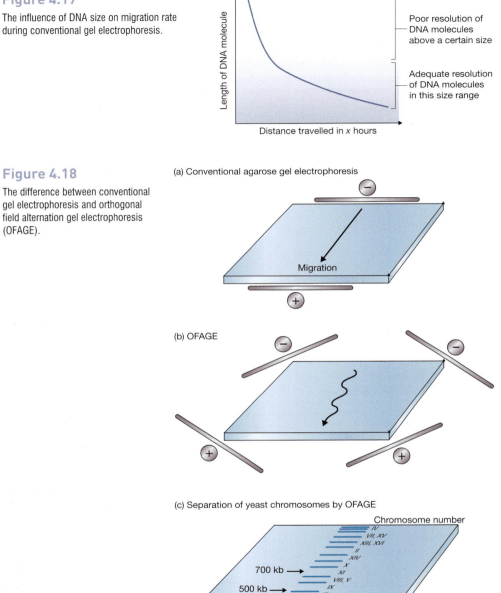

(a) Conventional agarose gel electrophoresis

Figure 4.18

The difference between conventional gel electrophoresis and orthogonal field alternation gel electrophoresis (OFAGE).

(b) OFAGE

(c) Separation of yeast chromosomes by OFAGE

$4^8 = 65,536$ bp. It is therefore unlikely that *Not*I fragments will be separated by standard gel electrophoresis.

The limitations of standard gel electrophoresis can be overcome if a more complex electric field is used. Several different systems have been designed, but the principle is best illustrated by **orthogonal field alternation gel electrophoresis (OFAGE)**. Instead of being applied directly along the length of the gel, as in the standard method (Figure 4.18a), the electric field now alternates between two pairs of electrodes, each pair set at an angle of 45° to the length of the gel (Figure 4.18b). The result is a pulsed field, with the DNA molecules in the gel having continually to change direction in accordance with the pulses. As the two fields alternate in a regular fashion, the net movement of the DNA molecules in the gel is still from one end to the other, in more or less a straight line. However, with every change in field direction, each DNA molecule has to realign through 90° before its migration can continue. This is the key point, because a short molecule can realign faster than a long one, allowing the short molecule to progress toward the bottom of the gel more quickly. This added dimension increases the resolving power of the gel quite dramatically, so that molecules up to several thousand kilobases in length can be separated.

This size range includes not only restriction fragments but also the intact chromosomal molecules of many lower eukaryotes, including yeast, several important filamentous fungi, and protozoans such as the malaria parasite *Plasmodium falciparum*. OFAGE and related techniques such as contour clamped homogeneous electric fields (CHEF) and field inversion gel electrophoresis (FIGE) can therefore be used to prepare gels showing the separated chromosomes of these organisms (Figure 4.18c), enabling DNA from these individual chromosomes to be purified.

4.3 Ligation – joining DNA molecules together

The final step in construction of a recombinant DNA molecule is the joining together of the vector molecule and the DNA to be cloned (Figure 4.19). This process is referred to as ligation, and the enzyme that catalyzes the reaction is called DNA ligase.

4.3.1 The mode of action of DNA ligase

All living cells produce DNA ligases, but the enzyme used in genetic engineering is usually purified from *E. coli* bacteria that have been infected with T4 phage. Within the cell the enzyme carries out the very important function of repairing any discontinuities that may arise in one of the strands of a double-stranded molecule (see Figure 4.4a). A discontinuity is quite simply a position where a phosphodiester bond between adjacent nucleotides is missing (contrast this with a nick, where one or more nucleotides are absent). Although discontinuities may arise by chance breakage of the cell's DNA molecules, they are also a natural result of processes such as DNA replication and recombination. Ligases therefore play several vital roles in the cell.

In the test tube, purified DNA ligases, as well as repairing single-strand discontinuities, can also join together individual DNA molecules or the two ends of the same molecule. The chemical reaction involved in ligating two molecules is exactly the same as discontinuity repair, except that two phosphodiester bonds must be made, one for each strand (Figure 4.20a).

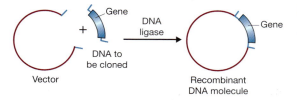

Figure 4.19

Ligation: the final step in construction of a recombinant DNA molecule.

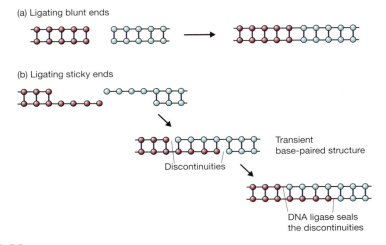

(a) Ligating blunt ends

(b) Ligating sticky ends

Transient
base-paired structure

Discontinuities

DNA ligase seals
the discontinuities

Figure 4.20

The different joining reactions catalysed by DNA ligase: (a) ligation of blunt-ended molecules; (b) ligation of sticky-ended molecules.

4.3.2 Sticky ends increase the efficiency of ligation

The ligation reaction in Figure 4.20a shows two blunt-ended fragments being joined together. Although this reaction can be carried out in the test tube, it is not very efficient. This is because the ligase is unable to "catch hold" of the molecule to be ligated, and has to wait for chance associations to bring the ends together. If possible, blunt end ligation should be performed at high DNA concentrations, to increase the chances of the ends of the molecules coming together in the correct way.

In contrast, ligation of complementary sticky ends is much more efficient. This is because compatible sticky ends can base pair with one another by hydrogen bonding (Figure 4.20b), forming a relatively stable structure for the enzyme to work on. If the phosphodiester bonds are not synthesized fairly quickly then the sticky ends fall apart again. These transient, base-paired structures do, however, increase the efficiency of ligation by increasing the length of time the ends are in contact with one another.

4.3.3 Putting sticky ends onto a blunt-ended molecule

For the reasons detailed in the preceding section, compatible sticky ends are desirable on the DNA molecules to be ligated together in a gene cloning experiment. Often these sticky ends can be provided by digesting both the vector and the DNA to be cloned with the same restriction endonuclease, or with different enzymes that produce the same sticky end, but it is not always possible to do this. A common situation is where the vector molecule has sticky ends, but the DNA fragments to be cloned are blunt-ended. Under these circumstances one of three methods can be used to put the correct sticky ends onto the DNA fragments.

Linkers

The first of these methods involves the use of linkers. These are short pieces of double-stranded DNA, of known nucleotide sequence, that are synthesized in the test tube.

(a) A typical linker

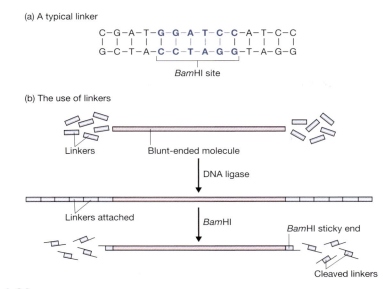

```
C-G-A-T-G-G-A-T-C-C-A-T-C-C
| | | | | | | | | | | | | |
G-C-T-A-C-C-T-A-G-G-T-A-G-G
```
                    ⎣_____⎦
                    *Bam*HI site

(b) The use of linkers

Linkers   Blunt-ended molecule

↓ DNA ligase

Linkers attached   *Bam*HI   *Bam*HI sticky end

                                       Cleaved linkers

Figure 4.21

Linkers and their use: (a) the structure of a typical linker; (b) the attachment of linkers to a blunt-ended molecule.

A typical linker is shown in Figure 4.21a. It is blunt-ended, but contains a restriction site, *Bam*HI in the example shown. DNA ligase can attach linkers to the ends of larger blunt-ended DNA molecules. Although a blunt end ligation, this particular reaction can be performed very efficiently because synthetic oligonucleotides, such as linkers, can be made in very large amounts and added into the ligation mixture at a high concentration.

More than one linker will attach to each end of the DNA molecule, producing the chain structure shown in Figure 4.21b. However, digestion with *Bam*HI cleaves the chains at the recognition sequences, producing a large number of cleaved linkers and the original DNA fragment, now carrying *Bam*HI sticky ends. This modified fragment is ready for ligation into a cloning vector restricted with *Bam*HI.

Adaptors

There is one potential drawback with the use of linkers. Consider what would happen if the blunt-ended molecule shown in Figure 4.21b contained one or more *Bam*HI recognition sequences. If this was the case, the restriction step needed to cleave the linkers and produce the sticky ends would also cleave the blunt-ended molecule (Figure 4.22). The resulting fragments will have the correct sticky ends, but that is no consolation if the gene contained in the blunt-ended fragment has now been broken into pieces.

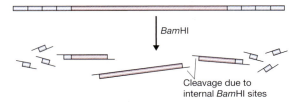

BamHI

Cleavage due to internal *Bam*HI sites

Figure 4.22

A possible problem with the use of linkers. Compare this situation with the desired result of *Bam*HI restriction, as shown in Figure 4.21(b).

Figure 4.23

Adaptors and the potential problem with their use. (a) A typical adaptor. (b) Two adaptors could ligate to one another to produce a molecule similar to a linker, so that (c) after ligation of adaptors a blunt-ended molecule is still blunt-ended and the restriction step is still needed.

(a) A typical adaptor

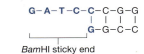

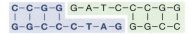

(b) Adaptors could ligate to one another

(c) The new DNA molecule is still blunt-ended

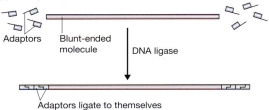

The second method of attaching sticky ends to a blunt-ended molecule is designed to avoid this problem. **Adaptors**, like linkers, are short synthetic oligonucleotides. But unlike linkers, an adaptor is synthesized so that it already has one sticky end (Figure 4.23a). The idea is of course to ligate the blunt end of the adaptor to the blunt ends of the DNA fragment, to produce a new molecule with sticky ends. This may appear to be a simple method but in practice a new problem arises. The sticky ends of individual adaptor molecules could base pair with each other to form dimers (Figure 4.23b), so that the new DNA molecule is still blunt-ended (Figure 4.23c). The sticky ends could be recreated by digestion with a restriction endonuclease, but that would defeat the purpose of using adaptors in the first place.

The answer to the problem lies in the precise chemical structure of the ends of the adaptor molecule. Normally the two ends of a polynucleotide strand are chemically distinct, a fact that is clear from a careful examination of the polymeric structure of DNA (Figure 4.24a). One end, referred to as the 5′ terminus, carries a phosphate group (5′-P); the other, the 3′ terminus, has a hydroxyl group (3′-OH). In the double helix the two strands are antiparallel (Figure 4.24b), so each end of a double-stranded molecule consists of one 5′-P terminus and one 3′-OH terminus. Ligation takes place between the 5′-P and 3′-OH ends (Figure 4.24c).

Adaptor molecules are synthesized so that the blunt end is the same as "natural" DNA, but the sticky end is different. The 3′-OH terminus of the sticky end is the same as usual, but the 5′-P terminus is modified: it lacks the phosphate group, and is in fact a 5′-OH terminus (Figure 4.25a). DNA ligase is unable to form a phosphodiester bridge between 5′-OH and 3′-OH ends. The result is that, although base pairing is always occurring between the sticky ends of adaptor molecules, the association is never stabilized by ligation (Figure 4.25b).

Adaptors can therefore be ligated to a blunt-ended DNA molecule but not to themselves. After the adaptors have been attached, the abnormal 5′-OH terminus is converted to the natural 5′-P form by treatment with the enzyme polynucleotide kinase (p. 50), producing a sticky-ended fragment that can be inserted into an appropriate vector.

(a) The structure of a polynucleotide strand showing the chemical distinction between the 5'-P and 3'-OH termini

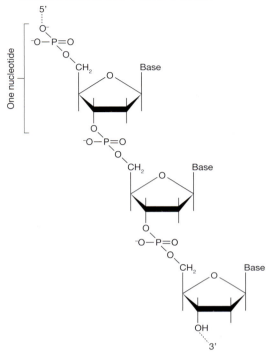

Figure 4.24

The distinction between the 5' and 3' termini of a polynucleotide.

(b) In the double helix the polynucleotide strands are antiparallel

(c) Ligation takes place between 5'-P and 3'-OH termini

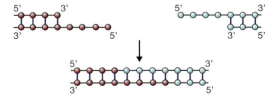

Producing sticky ends by homopolymer tailing

The technique of **homopolymer tailing** offers a radically different approach to the production of sticky ends on a blunt-ended DNA molecule. A homopolymer is simply a polymer in which all the subunits are the same. A DNA strand made up entirely of, say, deoxyguanosine is an example of a homopolymer, and is referred to as polydeoxyguanosine or poly(dG).

Tailing involves using the enzyme terminal deoxynucleotidyl transferase (p. 50) to add a series of nucleotides onto the 3'-OH termini of a double-stranded DNA molecule. If this reaction is carried out in the presence of just one deoxyribonucleotide, a homopolymer tail is produced (Figure 4.26a). Of course, to be able to ligate together

Figure 4.25

The use of adaptors: (a) the actual structure of an adaptor, showing the modified 5′-OH terminus; (b) conversion of blunt ends to sticky ends through the attachment of adaptors.

(a) The precise structure of an adaptor

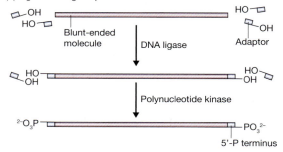

(b) Ligation using adaptors

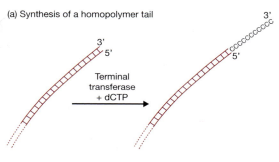

Figure 4.26

Homopolymer tailing: (a) synthesis of a homopolymer tail; (b) construction of a recombinant DNA molecule from a tailed vector plus tailed insert DNA; (c) repair of the recombinant DNA molecule.

(a) Synthesis of a homopolymer tail

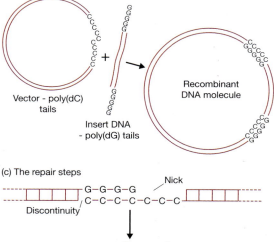

(b) Ligation of homopolymer tails

(c) The repair steps

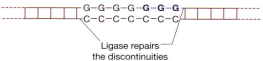

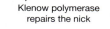
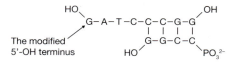

two tailed molecules, the homopolymers must be complementary. Frequently poly-deoxycytosine (poly(dC)) tails are attached to the vector and poly(dG) to the DNA to be cloned. Base pairing between the two occurs when the DNA molecules are mixed (Figure 4.26b).

In practice, the poly(dG) and poly(dC) tails are not usually exactly the same length, and the base-paired recombinant molecules that result have nicks as well as discontinuities (Figure 4.26c). Repair is therefore a two-step process, using Klenow polymerase to fill in the nicks followed by DNA ligase to synthesize the final phosphodiester bonds. This repair reaction does not always have to be performed in the test tube. If the complementary homopolymer tails are longer than about 20 nucleotides, then quite stable base-paired associations are formed. A recombinant DNA molecule, held together by base pairing although not completely ligated, is often stable enough to be introduced into the host cell in the next stage of the cloning experiment (see Figure 1.1). Once inside the host, the cell's own DNA polymerase and DNA ligase repair the recombinant DNA molecule, completing the construction begun in the test tube.

4.3.4 Blunt end ligation with a DNA topoisomerase

A more sophisticated, but easier and generally more efficient way of carrying out blunt end ligation, is to use a special type of enzyme called a DNA topoisomerase. In the cell, DNA topoisomerases are involved in processes that require turns of the double helix to be removed or added to a double-stranded DNA molecule. Turns are removed during DNA replication in order to unwind the helix and enable each polynucleotide to be replicated, and are added to newly synthesized circular molecules to introduce super-coiling. DNA topoisomerases are able to separate the two strands of a DNA molecule without actually rotating the double helix. They achieve this feat by causing transient single- or double-stranded breakages in the DNA backbone (Figure 4.27). DNA topoisomerases therefore have both nuclease and ligase activities.

To carry out blunt end ligation with a topoisomerase, a special type of cloning vector is needed. This is a plasmid that has been linearized by the nuclease activity of the DNA topoisomerase enzyme from vaccinia virus. The vaccinia topoisomerase cuts DNA at the sequence CCCTT, which is present just once in the plasmid. After cutting the

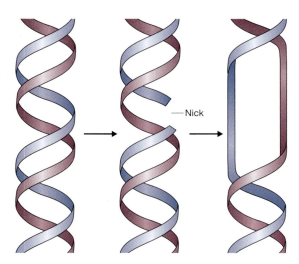

Figure 4.27

The mode of action of a Type 1 DNA topoisomerase, which removes or adds turns to a double helix by making a transient break in one of the strands.

Nick

(a) The ends of the vector resulting from topoisomerase cleavage

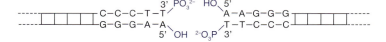

(b) Removal of terminal phosphates from the molecule to be cloned

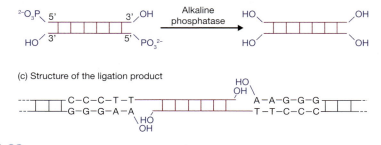

(c) Structure of the ligation product

Figure 4.28

Blunt end ligation with a DNA topoisomerase. (a) Cleavage of the vector with the topoisomerase leaves blunt ends with 5'-OH and 3'-P termini. (b) The molecule to be cloned must therefore be treated with alkaline phosphatase to convert its 5'-P ends into 5'-OH termini. (c) The topoisomerase ligates the 3'-P and 5'-OH ends, creating a double-stranded molecule with two discontinuities, which are repaired by cellular enzymes after introduction into the host bacteria.

plasmid, topoisomerase enzymes remain covalently bound to the resulting blunt ends. The reaction can be stopped at this point, enabling the vector to be stored until it is needed.

Cleavage by the topoisomerase results in 5'-OH and 3'-P termini (Figure 4.28a). If the blunt-ended molecules to be cloned have been produced from a larger molecule by cutting with a restriction enzyme, then they will have 5'-P and 3'-OH ends. Before mixing these molecules with the vector, their terminal phosphates must be removed to give 5'-OH ends that can ligate to the 3'-P termini of the vector. The molecules are therefore treated with alkaline phosphatase (Figure 4.28b).

Adding the phosphatased molecules to the vector reactivates the bound topoisomerases, which proceed to the ligation phase of their reaction. Ligation occurs between the 3'-P ends of the vectors and the 5'-OH ends of the phosphatased molecules. The blunt-ended molecules therefore become inserted into the vectors. Only one strand is ligated at each junction point (Figure 4.28c), but this is not a problem because the discontinuities will be repaired by cellular enzymes after the recombinant molecules have been introduced into the host bacteria.

Further reading

Deng, G. & Wu, R. (1981) An improved procedure for utilizing terminal transferase to add homopolymers to the 3′ termini of DNA. *Nucleic Acids Research*, 9, 4173–4188.

Helling, R.B., Goodman, H.M. & Boyer, H.W. (1974) Analysis of endonuclease R·*Eco*RI fragments of DNA from lambdoid bacteriophages and other viruses by agarose-gel electrophoresis. *Journal of Virology*, 14, 1235–1244.

Heyman, J.A., Cornthwaite, J., Foncerrada, L. et al. (1999) Genome-scale cloning and expression of individual open reading frames using topoisomerase I-mediated ligation. *Genome Research*, 9, 383–392. [A description of ligation using topoisomerase.]

Jacobsen, H., Klenow, H. & Overgaard-Hansen, K. (1974) The N-terminal amino acid sequences of DNA polymerase I from *Escherichia coli* and of the large and small fragments obtained by a limited proteolysis. *European Journal of Biochemistry*, 45, 623–627. [Production of the Klenow fragment of DNA polymerase I.]

Lehnman, I.R. (1974) DNA ligase: structure, mechanism, and function. *Science*, 186, 790–797.

REBASE: http://rebase.neb.com/rebase/ [A comprehensive list of all the known restriction endonucleases and their recognition sequences.]

Rothstein, R.J., Lau, L.F., Bahl, C.P., Narang, N.A. & Wu, R. (1979) Synthetic adaptors for cloning DNA. *Methods in Enzymology*, 68, 98–109.

Schwartz, D.C. & Cantor, C.R. (1984) Separation of yeast chromosome-sized DNAs by pulsed field gradient gel electrophoresis. *Cell*, 37, 67–75.

Smith, H.O. & Wilcox, K.W. (1970) A restriction enzyme from *Haemophilus influenzae*. *Journal of Molecular Biology*, 51, 379–391. [One of the first full descriptions of a restriction endonuclease.]

Zipper, H., Brunner, H., Bernhagen, J. & Vitzthum, F. (2004) Investigations on DNA intercalation and surface binding by SYBR Green I, its structure determination and methodological implications. *Nucleic Acids Research*, 32, e103. [Details of one the DNA dyes now used as an alternative to ethidium bromide for staining agarose gels.]

Chapter 5

Introduction of DNA into Living Cells

The manipulations described in Chapter 4 allow the molecular biologist to create novel recombinant DNA molecules. The next step in a gene cloning experiment is to introduce these molecules into living cells, usually bacteria, which then grow and divide to produce clones (see Figure 1.1). Strictly speaking, the word "cloning" refers only to the later stages of the procedure, and not to the construction of the recombinant DNA molecule itself.

Cloning serves two main purposes. First, it allows a large number of recombinant DNA molecules to be produced from a limited amount of starting material. At the outset only a few nanograms of recombinant DNA may be available, but each bacterium that takes up a plasmid subsequently divides numerous times to produce a colony, each cell of which contains multiple copies of the molecule. Several micrograms of recombinant DNA can usually be prepared from a single bacterial colony, representing a thousandfold increase over the starting amount (Figure 5.1). If the colony is used not as a source of DNA but as an inoculum for a liquid culture, the resulting cells may provide milligrams of DNA, a millionfold increase in yield. In this way cloning can supply the large amounts of DNA needed for molecular biological studies of gene structure and expression (Chapters 10 and 11).

The second important function of cloning can be described as purification. The manipulations that result in a recombinant DNA molecule can only rarely be controlled to the extent that no other DNA molecules are present at the end of the procedure. The

Gene Cloning and DNA Analysis: An Introduction. 6th edition. By T.A. Brown. Published 2010 by Blackwell Publishing.

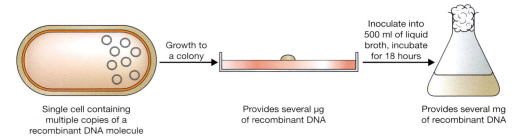

Figure 5.1

Cloning can supply large amounts of recombinant DNA.

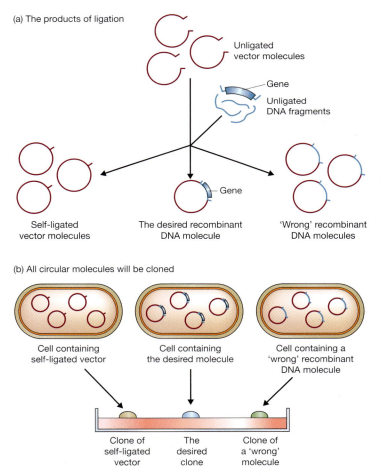

(a) The products of ligation

Unligated vector molecules

Gene

Unligated DNA fragments

Self-ligated vector molecules

The desired recombinant DNA molecule

Gene

'Wrong' recombinant DNA molecules

(b) All circular molecules will be cloned

Cell containing self-ligated vector

Cell containing the desired molecule

Cell containing a 'wrong' recombinant DNA molecule

Clone of self-ligated vector

The desired clone

Clone of a 'wrong' molecule

Figure 5.2

Cloning is analogous to purification. From a mixture of different molecules, clones containing copies of just one molecule can be obtained.

ligation mixture may contain, in addition to the desired recombinant molecule, any number of the following (Figure 5.2a):

- Unligated vector molecules
- Unligated DNA fragments
- Vector molecules that have recircularized without new DNA being inserted ("self-ligated" vector)
- Recombinant DNA molecules that carry the wrong inserted DNA fragment.

Unligated molecules rarely cause a problem because, even though they may be taken up by bacterial cells, only under exceptional circumstances will they be replicated. It is much more likely that enzymes within the host bacteria degrade these pieces of DNA. Self-ligated vector molecules and incorrect recombinant plasmids are more important because they are replicated just as efficiently as the desired molecule (Figure 5.2b). However, purification of the desired molecule can still be achieved through cloning because it is extremely unusual for any one cell to take up more than one DNA molecule. Each cell gives rise to a single colony, so each of the resulting clones consists of cells that all contain the same molecule. Of course, different colonies contain different molecules: some contain the desired recombinant DNA molecule, some have different recombinant molecules, and some contain self-ligated vector. The problem therefore becomes a question of identifying the colonies that contain the correct recombinant plasmids.

This chapter is concerned with the way in which plasmid and phage vectors, and recombinant molecules derived from them, are introduced into bacterial cells. During the course of the chapter it will become apparent that selection for colonies containing recombinant molecules, as opposed to colonies containing self-ligated vector, is relatively easy. The more difficult proposition of how to distinguish clones containing the correct recombinant DNA molecule from all the other recombinant clones will be tackled in Chapter 8.

5.1 Transformation—the uptake of DNA by bacterial cells

Most species of bacteria are able to take up DNA molecules from the medium in which they grow. Often a DNA molecule taken up in this way will be degraded, but occasionally it is able to survive and replicate in the host cell. In particular this happens if the DNA molecule is a plasmid with an origin of replication recognized by the host.

5.1.1 Not all species of bacteria are equally efficient at DNA uptake

In nature, transformation is probably not a major process by which bacteria obtain genetic information. This is reflected by the fact that in the laboratory only a few species (notably members of the genera *Bacillus* and *Streptococcus*) can be transformed with ease. Close study of these organisms has revealed that they possess sophisticated mechanisms for DNA binding and uptake.

Most species of bacteria, including *E. coli*, take up only limited amounts of DNA under normal circumstances. In order to transform these species efficiently, the bacteria have to undergo some form of physical and/or chemical treatment that enhances their ability to take up DNA. Cells that have undergone this treatment are said to be competent.

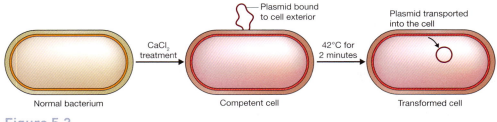

Figure 5.3

The binding and uptake of DNA by a competent bacterial cell.

5.1.2 Preparation of competent E. coli cells

As with many breakthroughs in recombinant DNA technology, the key development as far as transformation is concerned occurred in the early 1970s, when it was observed that *E. coli* cells that had been soaked in an ice cold salt solution were more efficient at DNA uptake than unsoaked cells. A solution of 50 mM calcium chloride ($CaCl_2$) is traditionally used, although other salts, notably rubidium chloride, are also effective.

Exactly why this treatment works is not understood. Possibly $CaCl_2$ causes the DNA to precipitate onto the outside of the cells, or perhaps the salt is responsible for some kind of change in the cell wall that improves DNA binding. In any case, soaking in $CaCl_2$ affects only DNA binding, and not the actual uptake into the cell. When DNA is added to treated cells, it remains attached to the cell exterior, and is not at this stage transported into the cytoplasm (Figure 5.3). The actual movement of DNA into competent cells is stimulated by briefly raising the temperature to 42°C. Once again, the exact reason why this heat shock is effective is not understood.

5.1.3 Selection for transformed cells

Transformation of competent cells is an inefficient procedure, however carefully the cells have been prepared. Although 1 ng of the plasmid vector called pUC8 (p. 92) can yield 1000–10,000 transformants, this represents the uptake of only 0.01% of all the available molecules. Furthermore, 10,000 transformants is only a very small proportion of the total number of cells that are present in a competent culture. This last fact means that some way must be found to distinguish a cell that has taken up a plasmid from the many thousands that have not been transformed.

Uptake and stable retention of a plasmid is usually detected by looking for expression of the genes carried by the plasmid. For example, *E. coli* cells are normally sensitive to the growth inhibitory effects of the antibiotics ampicillin and tetracycline. However, cells that contain the plasmid pBR322 (p. 89), which was one of the first cloning vectors to be developed back in the 1970s, are resistant to these antibiotics. This is because pBR322 carries two sets of genes, one gene that codes for a β-lactamase enzyme that modifies ampicillin into a form that is non-toxic to the bacterium, and a second set of genes that code for enzymes that detoxify tetracycline. After a transformation experiment with pBR322, only those *E. coli* cells that have taken up a plasmid are $amp^R tet^R$ and able to form colonies on an agar medium that contains ampicillin or tetracycline (Figure 5.4); non-transformants, which are still $amp^S tet^S$, do not produce colonies on the selective medium. Transformants and non-transformants are therefore easily distinguished.

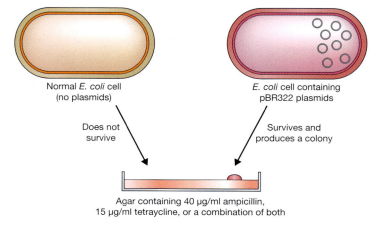

Normal *E. coli* cell
(no plasmids)

E. coli cell containing
pBR322 plasmids

Does not
survive

Survives and
produces a colony

Agar containing 40 μg/ml ampicillin,
15 μg/ml tetraycline, or a combination of both

Figure 5.4

Selecting cells that contain pBR322 plasmids by plating onto agar medium containing ampicillin and/or tetracycline.

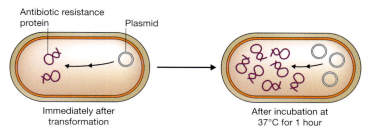

Antibiotic resistance
protein

Plasmid

Immediately after
transformation

After incubation at
37°C for 1 hour

Figure 5.5

Phenotypic expression. Incubation at 37°C for 1 hour before plating out improves the survival of the transformants on selective medium, because the bacteria have had time to begin synthesis of the antibiotic resistance enzymes.

Most plasmid cloning vectors carry at least one gene that confers antibiotic resistance on the host cells, with selection of transformants being achieved by plating onto an agar medium that contains the relevant antibiotic. Bear in mind, however, that resistance to the antibiotic is not due merely to the presence of the plasmid in the transformed cells. The resistance gene on the plasmid must also be expressed, so that the enzyme that detoxifies the antibiotic is synthesized. Expression of the resistance gene begins immediately after transformation, but it will be a few minutes before the cell contains enough of the enzyme to be able to withstand the toxic effects of the antibiotic. For this reason the transformed bacteria should not be plated onto the selective medium immediately after the heat shock treatment, but first placed in a small volume of liquid medium, in the absence of antibiotic, and incubated for a short time. Plasmid replication and expression can then get started, so that when the cells are plated out and encounter the antibiotic, they will already have synthesized sufficient resistance enzymes to be able to survive (Figure 5.5).

5.2 Identification of recombinants

Plating onto a selective medium enables transformants to be distinguished from non-transformants. The next problem is to determine which of the transformed colonies

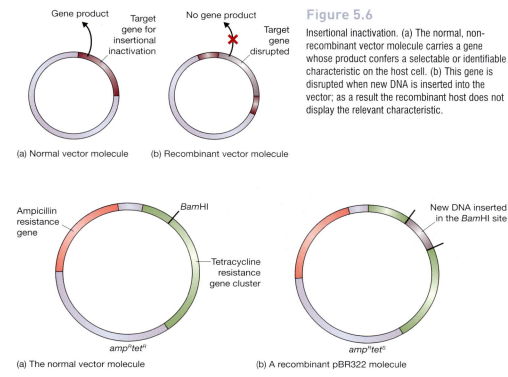

Gene product   Target gene for insertional inactivation   No gene product   Target gene disrupted

(a) Normal vector molecule   (b) Recombinant vector molecule

Figure 5.6

Insertional inactivation. (a) The normal, non-recombinant vector molecule carries a gene whose product confers a selectable or identifiable characteristic on the host cell. (b) This gene is disrupted when new DNA is inserted into the vector; as a result the recombinant host does not display the relevant characteristic.

Ampicillin resistance gene   BamHI   Tetracycline resistance gene cluster   New DNA inserted in the BamHI site

$amp^R tet^R$ $amp^R tet^S$

(a) The normal vector molecule   (b) A recombinant pBR322 molecule

Figure 5.7

The cloning vector pBR322: (a) the normal vector molecule; (b) a recombinant molecule containing an extra piece of DNA inserted into the BamHI site. For a more detailed map of pBR322, see Figure 6.1.

comprise cells that contain recombinant DNA molecules, and which contain self-ligated vector molecules (see Figure 5.2). With most cloning vectors, insertion of a DNA fragment into the plasmid destroys the integrity of one of the genes present on the molecule. Recombinants can therefore be identified because the characteristic coded by the inactivated gene is no longer displayed by the host cells (Figure 5.6). We will explore the general principles of insertional inactivation by looking at the different methods used with the two cloning vectors mentioned in the previous section—pBR322 and pUC8.

5.2.1 Recombinant selection with pBR322—insertional inactivation of an antibiotic resistance gene

pBR322 has several unique restriction sites that can be used to open up the vector before insertion of a new DNA fragment (Figure 5.7a). BamHI, for example, cuts pBR322 at just one position, within the cluster of genes that code for resistance to tetracycline. A recombinant pBR322 molecule, one that carries an extra piece of DNA in the BamHI site (Figure 5.7b), is no longer able to confer tetracycline resistance on its host, as one of the necessary genes is now disrupted by the inserted DNA. Cells containing this recombinant pBR322 molecule are still resistant to ampicillin, but sensitive to tetracycline ($amp^R tet^S$).

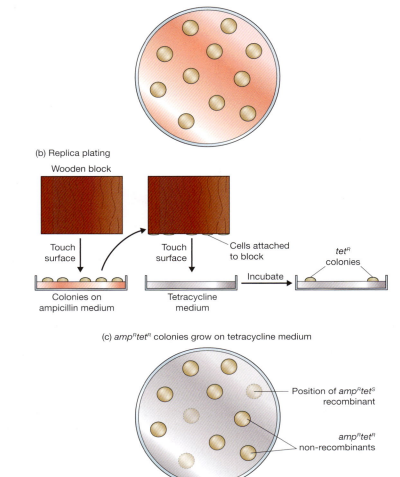

(a) Colonies on ampicillin medium

(b) Replica plating

Wooden block

Touch surface

Touch surface

Cells attached to block

Incubate

tet^R colonies

Colonies on ampicillin medium

Tetracycline medium

(c) $amp^R tet^R$ colonies grow on tetracycline medium

Position of $amp^R tet^S$ recombinant

$amp^R tet^R$ non-recombinants

Figure 5.8

Screening for pBR322 recombinants by insertional inactivation of the tetracycline resistance gene. (a) Cells are plated onto ampicillin agar: all the transformants produce colonies. (b) The colonies are replica plated onto tetracycline medium. (c) The colonies that grow on tetracycline medium are $amp^R tet^R$ and therefore non-recombinants. Recombinants ($amp^R tet^S$) do not grow, but their position on the ampicillin plate is now known.

Screening for pBR322 recombinants is performed in the following way. After transformation the cells are plated onto ampicillin medium and incubated until colonies appear (Figure 5.8a). All of these colonies are transformants (remember, untransformed cells are amp^S and so do not produce colonies on the selective medium), but only a few contain recombinant pBR322 molecules: most contain the normal, self-ligated plasmid. To identify the recombinants the colonies are replica plated onto agar medium that contains tetracycline (Figure 5.8b). After incubation, some of the original colonies regrow, but others do not (Figure 5.8c). Those that do grow consist of cells that carry the normal pBR322 with no inserted DNA and therefore a functional tetracycline resistance gene cluster ($amp^R tet^R$). The colonies that do not grow on tetracycline agar are recombinants ($amp^R tet^S$); once their positions are known, samples for further study can be recovered from the original ampicillin agar plate.

5.2.2 Insertional inactivation does not always involve antibiotic resistance

Although insertional inactivation of an antibiotic resistance gene provides an effective means of recombinant identification, the method is made inconvenient by the need to carry out two screenings, one with the antibiotic that selects for transformants, followed by the second screen, after replica-plating, with the antibiotic that distinguishes recombinants. Most modern plasmid vectors therefore make use of a different system. An example is pUC8 (Figure 5.9a), which carries the ampicillin resistance gene and a gene called *lacZ'*, which codes for part of the enzyme β-galactosidase. Cloning with pUC8 involves insertional inactivation of the *lacZ'* gene, with recombinants identified because of their inability to synthesize β-galactosidase (Figure 5.9b).

β-Galactosidase is one of a series of enzymes involved in the breakdown of lactose to glucose plus galactose. It is normally coded by the gene *lacZ*, which resides on the *E. coli* chromosome. Some strains of *E. coli* have a modified *lacZ* gene, one that lacks the segment referred to as *lacZ'* and coding for the α-peptide portion of β-galactosidase (Figure 5.10a). These mutants can synthesize the enzyme only when they harbor a plasmid, such as pUC8, that carries the missing *lacZ'* segment of the gene.

A cloning experiment with pUC8 involves selection of transformants on ampicillin agar followed by screening for β-galactosidase activity to identify recombinants. Cells

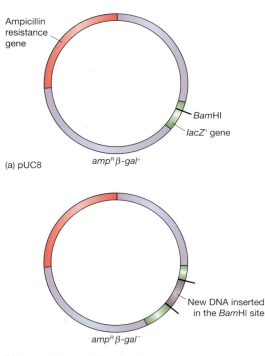

Ampicillin resistance gene

BamHI

lacZ' gene

(a) pUC8 $amp^R β$-gal$^+$

New DNA inserted in the BamHI site

$amp^R β$-gal$^-$

(b) A recombinant pUC8 molecule

Figure 5.9

The cloning vector pUC8: (a) the normal vector molecule; (b) a recombinant molecule containing an extra piece of DNA inserted into the *Bam*HI site. For a more detailed map of pUC8, see Figure 6.3.

(a) The role of the *lacZ'⁻* gene

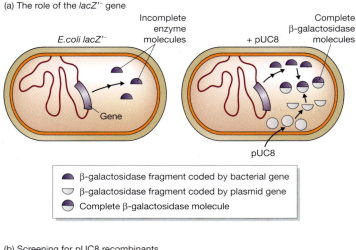

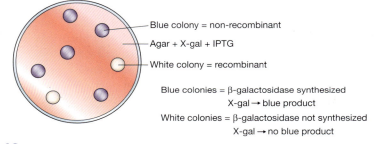

(b) Screening for pUC8 recombinants

Blue colony = non-recombinant

Agar + X-gal + IPTG

White colony = recombinant

Blue colonies = β-galactosidase synthesized
X-gal → blue product
White colonies = β-galactosidase not synthesized
X-gal → no blue product

Figure 5.10

The rationale behind insertional inactivation of the *lacZ'* gene carried by pUC8. (a) The bacterial and plasmid genes complement each other to produce a functional β-galactosidase molecule. (b) Recombinants are screened by plating onto agar containing X-gal and IPTG.

that harbor a normal pUC8 plasmid are amp^R and able to synthesize β-galactosidase (Figure 5.9a); recombinants are also amp^R but unable to make β-galactosidase (Figure 5.9b).

Screening for β-galactosidase presence or absence is in fact quite easy. Rather than assay for lactose being split to glucose and galactose, we test for a slightly different reaction that is also catalyzed by β-galactosidase. This involves a lactose analog called X-gal (5-bromo-4-chloro-3-indolyl-β-D-galactopyranoside) which is broken down by β-galactosidase to a product that is colored deep blue. If X-gal (plus an inducer of the enzyme such as isopropylthiogalactoside, IPTG) is added to the agar, along with ampicillin, then non-recombinant colonies, the cells of which synthesize β-galactosidase, will be colored blue, whereas recombinants with a disrupted *lacZ'* gene and unable to make β-galactosidase, will be white. This system, which is called Lac selection, is summarized in Figure 5.10b. Note that both ampicillin resistance and the presence or absence of β-galactosidase are tested for on a single agar plate. The two screenings are therefore carried out together and there is no need for the time-consuming replica-plating step that is necessary with plasmids such as pBR322.

5.3 Introduction of phage DNA into bacterial cells

There are two different methods by which a recombinant DNA molecule constructed with a phage vector can be introduced into a bacterial cell: transfection and *in vitro* packaging.

5.3.1 Transfection

Transfection is equivalent to transformation, the only difference being that phage DNA rather than a plasmid is involved. Just as with a plasmid, the purified phage DNA, or recombinant phage molecule, is mixed with competent *E. coli* cells and DNA uptake induced by heat shock. Transfection is the standard method for introducing the double-stranded RF form of an M13 cloning vector into *E. coli*.

5.3.2 In vitro *packaging of λ cloning vectors*

Transfection with λ DNA molecules is not a very efficient process when compared with the infection of a culture of cells with mature λ phage particles. It would therefore be useful if recombinant λ molecules could be packaged into their λ head-and-tail structures in the test tube.

   This may sound difficult but is actually relatively easy to achieve. Packaging requires a number of different proteins coded by the λ genome, but these can be prepared at a high concentration from cells infected with defective λ phage strains. Two different systems are in use. With the single strain system, the defective λ phage carries a mutation in the *cos* sites, so that these are not recognized by the endonuclease that normally cleaves the λ catenanes during phage replication (p. 21). This means that the defective phage cannot replicate, though it does direct synthesis of all the proteins needed for packaging. The proteins accumulate in the bacterium and can be purified from cultures of *E. coli* infected with the mutated λ. The protein preparation is then used for *in vitro* packaging of recombinant λ molecules (Figure 5.11a).

   With the second system two defective λ strains are needed. Both of these strains carry a mutation in a gene for one of the components of the phage protein coat: with one strain the mutation is in gene *D*, and with the second strain it is in gene *E* (see Figure 2.9). Neither strain is able to complete an infection cycle in *E. coli* because in the absence of the product of the mutated gene the complete capsid structure cannot be made. Instead the products of all the other coat protein genes accumulate (Figure 5.11b). An *in vitro* packaging mix can therefore be prepared by combining lysates of two cultures of cells, one infected with the λ D^- strain, the other infected with the E^- strain. The mixture now contains all the necessary components for *in vitro* packaging.

   With both systems, formation of phage particles is achieved simply by mixing the packaging proteins with λ DNA—assembly of the particles occurs automatically in the test tube (Figure 5.11c). The packaged λ DNA is then introduced into *E. coli* cells simply by adding the assembled phages to the bacterial culture and allowing the normal λ infective process to take place.

5.3.3 *Phage infection is visualized as plaques on an agar medium*

The final stage of the phage infection cycle is cell lysis (p. 18). If infected cells are spread onto a solid agar medium immediately after addition of the phage particles, or immediately

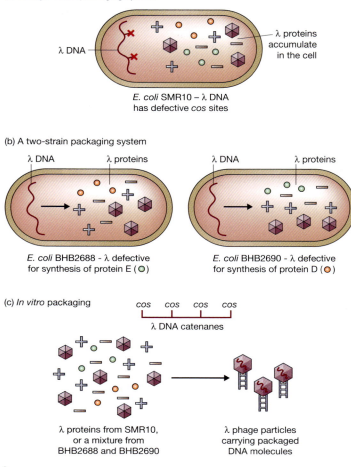

(a) A single-strain packaging system

λ DNA

λ proteins accumulate in the cell

E. coli SMR10 – λ DNA has defective *cos* sites

(b) A two-strain packaging system

λ DNA   λ proteins   λ DNA   λ proteins

E. coli BHB2688 - λ defective for synthesis of protein E (○)

E. coli BHB2690 - λ defective for synthesis of protein D (●)

(c) *In vitro* packaging

cos   cos   cos   cos

λ DNA catenanes

λ proteins from SMR10, or a mixture from BHB2688 and BHB2690

λ phage particles carrying packaged DNA molecules

Figure 5.11

In vitro packaging. (a) Synthesis of λ capsid proteins by *E. coli* strain SMR10, which carries a λ phage that has defective *cos* sites. (b) Synthesis of incomplete sets of λ capsid proteins by *E. coli* strains BHB2688 and BHB2690. (c) The cell lysates provide the complete set of capsid proteins and can package λ DNA molecules in the test tube.

after transfection with phage DNA, cell lysis can be visualized as **plaques** on a lawn of bacteria (Figure 5.12a). Each plaque is a zone of clearing produced as the phages lyse the cells and move on to infect and eventually lyse the neighboring bacteria (Figure 5.12b).

Both λ and M13 form plaques. With λ these are true plaques, produced by cell lysis. However, M13 plaques are slightly different as M13 does not lyse the host cells (p. 19). Instead M13 causes a decrease in the growth rate of infected cells, sufficient to produce a zone of relative clearing on a bacterial lawn. Although not true plaques, these zones of clearing are visually identical to normal phage plaques (Figure 5.12c).

The end result of a gene cloning experiment using a λ or M13 vector is therefore an agar plate covered in phage plaques. Each plaque is derived from a single transfected or infected cell and therefore contains identical phage particles. These may contain self-ligated vector molecules, or they may be recombinants.

(a) Plaques on a lawn of bacteria

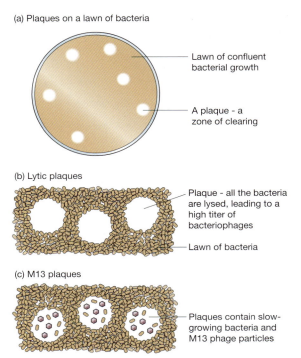

Lawn of confluent
bacterial growth

A plaque - a
zone of clearing

(b) Lytic plaques

Plaque - all the bacteria
are lysed, leading to a
high titer of
bacteriophages

Lawn of bacteria

(c) M13 plaques

Plaques contain slow-
growing bacteria and
M13 phage particles

Figure 5.12

Bacteriophage plaques. (a) The appearance of plaques on a lawn of bacteria. (b) Plaques produced by a phage that lyses the host cell (e.g., λ in the lytic infection cycle); the plaques contain lysed cells plus many phage particles. (c) Plaques produced by M13; these plaques contain slow-growing bacteria plus many M13 phage particles.

5.4 Identification of recombinant phages

A variety of ways of distinguishing recombinant plaques have been devised, the following being the most important.

5.4.1 Insertional inactivation of a *lacZ'* gene carried by the phage vector

All M13 cloning vectors (p. 94), as well as several λ vectors, carry a copy of the *lacZ'* gene. Insertion of new DNA into this gene inactivates β-galactosidase synthesis, just as with the plasmid vector pUC8. Recombinants are distinguished by plating cells onto X-gal agar: plaques comprising normal phages are blue; recombinant plaques are clear (Figure 5.13a).

5.4.2 Insertional inactivation of the λ *cI* gene

Several types of λ cloning vector have unique restriction sites in the *cI* gene (map position 38 on Figure 2.9). Insertional inactivation of this gene causes a change in plaque morphology. Normal plaques appear "turbid", whereas recombinants with a disrupted *cI* gene are "clear" (Figure 5.13b). The difference is readily apparent to the experienced eye.

5.4.3 Selection using the Spi phenotype

λ phages cannot normally infect *E. coli* cells that already possess an integrated form of a related phage called P2. λ is therefore said to be Spi+ (sensitive to P2 prophage

Figure 5.13

Strategies for the selection of recombinant phage.

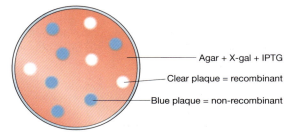

(a) Insertional activation of the *lacZ'* gene

Agar + X-gal + IPTG
Clear plaque = recombinant
Blue plaque = non-recombinant

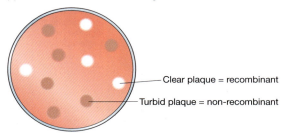

(b) Insertional activation of the λ *cl* gene

Clear plaque = recombinant
Turbid plaque = non-recombinant

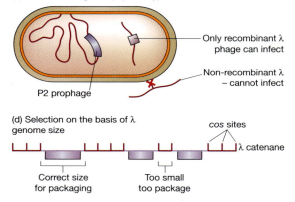

(c) Selection using the Spi phenotype

Only recombinant λ phage can infect
Non-recombinant λ – cannot infect
P2 prophage

(d) Selection on the basis of λ genome size

cos sites
λ catenane
Correct size for packaging
Too small too package

inhibition). Some λ cloning vectors are designed so that insertion of new DNA causes a change from Spi$^+$ to Spi$^-$, enabling the recombinants to infect cells that carry P2 prophages. Such cells are used as the host for cloning experiments with these vectors; only recombinants are Spi$^-$ so only recombinants form plaques (Figure 5.13c).

5.4.4 Selection on the basis of λ genome size

The λ packaging system, which assembles the mature phage particles, can only insert DNA molecules of between 37 and 52 kb into the head structure. Anything less than 37 kb is not packaged. Many λ vectors have been constructed by deleting large segments of the λ DNA molecule (p. 98) and so are less than 37 kb in length. These can only be packaged into mature phage particles after extra DNA has been inserted, bringing the total genome size up to 37 kb or more (Figure 5.13d). Therefore, with these vectors only recombinant phages are able to replicate.

5.5 Introduction of DNA into non-bacterial cells

Ways of introducing DNA into yeast, fungi, animals, and plants are also needed if these organisms are to be used as the hosts for gene cloning. Strictly speaking, these processes are not "transformation", as that term has a specific meaning that applies only to uptake of DNA by bacteria. However, molecular biologists have forgotten this over the years and "transformation" is now used to describe uptake of DNA by any organism.

In general terms, soaking cells in salt is effective only with a few species of bacteria, although treatment with lithium chloride or lithium acetate does enhance DNA uptake by yeast cells, and is frequently used in the transformation of *Saccharomyces cerevisiae*. However, for most higher organisms, more sophisticated methods are needed.

5.5.1 Transformation of individual cells

With most organisms the main barrier to DNA uptake is the cell wall. Cultured animal cells, which usually lack cell walls, are easily transformed, especially if the DNA is precipitated onto the cell surface with calcium phosphate (Figure 5.14a) or enclosed in liposomes that fuse with the cell membrane (Figure 5.14b). For other types of cell the answer is often to remove the cell wall. Enzymes that degrade yeast, fungal, and plant cell walls are available, and under the right conditions intact protoplasts can be obtained (Figure 5.14c). Protoplasts generally take up DNA quite readily, but transformation can be stimulated by special techniques such as electroporation, during which the cells are subjected to a short electrical pulse, thought to induce the transient formation of pores in the cell membrane, through which DNA molecules are able to enter the cell. After transformation the protoplasts are washed to remove the degradative enzymes and the cell wall spontaneously re-forms.

In contrast to the transformation systems described so far, there are two physical methods for introducing DNA into cells. The first of these is microinjection, which makes use of a very fine pipette to inject DNA molecules directly into the nucleus of the cells to be transformed (Figure 5.15a). This technique was initially applied to animal cells but has subsequently been successful with plant cells. The second method involves bombardment of the cells with high velocity microprojectiles, usually particles of gold or tungsten that have been coated with DNA. These microprojectiles are fired at the cells from a particle gun (Figure 5.15b). This unusual technique is termed biolistics and has been used with a number of different types of cell.

5.5.2 Transformation of whole organisms

With animals and plants the desired end product might not be transformed cells, but a transformed organism. Plants are relatively easy to regenerate from cultured cells, though problems have been experienced in developing regeneration procedures for monocotyledonous species such as cereals and grasses. A single transformed plant cell can therefore give rise to a transformed plant, which carries the cloned DNA in every cell, and passes the cloned DNA on to its progeny following flowering and seed formation (see Figure 7.13). Animals, of course, cannot be regenerated from cultured cells, so obtaining transformed animals requires a rather more subtle approach. One technique with mammals such as mice is to remove fertilized eggs from the oviduct, to microinject DNA, and then to reimplant the transformed cells into the mother's reproductive tract. We will look more closely at these methods for obtaining transformed animals in Chapter 13.

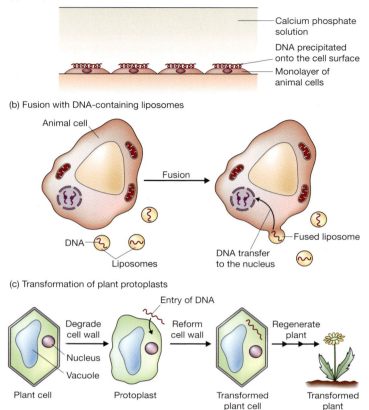

(a) Precipitation of DNA on to animal cells

Calcium phosphate solution

DNA precipitated onto the cell surface

Monolayer of animal cells

(b) Fusion with DNA-containing liposomes

Animal cell

Fusion

DNA

Liposomes

Fused liposome

DNA transfer to the nucleus

(c) Transformation of plant protoplasts

Entry of DNA

Degrade cell wall

Reform cell wall

Regenerate plant

Nucleus

Vacuole

Plant cell

Protoplast

Transformed plant cell

Transformed plant

Figure 5.14

Strategies for introducing new DNA into animal and plant cells: (a) precipitation of DNA on to animal cells; (b) introduction of DNA into animal cells by liposome fusion; (c) transformation of plant protoplasts.

Figure 5.15

Two physical methods for introducing DNA into cells.

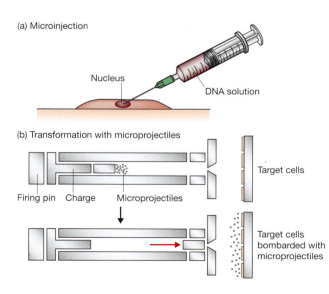

(a) Microinjection

Nucleus

DNA solution

(b) Transformation with microprojectiles

Target cells

Firing pin   Charge   Microprojectiles

Target cells bombarded with microprojectiles

Further reading

Calvin, N.M. & Hanawalt, P.C. (1988) High efficiency transformation of bacterial cells by electroporation. *Journal of Bacteriology*, 170, 2796–2801.

Capecchi, M.R. (1980) High efficiency transformation by direct microinjection of DNA into cultured mammalian cells. *Cell*, 22, 479–488.

Hammer, R.E., Pursel, V.G., Rexroad, C.E. et al. (1985) Production of transgenic rabbits, sheep and pigs by microinjection. *Nature*, 315, 680–683.

Hohn, B. (1979) *In vitro* packaging of lambda and cosmid DNA. *Methods in Enzymology*, 68, 299–309.

Klein, T.M., Wolf, E.D., Wu, R. & Sanford, J.C. (1987) High velocity microprojectiles for delivering nucleic acids into living cells. *Nature*, 327, 70–73. [Biolistics.]

Lederberg, J. & Lederberg, E.M. (1952) Replica plating and indirect selection of bacterial mutants. *Journal of Bacteriology*, 63, 399–406.

Mandel, M. & Higa, A. (1970) Calcium-dependent bacteriophage DNA infection. *Journal of Molecular Biology*, 53, 159–162. [The first description of the use of calcium chloride to prepare competent *E. coli* cells.]

Chapter 6

Cloning Vectors for *E. coli*

The basic experimental techniques involved in gene cloning have now been described. In Chapters 3, 4, and 5 we have seen how DNA is purified from cell extracts, how recombinant DNA molecules are constructed in the test tube, how DNA molecules are reintroduced into living cells, and how recombinant clones are distinguished. Now we must look more closely at the cloning vector itself, in order to consider the range of vectors available to the molecular biologist, and to understand the properties and uses of each individual type.

The greatest variety of cloning vectors exists for use with *E. coli* as the host organism. This is not surprising in view of the central role that this bacterium has played in basic research over the past 50 years. The tremendous wealth of information that exists concerning the microbiology, biochemistry, and genetics of *E. coli* has meant that virtually all fundamental studies of gene structure and function were initially carried out with this bacterium as the experimental organism. Even when a eukaryote is being studied, *E. coli* is still used as the workhorse for preparation of cloned DNA for sequencing, and for construction of recombinant genes that will subsequently be placed back in the eukaryotic host in order to study their function and expression. In recent years, gene cloning and molecular biological research have become mutually synergistic—breakthroughs in gene cloning have acted as a stimulus to research, and the needs of research have spurred on the development of new, more sophisticated cloning vectors.

Gene Cloning and DNA Analysis: An Introduction. 6th edition. By T.A. Brown. Published 2010 by Blackwell Publishing.

In this chapter the most important types of *E. coli* cloning vector will be described, and their specific uses outlined. In Chapter 7, cloning vectors for yeast, fungi, plants, and animals will be considered.

6.1 Cloning vectors based on *E. coli* plasmids

The simplest cloning vectors, and the ones most widely used in gene cloning, are those based on small bacterial plasmids. A large number of different plasmid vectors are available for use with *E. coli*, many obtainable from commercial suppliers. They combine ease of purification with desirable properties such as high transformation efficiency, convenient selectable markers for transformants and recombinants, and the ability to clone reasonably large (up to about 8 kb) pieces of DNA. Most "routine" gene cloning experiments make use of one or other of these plasmid vectors.

One of the first vectors to be developed was pBR322, which was introduced in Chapter 5 to illustrate the general principles of transformant selection and recombinant identification (p. 77). Although pBR322 lacks the more sophisticated features of the newest cloning vectors, and so is no longer used extensively in research, it still illustrates the important, fundamental properties of any plasmid cloning vector. We will therefore begin our study of *E. coli* vectors by looking more closely at pBR322.

6.1.1 The nomenclature of plasmid cloning vectors

The name "pBR322" conforms with the standard rules for vector nomenclature:

- "p" indicates that this is indeed a plasmid.
- "BR" identifies the laboratory in which the vector was originally constructed (BR stands for Bolivar and Rodriguez, the two researchers who developed pBR322).
- "322" distinguishes this plasmid from others developed in the same laboratory (there are also plasmids called pBR325, pBR327, pBR328, etc.).

6.1.2 The useful properties of pBR322

The genetic and physical map of pBR322 (Figure 6.1) gives an indication of why this plasmid was such a popular cloning vector.

The first useful feature of pBR322 is its size. In Chapter 2 it was stated that a cloning vector ought to be less than 10 kb in size, to avoid problems such as DNA breakdown

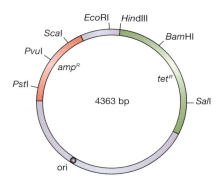

Figure 6.1

A map of pBR322 showing the positions of the ampicillin resistance (*amp^R*) and tetracycline resistance (*tet^R*) genes, the origin of replication (ori) and some of the most important restriction sites.

during purification. pBR322 is 4363 bp, which means that not only can the vector itself be purified with ease, but so can recombinant DNA molecules constructed with it. Even with 6 kb of additional DNA, a recombinant pBR322 molecule is still a manageable size.

The second feature of pBR322 is that, as described in Chapter 5, it carries two sets of antibiotic resistance genes. Either ampicillin or tetracycline resistance can be used as a selectable marker for cells containing the plasmid, and each marker gene includes unique restriction sites that can be used in cloning experiments. Insertion of new DNA into pBR322 that has been restricted with *Pst*I, *Pvu*I, or *Sca*I inactivates the *amp*^{R} gene, and insertion using any one of eight restriction endonucleases (notably *Bam*HI and *Hin*dIII) inactivates tetracycline resistance. This great variety of restriction sites that can be used for insertional inactivation means that pBR322 can be used to clone DNA fragments with any of several kinds of sticky end.

A third advantage of pBR322 is that it has a reasonably high copy number. Generally there are about 15 molecules present in a transformed *E. coli* cell, but this number can be increased, up to 1000–3000, by plasmid amplification in the presence of a protein synthesis inhibitor such as chloramphenicol (p. 39). An *E. coli* culture therefore provides a good yield of recombinant pBR322 molecules.

6.1.3 The pedigree of pBR322

The remarkable convenience of pBR322 as a cloning vector did not arise by chance. The plasmid was in fact designed in such a way that the final construct would possess these desirable properties. An outline of the scheme used to construct pBR322 is shown in Figure 6.2a. It can be seen that its production was a tortuous business that required full and skilfull use of the DNA manipulative techniques described in Chapter 4. A summary of the result of these manipulations is provided in Figure 6.2b, from which it can be seen that pBR322 comprises DNA derived from three different naturally occurring plasmids. The *amp*^{R} gene originally resided on the plasmid R1, a typical antibiotic resistance plasmid that occurs in natural populations of *E. coli* (p. 17). The *tet*^{R} gene is derived from R6-5, a second antibiotic-resistant plasmid. The replication origin of pBR322, which directs multiplication of the vector in host cells, is originally from pMB1, which is closely related to the colicin-producing plasmid ColE1 (p. 17).

6.1.4 More sophisticated E. coli plasmid cloning vectors

pBR322 was developed in the late 1970s, the first research paper describing its use being published in 1977. Since then many other plasmid cloning vectors have been constructed, the majority of these derived from pBR322 by manipulations similar to those summarized in Figure 6.2a. One of the first of these was pBR327, which was produced by removing a 1089 bp segment from pBR322. This deletion left the *amp*^{R} and *tet*^{R} genes intact, but changed the replicative and conjugative abilities of the resulting plasmid. As a result, pBR327 differs from pBR322 in two important ways:

- pBR327 has a higher copy number than pBR322, being present at about 30–45 molecules per *E. coli* cell. This is not of great relevance as far as plasmid yield is concerned, as both plasmids can be amplified to copy numbers greater than 1000. However, the higher copy number of pBR327 in normal cells makes this vector more suitable if the aim of the experiment is to study the function of the cloned gene. In these cases gene dosage becomes important, because the more copies there

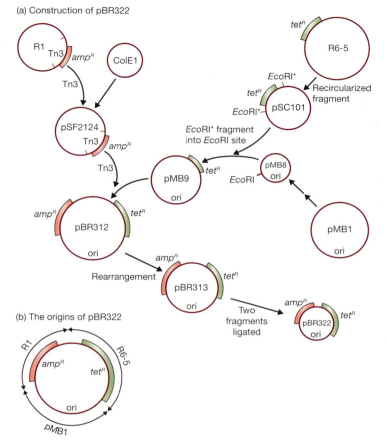

Figure 6.2

The pedigree of pBR322. (a) The manipulations involved in construction of pBR322. (b) A summary of the origins of pBR322.

are of a cloned gene, the more likely it is that the effect of the cloned gene on the host cell will be detectable. pBR327, with its high copy number, is therefore a better choice than pBR322 for this kind of work.

- The deletion also destroys the conjugative ability of pBR322, making pBR327 a non-conjugative plasmid that cannot direct its own transfer to other *E. coli* cells. This is important for **biological containment**, averting the possibility of a recombinant pBR327 molecule escaping from the test tube and colonizing bacteria in the gut of a careless molecular biologist. In contrast, pBR322 could theoretically be passed to natural populations of *E. coli* by conjugation, though in fact pBR322 also has safeguards (though less sophisticated ones) to minimize the chances of this happening. pBR327 is, however, preferable if the cloned gene is potentially harmful should an accident occur.

Although pBR327, like pBR322, is no longer widely used, its properties have been inherited by most of today's modern plasmid vectors. There are a great number of these, and it would be pointless to attempt to describe them all. Two additional examples will suffice to illustrate the most important features.

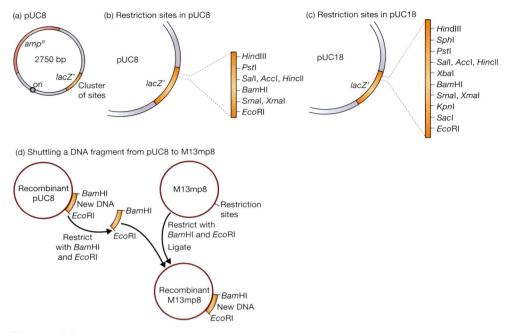

Figure 6.3

The pUC plasmids. (a) The structure of pUC8. (b) The restriction site cluster in the *lacZ'* gene of pUC8. (c) The restriction site cluster in pUC18. (d) Shuttling a DNA fragment from pUC8 to M13mp8.

pUC8—a Lac selection plasmid

This vector was mentioned in Chapter 5 when identification of recombinants by insertional inactivation of the β-galactosidase gene was described (p. 79). pUC8 (Figure 6.3a) is descended from pBR322, although only the replication origin and the *amp^R* gene remain. The nucleotide sequence of the *amp^R* gene has been changed so that it no longer contains the unique restriction sites: all these cloning sites are now clustered into a short segment of the *lacZ'* gene carried by pUC8.

pUC8 has three important advantages that have led to it becoming one of the most popular *E. coli* cloning vectors. The first of these is fortuitous: the manipulations involved in construction of pUC8 were accompanied by a chance mutation, within the origin of replication, which results in the plasmid having a copy number of 500–700 even before amplification. This has a significant effect on the yield of cloned DNA obtainable from *E. coli* cells transformed with recombinant pUC8 plasmids.

The second advantage is that identification of recombinant cells can be achieved by a single step process, by plating onto agar medium containing ampicillin plus X-gal (p. 79). With both pBR322 and pBR327, selection of recombinants is a two-step procedure, requiring replica plating from one antibiotic medium to another (p. 78). A cloning experiment with pUC8 can therefore be carried out in half the time needed with pBR322 or pBR327.

The third advantage of pUC8 lies with the clustering of the restriction sites, which allows a DNA fragment with two different sticky ends (say *Eco*RI at one end and *Bam*HI at the other) to be cloned without resorting to additional manipulations such as linker attachment (Figure 6.3b). Other pUC vectors carry different combinations of restriction

(a) pGEM3Z

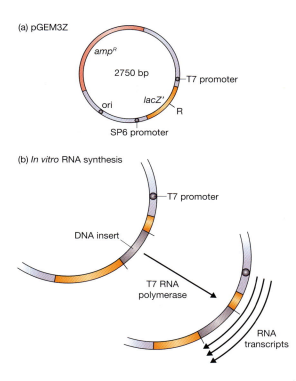

SP6 promoter

Figure 6.4

pGEM3Z. (a) Map of the vector. (b) *In vitro* RNA synthesis. R = cluster of restriction sites for *Eco*RI, *Sac*I, *Kpn*I, *Ava*I, *Sma*I, *Bam*HI, *Xba*I, *Sal*I, *Acc*I, *Hinc*II, *Pst*I, *Sph*I, and *Hind*III.

(b) *In vitro* RNA synthesis

sites and provide even greater flexibility in the types of DNA fragment that can be cloned (Figure 6.3c). Furthermore, the restriction site clusters in these vectors are the same as the clusters in the equivalent M13mp series of vectors (p. 95). DNA cloned into a member of the pUC series can therefore be transferred directly to its M13mp counterpart, enabling the cloned gene to be obtained as single-stranded DNA (Figure 6.3d).

pGEM3Z—in vitro transcription of cloned DNA

pGEM3Z (Figure 6.4a) is very similar to a pUC vector: it carries the *amp*^R and *lacZ′* genes, the latter containing a cluster of restriction sites, and it is almost exactly the same size. The distinction is that pGEM3Z has two additional short pieces of DNA, each of which acts as the recognition site for attachment of an RNA polymerase enzyme. These two **promoter** sequences lie on either side of the cluster of restriction sites used for introduction of new DNA into the pGEM3Z molecule. This means that if a recombinant pGEM3Z molecule is mixed with purified RNA polymerase in the test tube, transcription occurs and RNA copies of the cloned fragment are synthesized (Figure 6.4b). The RNA that is produced could be used as a hybridization probe (p. 133), or might be required for experiments aimed at studying RNA processing (e.g., the removal of introns) or protein synthesis.

The promoters carried by pGEM3Z and other vectors of this type are not the standard sequences recognized by the *E. coli* RNA polymerase. Instead, one of the promoters is specific for the RNA polymerase coded by T7 bacteriophage and the other for the RNA polymerase of SP6 phage. These RNA polymerases are synthesized during infection of *E. coli* with one or other of the phages and are responsible for transcribing the phage genes. They are chosen for use in *in vitro* transcription as they are very active enzymes – remember that the entire lytic infection cycle takes only 20 minutes (p. 18), so the phage genes must be transcribed very quickly. These polymerases are able to synthesize 1–2 mg of RNA per minute, substantially more than can be produced by the standard *E. coli* enzyme.

6.2 Cloning vectors based on M13 bacteriophage

The most essential requirement for any cloning vector is that it has a means of replicating in the host cell. For plasmid vectors this requirement is easy to satisfy, as relatively short DNA sequences are able to act as plasmid origins of replication, and most, if not all, of the enzymes needed for replication are provided by the host cell. Elaborate manipulations, such as those that resulted in pBR322 (see Figure 6.2a), are therefore possible so long as the final construction has an intact, functional replication origin.

With bacteriophages such as M13 and λ, the situation as regards replication is more complex. Phage DNA molecules generally carry several genes that are essential for replication, including genes coding for components of the phage protein coat and phage-specific DNA replicative enzymes. Alteration or deletion of any of these genes will impair or destroy the replicative ability of the resulting molecule. There is therefore much less freedom to modify phage DNA molecules, and generally phage cloning vectors are only slightly different from the parent molecule.

6.2.1 How to construct a phage cloning vector

The problems in constructing a phage cloning vector are illustrated by considering M13. The normal M13 genome is 6.4 kb in length, but most of this is taken up by ten closely packed genes (Figure 6.5), each essential for the replication of the phage. There is only a single 507-nucleotide intergenic sequence into which new DNA could be inserted without disrupting one of these genes, and this region includes the replication origin which must itself remain intact. Clearly there is only limited scope for modifying the M13 genome.

The first step in construction of an M13 cloning vector was to introduce the *lacZ'* gene into the intergenic sequence. This gave rise to M13mp1, which forms blue plaques on X-gal agar (Figure 6.6a). M13mp1 does not possess any unique restriction sites in the *lacZ'* gene. It does, however, contain the hexanucleotide GGATTC near the start of the gene. A single nucleotide change would make this GAATTC, which is an *Eco*RI site. This alteration was carried out using *in vitro* mutagenesis (p. 200), resulting in M13mp2

Figure 6.5

The M13 genome, showing the positions of genes I to X.

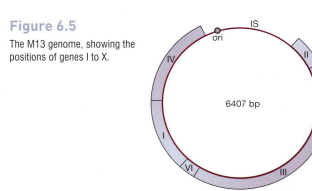

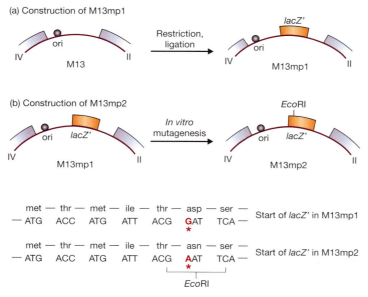

met — thr — met — ile — thr — asp — ser —   Start of *lacZ'* in M13mp1
— ATG   ACC   ATG   ATT   ACG   **G**AT   TCA —
                                          *

met — thr — met — ile — thr — asn — ser —   Start of *lacZ'* in M13mp2
— ATG   ACC   ATG   ATT   ACG   **A**AT   TCA —
                                          *
                                    └─────┬─────┘
                                       *Eco*RI

Figure 6.6

Construction of (a) M13mp1, and (b) M13mp2 from the wild-type M13 genome.

(Figure 6.6b). M13mp2 has a slightly altered *lacZ'* gene (the sixth codon now specifies asparagine instead of aspartic acid), but the β-galactosidase enzyme produced by cells infected with M13mp2 is still perfectly functional.

The next step in the development of M13 vectors was to introduce additional restriction sites into the *lacZ'* gene. This was achieved by synthesizing in the test tube a short oligonucleotide, called a **polylinker**, which consists of a series of restriction sites and has *Eco*RI sticky ends (Figure 6.7a). This polylinker was inserted into the *Eco*RI site of M13mp2, to give M13mp7 (Figure 6.7b), a more complex vector with four possible cloning sites (*Eco*RI, *Bam*HI, *Sal*I, and *Pst*I). The polylinker is designed so that it does

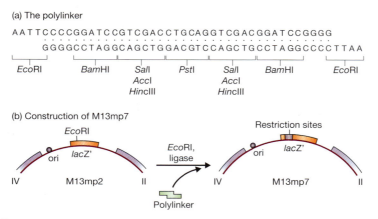

Figure 6.7

Construction of M13mp7: (a) the polylinker, and (b) its insertion into the *Eco*RI site of M13mp2. Note that the *Sal*I restriction sites are also recognized by *Acc*I and *Hinc*II.

not totally disrupt the *lacZ'* gene: a reading frame is maintained throughout the polylinker, and a functional, though altered, β-galactosidase enzyme is still produced.

The most sophisticated M13 vectors have more complex polylinkers inserted into the *lacZ'* gene. An example is M13mp8, which has the same series of restriction sites as the plasmid pUC8 (p. 92). As with the plasmid vector, one advantage of M13mp8 is its ability to take DNA fragments with two different sticky ends.

6.2.2 Hybrid plasmid–M13 vectors

Although M13 vectors are very useful for the production of single-stranded versions of cloned genes, they do suffer from one disadvantage. There is a limit to the size of DNA fragment that can be cloned with an M13 vector, with 1500 bp generally being looked on as the maximum capacity, though fragments up to 3 kb have occasionally been cloned. To get around this problem a number of hybrid vectors ("phagemids") have been developed by combining a part of the M13 genome with plasmid DNA.

An example is provided by pEMBL8 (Figure 6.8a), which was made by transferring into pUC8 a 1300 bp fragment of the M13 genome. This piece of M13 DNA contains

Figure 6.8

pEMBL8: a hybrid plasmid–M13 vector that can be converted into single-stranded DNA.

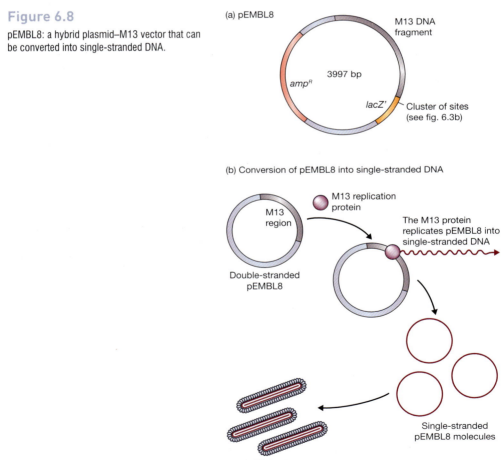

(a) pEMBL8

M13 DNA fragment

amp^R

3997 bp

lacZ'

Cluster of sites (see fig. 6.3b)

(b) Conversion of pEMBL8 into single-stranded DNA

M13 region

Double-stranded pEMBL8

M13 replication protein

The M13 protein replicates pEMBL8 into single-stranded DNA

Single-stranded pEMBL8 molecules

pEMBL8 'phage' particles

the signal sequence recognized by the enzymes that convert the normal double-stranded M13 molecule into single-stranded DNA before secretion of new phage particles. This signal sequence is still functional even though detached from the rest of the M13 genome, so pEMBL8 molecules are also converted into single-stranded DNA and secreted as defective phage particles (Figure 6.8b). All that is necessary is that the *E. coli* cells used as hosts for a pEMBL8 cloning experiment are subsequently infected with normal M13 to act as a helper phage, providing the necessary replicative enzymes and phage coat proteins. pEMBL8, being derived from pUC8, has the polylinker cloning sites within the *lacZ'* gene, so recombinant plaques can be identified in the standard way on agar containing X-gal. With pEMBL8, single-stranded versions of cloned DNA fragments up to 10 kb in length can be obtained, greatly extending the range of the M13 cloning system.

6.3 Cloning vectors based on λ bacteriophage

Two problems had to be solved before λ-based cloning vectors could be developed:

- The λ DNA molecule can be increased in size by only about 5%, representing the addition of only 3 kb of new DNA. If the total size of the molecule is more than 52 kb, then it cannot be packaged into the λ head structure and infective phage particles are not formed. This severely limits the size of a DNA fragment that can be inserted into an unmodified λ vector (Figure 6.9a).

- The λ genome is so large that it has more than one recognition sequence for virtually every restriction endonuclease. Restriction cannot be used to cleave the normal λ molecule in a way that will allow insertion of new DNA, because the molecule would be cut into several small fragments that would be very unlikely to re-form a viable λ genome on religation (Figure 6.9b).

In view of these difficulties it is perhaps surprising that a wide variety of λ cloning vectors have been developed, their primary use being to clone large pieces of DNA, from 5 to 25 kb, much too big to be handled by plasmid or M13 vectors.

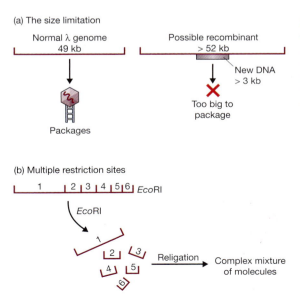

(a) The size limitation

Normal λ genome
49 kb

Possible recombinant
> 52 kb

New DNA
> 3 kb

Too big to package

Packages

(b) Multiple restriction sites

1 2 3 4 5 6 *Eco*RI

*Eco*RI

Religation → Complex mixture of molecules

Figure 6.9

The two problems that had to be solved before λ cloning vectors could be developed. (a) The size limitation placed on the λ genome by the need to package it into the phage head. (b) λ DNA has multiple recognition sites for almost all restriction endonucleases.

Figure 6.10

The λ genetic map, showing the position of the main non-essential region that can be deleted without affecting the ability of the phage to follow the lytic infection cycle. There are other, much shorter non-essential regions in other parts of the genome.

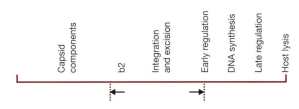

6.3.1 Segments of the λ genome can be deleted without impairing viability

The way forward for the development of λ cloning vectors was provided by the discovery that a large segment in the central region of the λ DNA molecule can be removed without affecting the ability of the phage to infect *E. coli* cells. Removal of all or part of this non-essential region, between positions 20 and 35 on the map shown in Figure 2.9, decreases the size of the resulting λ molecule by up to 15 kb. This means that as much as 18 kb of new DNA can now be added before the cut-off point for packaging is reached (Figure 6.10).

This "non-essential" region in fact contains most of the genes involved in integration and excision of the λ prophage from the *E. coli* chromosome. A deleted λ genome is therefore non-lysogenic and can follow only the lytic infection cycle. This in itself is desirable for a cloning vector as it means induction is not needed before plaques are formed (p. 40).

6.3.2 Natural selection can be used to isolate modified λ that lack certain restriction sites

Even a deleted λ genome, with the non-essential region removed, has multiple recognition sites for most restriction endonucleases. This is a problem that is often encountered when a new vector is being developed. If just one or two sites need to be removed, then the technique of *in vitro* mutagenesis (p. 200) can be used. For example, an *Eco*RI site, GAATTC, could be changed to GGATTC, which is not recognized by the enzyme. However, *in vitro* mutagenesis was in its infancy when the first λ vectors were under development, and even today would not be an efficient means of changing more than a few sites in a single molecule.

Instead, natural selection was used to provide strains of λ that lack the unwanted restriction sites. Natural selection can be brought into play by using as a host an *E. coli* strain that produces *Eco*RI. Most λ DNA molecules that invade the cell are destroyed by this restriction endonuclease, but a few survive and produce plaques. These are mutant phages, from which one or more *Eco*RI sites have been lost spontaneously (Figure 6.11). Several cycles of infection will eventually result in λ molecules that lack all or most of the *Eco*RI sites.

6.3.3 Insertion and replacement vectors

Once the problems posed by packaging constraints and by the multiple restriction sites had been solved, the way was open for the development of different types of λ-based cloning vectors. The first two classes of vector to be produced were λ **insertion** and λ **replacement** (or substitution) vectors.

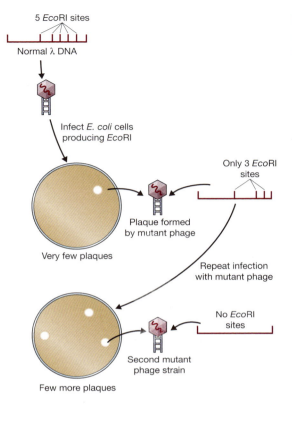

Figure 6.11

Using natural selection to isolate λ phage lacking *Eco*RI restriction sites.

5 *Eco*RI sites

Normal λ DNA

Infect *E. coli* cells
producing *Eco*RI

Only 3 *Eco*RI
sites

Plaque formed
by mutant phage

Very few plaques

Repeat infection
with mutant phage

No *Eco*RI
sites

Second mutant
phage strain

Few more plaques

(a) Construction of a λ insertion vector

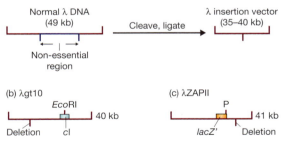

Figure 6.12

λ insertion vectors. P = polylinker in the *lacZ′* gene of λZAPII, containing unique restriction sites for *Sac*I, *Not*I, *Xba*I, *Spe*I, *Eco*RI, and *Xho*I.

Normal λ DNA
(49 kb)

Cleave, ligate

λ insertion vector
(35–40 kb)

Non-essential
region

(b) λgt10

*Eco*RI

Deletion   cI

40 kb

(c) λZAPII

P

lacZ′   Deletion

41 kb

Insertion vectors

With an insertion vector (Figure 6.12a), a large segment of the non-essential region has been deleted, and the two arms ligated together. An insertion vector possesses at least one unique restriction site into which new DNA can be inserted. The size of the DNA fragment that an individual vector can carry depends, of course, on the extent to which the non-essential region has been deleted. Two popular insertion vectors are:

- **λgt10** (Figure 6.12b), which can carry up to 8 kb of new DNA, inserted into a unique *Eco*RI site located in the *c*I gene. Insertional inactivation of this gene means that recombinants are distinguished as clear rather than turbid plaques (p. 83).

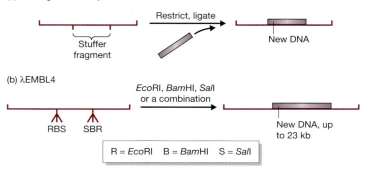

(a) Cloning with a λ replacement vector

Restrict, ligate

Stuffer fragment

New DNA

(b) λEMBL4

EcoRI, BamHI, SalI or a combination

RBS   SBR

New DNA, up to 23 kb

R = EcoRI   B = BamHI   S = SalI

Figure 6.13

λ replacement vectors. (a) Cloning with a λ replacement vector. (b) Cloning with λEMBL4.

- **λZAPII** (Figure 6.12c), with which insertion of up to 10 kb DNA into any of 6 restriction sites within a polylinker inactivates the *lacZ'* gene carried by the vector. Recombinants give clear rather than blue plaques on X-gal agar.

Replacement vectors

A λ replacement vector has two recognition sites for the restriction endonuclease used for cloning. These sites flank a segment of DNA that is replaced by the DNA to be cloned (Figure 6.13a). Often the replaceable fragment (or **"stuffer fragment"** in cloning jargon) carries additional restriction sites that can be used to cut it up into small pieces, so that its own re-insertion during a cloning experiment is very unlikely. Replacement vectors are generally designed to carry larger pieces of DNA than insertion vectors can handle. Recombinant selection is often on the basis of size, with non-recombinant vectors being too small to be packaged into λ phage heads (p. 84).

An example of a replacement vectors is:

- **λEMBL4** (Figure 6.13b) can carry up to 20 kb of inserted DNA by replacing a segment flanked by pairs of *EcoRI*, *BamHI*, and *SalI* sites. Any of these three restriction endonucleases can be used to remove the stuffer fragment, so DNA fragments with a variety of sticky ends can be cloned. Recombinant selection with λEMBL4 can be on the basis of size, or can utilize the Spi phenotype (p. 83).

6.3.4 Cloning experiments with λ insertion or replacement vectors

A cloning experiment with a λ vector can proceed along the same lines as with a plasmid vector—the λ molecules are restricted, new DNA is added, the mixture is ligated, and the resulting molecules used to transfect a competent *E. coli* host (Figure 6.14a). This type of experiment requires that the vector be in its circular form, with the *cos* sites hydrogen bonded to each other.

Although satisfactory for many purposes, a procedure based on transfection is not particularly efficient. A greater number of recombinants will be obtained if one or two refinements are introduced. The first is to use the linear form of the vector. When the linear form of the vector is digested with the relevant restriction endonuclease, the left and right arms are released as separate fragments. A recombinant molecule can

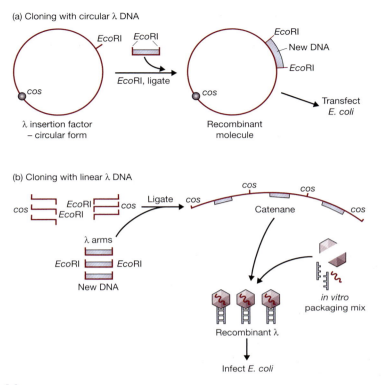

(a) Cloning with circular λ DNA

(b) Cloning with linear λ DNA

Figure 6.14

Different strategies for cloning with a λ vector. (a) Using the circular form of λ as a plasmid. (b) Using left and right arms of the λ genome, plus *in vitro* packaging, to achieve a greater number of recombinant plaques.

be constructed by mixing together the DNA to be cloned with the vector arms (Figure 6.14b). Ligation results in several molecular arrangements, including catenanes comprising left arm–DNA–right arm repeated many times (Figure 6.14b). If the inserted DNA is the correct size, then the *cos* sites that separate these structures will be the right distance apart for *in vitro* packaging (p. 81). Recombinant phage are therefore produced in the test tube and can be used to infect an *E. coli* culture. This strategy, in particular the use of *in vitro* packaging, results in a large number of recombinant plaques.

6.3.5 Long DNA fragments can be cloned using a cosmid

The final and most sophisticated type of λ-based vector is the **cosmid**. Cosmids are hybrids between a phage DNA molecule and a bacterial plasmid, and their design centers on the fact that the enzymes that package the λ DNA molecule into the phage protein coat need only the *cos* sites in order to function (p. 21). The *in vitro* packaging reaction works not only with λ genomes, but also with any molecule that carries *cos* sites separated by 37–52 kb of DNA.

A cosmid is basically a plasmid that carries a *cos* site (Figure 6.15a). It also needs a selectable marker, such as the ampicillin resistance gene, and a plasmid origin of replication, as cosmids lack all the λ genes and so do not produce plaques. Instead colonies are formed on selective media, just as with a plasmid vector.

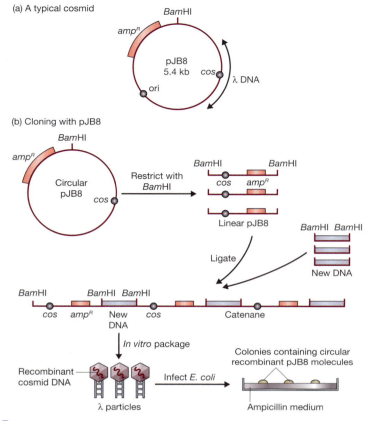

Figure 6.15

A typical cosmid and the way it is used to clone long fragments of DNA.

A cloning experiment with a cosmid is carried out as follows (Figure 6.15b). The cosmid is opened at its unique restriction site and new DNA fragments inserted. These fragments are usually produced by partial digestion with a restriction endonuclease, as total digestion almost invariably results in fragments that are too small to be cloned with a cosmid. Ligation is carried out so that catenanes are formed. Providing the inserted DNA is the right size, *in vitro* packaging cleaves the *cos* sites and places the recombinant cosmids in mature phage particles. These λ phage are then used to infect an *E. coli* culture, though of course plaques are not formed. Instead, infected cells are plated onto a selective medium and antibiotic-resistant colonies are grown. All colonies are recombinants, as non-recombinant linear cosmids are too small to be packaged into λ heads.

6.4 λ and other high-capacity vectors enable genomic libraries to be constructed

The main use of all λ-based vectors is to clone DNA fragments that are too long to be handled by plasmid or M13 vectors. A replacement vector, such as λEMBL4, can carry up to 20 kb of new DNA, and some cosmids can manage fragments up to 40 kb. This

Table 6.1

Number of clones needed for genomic libraries of a variety of organisms.

SPECIES	GENOME SIZE (bp)	NUMBER OF CLONES*	
		17 kb FRAGMENTS[†]	35 kb FRAGMENTS[‡]
E. coli	4.6×10^6	820	410
Saccharomyces cerevisiae	1.8×10^7	3225	1500
Drosophila melanogaster	1.2×10^8	21,500	10,000
Rice	5.7×10^8	100,000	49,000
Human	3.2×10^9	564,000	274,000
Frog	2.3×10^{10}	4,053,000	1,969,000

*Calculated for a probability (p) of 95% that any particular gene will be present in the library.

[†]Fragments suitable for a replacement vector such as λEMBL4.

[‡]Fragments suitable for a cosmid.

compares with a maximum insert size of about 8 kb for most plasmids and less than 3 kb for M13 vectors.

The ability to clone such long DNA fragments means that **genomic libraries** can be generated. A genomic library is a set of recombinant clones that contains all of the DNA present in an individual organism. An *E. coli* genomic library, for example, contains all the *E. coli* genes, so any desired gene can be withdrawn from the library and studied. Genomic libraries can be retained for many years, and propagated so that copies can be sent from one research group to another.

The big question is how many clones are needed for a genomic library? The answer can be calculated with the formula:

$$N = \frac{\ln(1 - p)}{\ln\left(1 - \dfrac{a}{b}\right)}$$

where N is the number of clones that are required, p is probability that any given gene will be present, a is the average size of the DNA fragments inserted into the vector, and b is the total size of the genome.

Table 6.1 shows the number of clones needed for genomic libraries of a variety of organisms, constructed using a λ replacement vector or a cosmid. For humans and other mammals, several hundred thousand clones are required. It is by no means impossible to obtain several hundred thousand clones, and the methods used to identify a clone carrying a desired gene (Chapter 8) can be adapted to handle such large numbers, so genomic libraries of these sizes are by no means unreasonable. However, ways of reducing the number of clones needed for mammalian genomic libraries are continually being sought.

One solution is to develop new cloning vectors able to handle longer DNA inserts. The most popular of these vectors are **bacterial artificial chromosomes (BACs)**, which are based on the F plasmid (p. 16). The F plasmid is relatively large and vectors derived from it have a higher capacity than normal plasmid vectors. BACs can handle DNA inserts up to 300 kb in size, reducing the size of the human genomic library to just 30,000 clones. Other high-capacity vectors have been constructed from bacteriophage **P1**, which has the advantage over λ of being able to squeeze 110 kb of DNA into its capsid structure. Cosmid-type vectors based on P1 have been designed and used to clone DNA fragments ranging in size from 75 to 100 kb. Vectors that combine the features of P1 vectors and BACs, called **P1-derived artificial chromosomes (PACs)**, also have a capacity of up to 300 kb.

6.5 Vectors for other bacteria

Cloning vectors have also been developed for several other species of bacteria, including *Streptomyces*, *Bacillus*, and *Pseudomonas*. Some of these vectors are based on plasmids specific to the host organism, and some on broad host range plasmids able to replicate in a variety of bacterial hosts. A few are derived from bacteriophages specific to these organisms. Antibiotic resistance genes are generally used as the selectable markers. Most of these vectors are very similar to *E. coli* vectors in terms of their general purposes and uses.

Further reading

Bolivar, F., Rodriguez, R.L., Green, P.J. et al. (1977) Construction and characterization of new cloning vectors. II. A multipurpose cloning system. *Gene*, 2, 95–113. [pBR322.]

Frischauf, A.-M., Lehrach, H., Poustka, A. & Murray, N. (1983) Lambda replacement vectors carrying polylinker sequences. *Journal of Molecular Biology*, 170, 827–842. [The λEMBL vectors.]

Iouannou, P.A., Amemiya, C.T., Garnes, J. et al. (1994) P1-derived vector for the propagation of large human DNA fragments. *Nature Genetics*, 6, 84–89.

Melton, D.A., Krieg, P.A., Rebagliati, M.R., Maniatis, T., Zinn, K. & Green, M.R. (1984) Efficient *in vitro* synthesis of biologically active RNA and RNA hybridization probes from plasmids containing a bacteriophage SP6 promoter. *Nucleic Acids Research*, 12, 7035–7056. [RNA synthesis from DNA cloned in a plasmid such as pGEM3Z.]

Sanger, F., Coulson, A.R., Barrell, B.G. et al. (1980) Cloning in single-stranded bacteriophage as an aid to rapid DNA sequencing. *Journal of Molecular Biology*, 143, 161–178. [M13 vectors.]

Shiyuza, H., Birren, B., Kim, U.J. et al. (1992) Cloning and stable maintenance of 300 kilobase-pair fragments of human DNA in *Escherichia coli* using an F-factor-based vector. *Proceedings of the National Academy of Sciences of the USA*, 89, 8794–8797. [The first description of a BAC.]

Sternberg, N. (1992) Cloning high molecular weight DNA fragments by the bacteriophage P1 system. *Trends in Genetics*, 8, 11–16.

Yanisch-Perron, C., Vieira, J. & Messing, J. (1985) Improved M13 phage cloning vectors and host strains: nucleotide sequences of the M13mp18 and pUC19 vectors. *Gene*, 33, 103–119.

Chapter 7

Cloning Vectors for Eukaryotes

Most cloning experiments are carried out with *E. coli* as the host, and the widest variety of cloning vectors are available for this organism. *E. coli* is particularly popular when the aim of the cloning experiment is to study the basic features of molecular biology such as gene structure and function. However, under some circumstances it may be desirable to use a different host for a gene cloning experiment. This is especially true in biotechnology (Chapter 13), where the aim may not be to study a gene, but to use cloning to obtain large amounts of an important pharmaceutical protein (e.g., a hormone such as insulin), or to change the properties of the organism (e.g., to introduce herbicide resistance into a crop plant). We must therefore consider cloning vectors for organisms other than *E. coli*.

7.1 Vectors for yeast and other fungi

The yeast *Saccharomyces cerevisiae* is one of the most important organisms in biotechnology. As well as its role in brewing and breadmaking, yeast has been used as a host organism for the production of important pharmaceuticals from cloned genes (p. 237). Development of cloning vectors for yeast was initially stimulated by the discovery of a plasmid that is present in most strains of *S. cerevisiae* (Figure 7.1). The 2 μm plasmid, as it is called, is one of only a very limited number of plasmids found in eukaryotic cells.

Gene Cloning and DNA Analysis: An Introduction. 6th edition. By T.A. Brown. Published 2010 by Blackwell Publishing.

Figure 7.1

The yeast 2 μm plasmid. *REP1* and *REP2* are involved in replication of the plasmid, and *FLP* codes for a protein that can convert the A form of the plasmid (shown here) to the B form, in which the gene order has been rearranged by intramolecular recombination. The function of *D* is not exactly known.

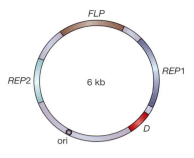

7.1.1 Selectable markers for the 2 μm plasmid

The 2 μm plasmid is an excellent basis for a cloning vector. It is 6 kb in size, which is ideal for a vector, and exists in the yeast cell at a copy number of between 70 and 200. Replication makes use of a plasmid origin, several enzymes provided by the host cell, and the proteins coded by the *REP1* and *REP2* genes carried by the plasmid.

However, all is not perfectly straightforward in using the 2 μm plasmid as a cloning vector. First, there is the question of a selectable marker. Some yeast cloning vectors carry genes conferring resistance to inhibitors such as methotrexate and copper, but most of the popular yeast vectors make use of a radically different type of selection system. In practice, a normal yeast gene is used, generally one that codes for an enzyme involved in amino acid biosynthesis. An example is the gene *LEU2*, which codes for β-isopropyl-malate dehydrogenase, one of the enzymes involved in the conversion of pyruvic acid to leucine.

In order to use *LEU2* as a selectable marker, a special kind of host organism is needed. The host must be an **auxotrophic** mutant that has a non-functional *LEU2* gene. Such a *leu2⁻* yeast is unable to synthesize leucine and can survive only if this amino acid is supplied as a nutrient in the growth medium (Figure 7.2a). Selection is possible because transformants contain a plasmid-borne copy of the *LEU2* gene, and so are able to grow in the absence of the amino acid. In a cloning experiment, cells are plated out onto **minimal medium**, which contains no added amino acids. Only transformed cells are able to survive and form colonies (Figure 7.2b).

7.1.2 Vectors based on the 2 μm plasmid—yeast episomal plasmids

Vectors derived from the 2 μm plasmid are called **yeast episomal plasmids (YEps)**. Some YEps contain the entire 2 μm plasmid, others include just the 2 μm origin of replication. An example of the latter type is YEp13 (Figure 7.3).

YEp13 illustrates several general features of yeast cloning vectors. First, it is a **shuttle vector**. As well as the 2 μm origin of replication and the selectable *LEU2* gene, YEp13 also includes the entire pBR322 sequence, and can therefore replicate and be selected for in both yeast and *E. coli*. There are several lines of reasoning behind the use of shuttle vectors. One is that it might be difficult to recover the recombinant DNA molecule from a transformed yeast colony. This is not such a problem with YEps, which are present in yeast cells primarily as plasmids, but with other yeast vectors, which may integrate into one of the yeast chromosomes (p. 108), purification might be impossible. This is a disadvantage because in many cloning experiments purification of

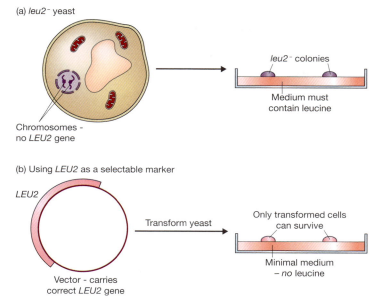

(a) *leu2⁻* yeast

leu2⁻ colonies

Medium must contain leucine

Chromosomes - no *LEU2* gene

(b) Using *LEU2* as a selectable marker

LEU2

Transform yeast

Only transformed cells can survive

Minimal medium – *no* leucine

Vector - carries correct *LEU2* gene

Figure 7.2

Using the *LEU2* gene as a selectable marker in a yeast cloning experiment.

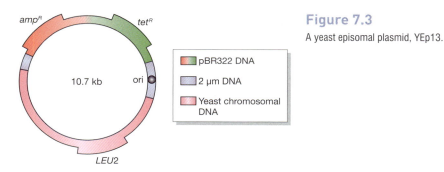

amp^R

tet^R

10.7 kb

ori

LEU2

■ pBR322 DNA

□ 2 μm DNA

▨ Yeast chromosomal DNA

Figure 7.3

A yeast episomal plasmid, YEp13.

recombinant DNA is essential in order for the correct construct to be identified by, for example, DNA sequencing.

The standard procedure when cloning in yeast is therefore to perform the initial cloning experiment with *E. coli*, and to select recombinants in this organism. Recombinant plasmids can then be purified, characterized, and the correct molecule introduced into yeast (Figure 7.4).

7.1.3 A YEp may insert into yeast chromosomal DNA

The word "episomal" indicates that a YEp can replicate as an independent plasmid, but also implies that integration into one of the yeast chromosomes can occur (see the definition of "episome" on p. 14). Integration occurs because the gene carried on the vector as a selectable marker is very similar to the mutant version of the gene present in the yeast chromosomal DNA. With YEp13, for example, homologous recombination can occur between the plasmid *LEU2* gene and the yeast mutant *LEU2* gene, resulting

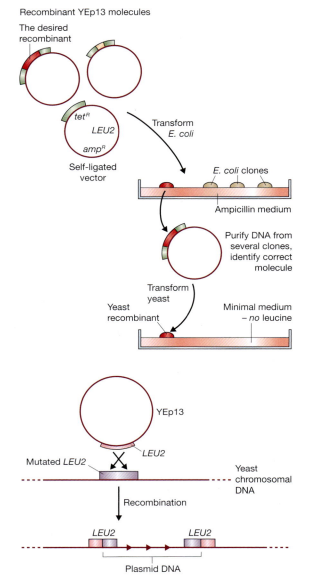

Figure 7.4

Cloning with an *E. coli*–yeast shuttle vector such as YEp13.

Figure 7.5

Recombination between plasmid and chromosomal *LEU2* genes can integrate YEp13 into yeast chromosomal DNA. After integration there are two copies of the *LEU2* gene; usually one is functional, and the other mutated.

in insertion of the entire plasmid into one of the yeast chromosomes (Figure 7.5). The plasmid may remain integrated, or a later recombination event may result in it being excised again.

7.1.4 Other types of yeast cloning vector

In addition to YEps, there are several other types of cloning vector for use with *S. cerevisiae*. Two important ones are as follows:

- **Yeast integrative plasmids (YIps)** are basically bacterial plasmids carrying a yeast gene. An example is YIp5, which is pBR322 with an inserted *URA3* gene

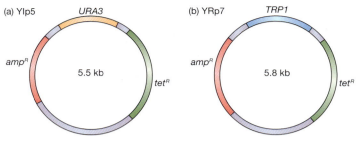

Figure 7.6

A YIp and a YRp.

(Figure 7.6a). This gene codes for orotidine-5′-phosphate decarboxylase (an enzyme that catalyzes one of the steps in the biosynthesis pathway for pyrimidine nucleotides) and is used as a selectable marker in exactly the same way as *LEU2*. A YIp cannot replicate as a plasmid as it does not contain any parts of the 2 μm plasmid, and instead depends for its survival on integration into yeast chromosomal DNA. Integration occurs just as described for a YEp (Figure 7.5).

● **Yeast replicative plasmids (YRps)** are able to multiply as independent plasmids because they carry a chromosomal DNA sequence that includes an origin of replication. Replication origins are known to be located very close to several yeast genes, including one or two which can be used as selectable markers. YRp7 (Figure 7.6b) is an example of a replicative plasmid. It is made up of pBR322 plus the yeast gene *TRP1*. This gene, which is involved in tryptophan biosynthesis, is located adjacent to a chromosomal origin of replication. The yeast DNA fragment present in YRp7 contains both *TRP1* and the origin.

Three factors come into play when deciding which type of yeast vector is most suit-able for a particular cloning experiment. The first of these is **transformation frequency**, a measure of the number of transformants that can be obtained per microgram of plasmid DNA. A high transformation frequency is necessary if a large number of recom-binants are needed, or if the starting DNA is in short supply. YEps have the highest transformation frequency, providing between 10,000 and 100,000 transformed cells per μg. YRps are also quite productive, giving between 1000 and 10,000 transformants per μg, but a YIp yields less than 1000 transformants per μg, and only 1–10 unless special procedures are used. The low transformation frequency of a YIp reflects the fact that the rather rare chromosomal integration event is necessary before the vector can be retained in a yeast cell.

The second important factor is copy number. YEps and YRps have the highest copy numbers: 20–50 and 5–100, respectively. In contrast, a YIp is usually present at just one copy per cell. These figures are important if the objective is to obtain protein from the cloned gene, as the more copies there are of the gene the greater the expected yield of the protein product.

So why would one ever wish to use a YIp? The answer is because YIps produce very stable recombinants, as loss of a YIp that has become integrated into a chromosome occurs at only a very low frequency. In contrast, YRp recombinants are extremely unstable, the plasmids tending to congregate in the mother cell when a daughter cell buds off, so the daughter cell is non-recombinant. YEp recombinants suffer from similar problems, though an improved understanding of the biology of the 2 μm plasmid has enabled

Figure 7.7

Chromosome structure.

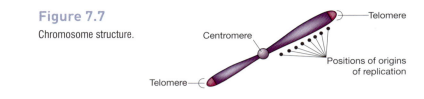

more stable YEps to be developed in recent years. Nevertheless, a YIp is the vector of choice if the needs of the experiment dictate that the recombinant yeast cells must retain the cloned gene for long periods in culture.

7.1.5 Artificial chromosomes can be used to clone long pieces of DNA in yeast

The final type of yeast cloning vector to consider is the **yeast artificial chromosome (YAC)**, which presents a totally different approach to gene cloning. The development of YACs was a spin-off from fundamental research into the structure of eukaryotic chromosomes, work that has identified the key components of a chromosome as being (Figure 7.7):

- The centromere, which is required for the chromosome to be distributed correctly to daughter cells during cell division;
- Two telomeres, the structures at the ends of a chromosome, which are needed in order for the ends to be replicated correctly and which also prevent the chromosome from being nibbled away by exonucleases;
- The origins of replication, which are the positions along the chromosome at which DNA replication initiates, similar to the origin of replication of a plasmid.

Once chromosome structure had been defined in this way, the possibility arose that the individual components might be isolated by recombinant DNA techniques and then joined together again in the test tube, creating an artificial chromosome. As the DNA molecules present in natural yeast chromosomes are several hundred kilobases in length, it might be possible with an artificial chromosome to clone long pieces of DNA.

The structure and use of a YAC vector

Several YAC vectors have been developed but each one is constructed along the same lines, with pYAC3 being a typical example (Figure 7.8a). At first glance, pYAC3 does not look much like an artificial chromosome, but on closer examination its unique features become apparent. pYAC3 is essentially a pBR322 plasmid into which a number of yeast genes have been inserted. Two of these genes, *URA3* and *TRP1*, have been encountered already as the selectable markers for YIp5 and YRp7, respectively. As in YRp7, the DNA fragment that carries *TRP1* also contains an origin of replication, but in pYAC3 this fragment is extended even further to include the sequence called *CEN4*, which is the DNA from the centromere region of chromosome 4. The *TRP1*–origin–*CEN4* fragment therefore contains two of the three components of the artificial chromosome.

The third component, the telomeres, is provided by the two sequences called *TEL*. These are not themselves complete telomere sequences, but once inside the yeast nucleus they act as seeding sequences onto which telomeres will be built. This just leaves one other part of pYAC3 that has not been mentioned: *SUP4*, which is the selectable marker into which new DNA is inserted during the cloning experiment.

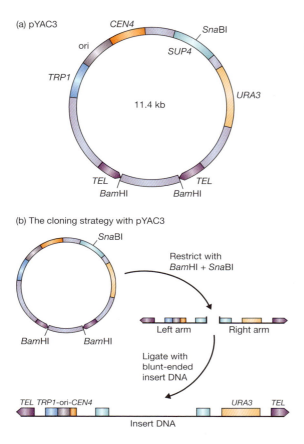

Figure 7.8

A YAC vector and the way it is used to clone large pieces of DNA.

The cloning strategy with pYAC3 is as follows (Figure 7.8b). The vector is first restricted with a combination of *Bam*HI and *Sna*BI, cutting the molecule into three fragments. The fragment flanked by *Bam*HI sites is discarded, leaving two arms, each bounded by one *TEL* sequence and one *Sna*BI site. The DNA to be cloned, which must have blunt ends (*Sna*BI is a blunt end cutter, recognizing the sequence TACGTA), is ligated between the two arms, producing the artificial chromosome. Protoplast trans-formation (p. 85) is then used to introduce the artificial chromosome into *S. cerevisiae*. The yeast strain that is used is a double auxotrophic mutant, *trp1⁻ ura3⁻*, which is con-verted to *trp1⁺ ura3⁺* by the two markers on the artificial chromosome. Transformants are therefore selected by plating onto minimal medium, on which only cells containing a correctly constructed artificial chromosome are able to grow. Any cell transformed with an incorrect artificial chromosome, containing two left or two right arms rather than one of each, is not able to grow on minimal medium as one of the markers is absent. The presence of the insert DNA in the vector can be checked by testing for inser-tional inactivation of *SUP4*, which is carried out by a simple color test: white colonies are recombinants, red colonies are not.

Applications for YAC vectors

The initial stimulus in designing artificial chromosomes came from yeast geneticists who wanted to use them to study various aspects of chromosome structure and behavior, for instance to examine the segregation of chromosomes during meiosis. These experiments established that artificial chromosomes are stable during propagation in yeast cells and raised the possibility that they might be used as vectors for genes that are too long to be cloned as a single fragment in an *E. coli* vector. Several important mammalian genes are greater than 100 kb in length (e.g., the human cystic fibrosis gene is 250 kb), beyond the capacity of all but the most sophisticated *E. coli* cloning systems (p. 102), but well within the range of a YAC vector. Yeast artificial chromosomes therefore opened the way to studies of the functions and modes of expression of genes that had previously been intractable to analysis by recombinant DNA techniques. A new dimension to these experiments was provided by the discovery that under some circumstances YACs can be propagated in mammalian cells, enabling the functional analysis to be carried out in the organism in which the gene normally resides.

Yeast artificial chromosomes are equally important in the production of gene libraries. Recall that with fragments of 300 kb, the maximum insert size for the highest capacity *E. coli* vector, some 30,000 clones are needed for a human gene library (p. 103). However, YAC vectors are routinely used to clone 600 kb fragments, and special types are able to handle DNA up to 1400 kb in length, the latter bringing the size of a human gene library down to just 6500 clones. Unfortunately these "mega-YACs" have run into problems with insert stability, the cloned DNA sometimes becoming rearranged by intramolecular recombination. Nevertheless, YACs have been of immense value in providing long pieces of cloned DNA for use in large-scale DNA sequencing projects.

7.1.6 Vectors for other yeasts and fungi

Cloning vectors for other species of yeast and fungi are needed for basic studies of the molecular biology of these organisms and to extend the possible uses of yeasts and fungi in biotechnology. Episomal plasmids based on the *S. cerevisiae* 2 μm plasmid are able to replicate in a few other types of yeast, but the range of species is not broad enough for 2 μm vectors to be of general value. In any case, the requirements of biotechnology are better served by integrative plasmids, equivalent to YIps, as these provide stable recombinants that can be grown for long periods in bioreactors (p. 225). Efficient integrative vectors are now available for a number of species, including yeasts such as *Pichia pastoris* and *Kluveromyces lactis*, and the filamentous fungi such as *Aspergillus nidulans* and *Neurospora crassa*.

7.2 Cloning vectors for higher plants

Cloning vectors for higher plants were developed in the 1980s and their use has led to the genetically modified (GM) crops that are in the headlines today. We will examine the genetic modification of crops and other plants in Chapter 15. Here we look at the cloning vectors and how they are used.

Three types of vector system have been used with varying degrees of success with higher plants:

- Vectors based on naturally occurring plasmids of *Agrobacterium*;
- Direct gene transfer using various types of plasmid DNA;
- Vectors based on plant viruses.

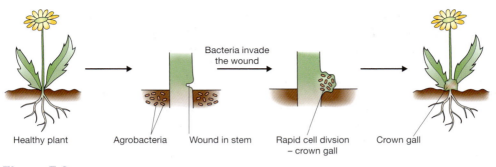

Figure 7.9

Crown gall disease.

7.2.1 Agrobacterium tumefaciens—nature's smallest genetic engineer

Although no naturally occurring plasmids are known in higher plants, one bacterial plasmid, the Ti plasmid of *Agrobacterium tumefaciens*, is of great importance.

A. tumefaciens is a soil microorganism that causes crown gall disease in many species of dicotyledonous plants. Crown gall occurs when a wound on the stem allows *A. tumefaciens* bacteria to invade the plant. After infection the bacteria cause a cancerous proliferation of the stem tissue in the region of the crown (Figure 7.9).

The ability to cause crown gall disease is associated with the presence of the Ti (tumor inducing) plasmid within the bacterial cell. This is a large (greater than 200 kb) plasmid that carries numerous genes involved in the infective process (Figure 7.10a). A remarkable feature of the Ti plasmid is that, after infection, part of the molecule is integrated into the plant chromosomal DNA (Figure 7.10b). This segment, called the T-DNA, is between 15 and 30 kb in size, depending on the strain. It is maintained in a stable form in the plant cell and is passed on to daughter cells as an integral part of the chromosomes. But the most remarkable feature of the Ti plasmid is that the T-DNA contains eight or so genes that are expressed in the plant cell and are responsible for the cancerous properties of the transformed cells. These genes also direct synthesis of unusual compounds, called opines, that the bacteria use as nutrients (Figure 7.10c). In short, *A. tumefaciens* genetically engineers the plant cell for its own purposes.

Using the Ti plasmid to introduce new genes into a plant cell

It was realized very quickly that the Ti plasmid could be used to transport new genes into plant cells. All that would be necessary would be to insert the new genes into the T-DNA and then the bacterium could do the hard work of integrating them into the plant chromosomal DNA. In practice this has proved a tricky proposition, mainly because the large size of the Ti plasmid makes manipulation of the molecule very difficult.

The main problem is, of course, that a unique restriction site is an impossibility with a plasmid 200 kb in size. Novel strategies have to be developed for inserting new DNA into the plasmid. Two are in general use:

- **The binary vector strategy** (Figure 7.11) is based on the observation that the T-DNA does not need to be physically attached to the rest of the Ti plasmid.

Figure 7.10

The Ti plasmid and its integration into the plant chromosomal DNA after *A. tumefaciens* infection.

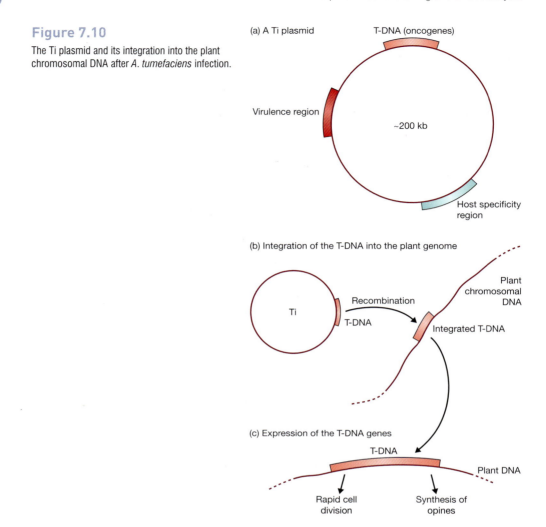

A two-plasmid system, with the T-DNA on a relatively small molecule, and the rest of the plasmid in normal form, is just as effective at transforming plant cells. In fact, some strains of *A. tumefaciens*, and related agrobacteria, have natural binary plasmid systems. The T-DNA plasmid is small enough to have a unique restriction site and to be manipulated using standard techniques.

• **The co-integration strategy** (Figure 7.12) uses an entirely new plasmid, based on an *E. coli* vector, but carrying a small portion of the T-DNA. The homology between the new molecule and the Ti plasmid means that if both are present in the same *A. tumefaciens* cell, recombination can integrate the *E. coli* plasmid into the T-DNA region. The gene to be cloned is therefore inserted into a unique restriction site on the small *E. coli* plasmid, introduced into *A. tumefaciens* cells carrying a Ti plasmid, and the natural recombination process left to integrate the new gene into the T-DNA. Infection of the plant leads to insertion of the new gene, along with the rest of the T-DNA, into the plant chromosomes.

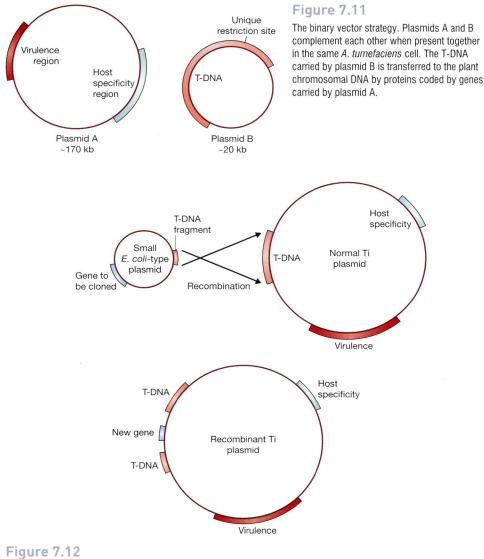

Figure 7.11

The binary vector strategy. Plasmids A and B complement each other when present together in the same *A. tumefaciens* cell. The T-DNA carried by plasmid B is transferred to the plant chromosomal DNA by proteins coded by genes carried by plasmid A.

Figure 7.12

The cointegration strategy.

Production of transformed plants with the Ti plasmid

If *A. tumefaciens* bacteria that contain an engineered Ti plasmid are introduced into a plant in the natural way, by infection of a wound in the stem, then only the cells in the resulting crown gall will possess the cloned gene (Figure 7.13a). This is obviously of little value to the biotechnologist. Instead a way of introducing the new gene into every cell in the plant is needed.

There are several solutions, the simplest being to infect not the mature plant but a culture of plant cells or protoplasts (p. 85) in liquid medium (Figure 7.13b). Plant cells and protoplasts whose cell walls have re-formed can be treated in the same way as microorganisms: for example, they can be plated onto a selective medium in order to

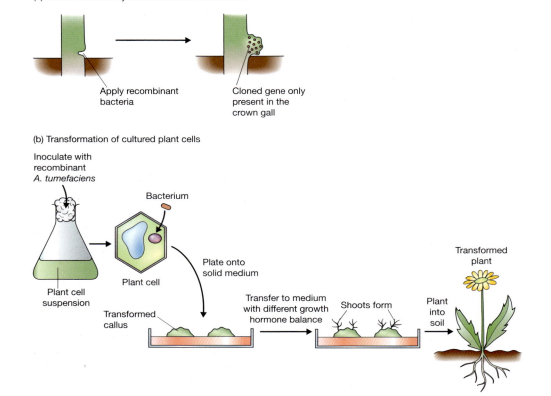

(a) Wound infection by recombinant *A. tumefaciens*

Apply recombinant bacteria

Cloned gene only present in the crown gall

(b) Transformation of cultured plant cells

Inoculate with recombinant *A. tumefaciens*

Bacterium

Plant cell

Plate onto solid medium

Transformed plant

Plant cell suspension

Transformed callus

Transfer to medium with different growth hormone balance

Shoots form

Plant into soil

Figure 7.13

Transformation of plant cells by recombinant *A. tumefaciens*. (a) Infection of a wound: transformed plant cells are present only in the crown gall. (b) Transformation of a cell suspension: all the cells in the resulting plant are transformed.

isolate transformants. A mature plant regenerated from transformed cells will contain the cloned gene in every cell and will pass the cloned gene to its offspring. However, regeneration of a transformed plant can occur only if the Ti vector has been "disarmed" so that the transformed cells do not display cancerous properties. Disarming is possible because the cancer genes, all of which lie in the T-DNA, are not needed for the infection process, infectivity being controlled mainly by the virulence region of the Ti plasmid. In fact, the only parts of the T-DNA that are involved in infection are two 25 bp repeat sequences found at the left and right borders of the region integrated into the plant DNA. Any DNA placed between these two repeat sequences will be treated as "T-DNA" and transferred to the plant. It is therefore possible to remove all the cancer genes from the normal T-DNA, and replace them with an entirely new set of genes, without disturbing the infection process.

A number of disarmed Ti cloning vectors are now available, a typical example being the binary vector pBIN19 (Figure 7.14). The left and right T-DNA borders present in this vector flank a copy of the *lacZ'* gene, containing a number of cloning sites, and a kanamycin resistance gene that functions after integration of the vector sequences into the plant chromosome. As with a yeast shuttle vector (p. 106), the initial manipulations

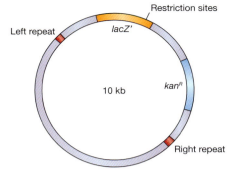

Figure 7.14
The binary Ti vector pBIN19. *kan*^{R} = kanamycin resistance gene.

that result in insertion of the gene to be cloned into pBIN19 are carried out in *E. coli*, the correct recombinant pBIN19 molecule then being transferred to *A. tumefaciens* and thence into the plant. Transformed plant cells are selected by plating onto agar medium containing kanamycin.

The Ri plasmid

Over the years there has also been interest in developing plant cloning vectors based on the Ri plasmid of *Agrobacterium rhizogenes*. Ri and Ti plasmids are very similar, the main difference being that transfer of the T-DNA from an Ri plasmid to a plant results not in a crown gall but in hairy root disease, typified by a massive proliferation of a highly branched root system. The possibility of growing transformed roots at high density in liquid culture has been explored by biotechnologists as a potential means of obtaining large amounts of protein from genes cloned in plants (p. 242).

Limitations of cloning with Agrobacterium plasmids

Higher plants are divided into two broad categories, the monocots and the dicots. Several factors have combined to make it much easier to clone genes in dicots such as tomato, tobacco, potato, peas, and beans, but much more difficult to obtain the same results with monocots. This has been frustrating because monocots include wheat, barley, rice, and maize, which are the most important crop plants and hence the most desirable targets for genetic engineering projects.

The main difficulty stems from the fact that in nature *A. tumefaciens* and *A. rhizogenes* infect only dicotyledonous plants; monocots are outside of the normal host range. For some time it was thought that this natural barrier was insurmountable and that monocots were totally resistant to transformation with Ti and Ri vectors, but eventually artificial techniques for achieving T-DNA transfer were devised. However, this was not the end of the story. Transformation with an *Agrobacterium* vector normally involves regeneration of an intact plant from a transformed protoplast, cell, or callus culture. The ease with which a plant can be regenerated depends very much on the particular species involved and, once again, the most difficult plants are the monocots. Attempts to circumvent this problem have centered on the use of biolistics—bombardment with microprojectiles (p. 85)—to introduce plasmid DNA directly into plant embryos. Although this is a fairly violent transformation procedure it does not appear to be too damaging for the embryos, which still continue their normal development program to produce mature plants. The approach has been successful with maize and several other important monocots.

Figure 7.15
Direct gene transfer.

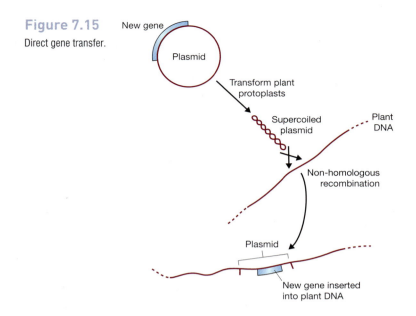

7.2.2 Cloning genes in plants by direct gene transfer

Biolistics circumvents the need to use *Agrobacterium* as the means of transferring DNA into the plant cells. **Direct gene transfer** takes the process one step further and dispenses with the Ti plasmid altogether.

Direct gene transfer into the nucleus

Direct gene transfer is based on the observation, first made in 1984, that a supercoiled bacterial plasmid, although unable to replicate in a plant cell on its own, can become integrated by recombination into one of the plant chromosomes. The recombination event is poorly understood but is almost certainly distinct from the processes responsible for T-DNA integration. It is also distinct from the chromosomal integration of a yeast vector (p. 107), as there is no requirement for a region of similarity between the bacterial plasmid and the plant DNA. In fact, integration appears to occur randomly at any position in any of the plant chromosomes (Figure 7.15).

One method involves resuspending protoplasts in a viscous solution of polyethylene glycol, a polymeric, negatively charged compound that is thought to precipitate DNA onto the surfaces of the protoplasts and to induce uptake by endocytosis (Figure 7.16). Electroporation (p. 85) is also sometimes used to increase transformation frequency. After treatment, protoplasts are left for a few days in a solution that encourages regeneration of the cell walls. The cells are then spread onto selective medium to identify transformants and to provide callus cultures from which intact plants can be grown (exactly as described for the *Agrobacterium* system, Figure 7.13b).

Direct gene transfer therefore makes use of supercoiled plasmid DNA, possibly a simple bacterial plasmid, into which an appropriate selectable marker (e.g., a kanamycin resistance gene) and the gene to be cloned have been inserted. Biolistics is frequently used to introduce the plasmid DNA into plant embryos, but if the species being engineered can be regenerated from protoplasts or single cells, then other strategies, possibly more efficient than biolistics, are possible.

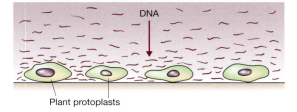

Figure 7.16

Direct gene transfer by precipitation of DNA onto the surfaces of protoplasts.

Plant protoplasts

Transfer of genes into the chloroplast genome

If biolistics is used to introduce DNA in a plant embryo, then some particles may penetrate one or more of the chloroplasts present in the cells. Chloroplasts contain their own genomes, distinct from (and much shorter) than the DNA molecules in the nucleus, and under some circumstances plasmid DNA can become integrated into this chloroplast genome. Unlike the integration of DNA into nuclear chromosomes, integration into the chloroplast genome will not occur randomly. Instead the DNA to be cloned must be flanked by sequences similar to the region of the chloroplast genome into which the DNA is to be inserted, so that insertion can take place by homologous recombination (see p. 107). Each of these flanking sequences must be 500 bp or so in length. A low level of chloroplast transformation can also be achieved after PEG-induced DNA delivery into protoplasts if the plasmid that is taken up carries these flanking sequences.

A plant cell contains tens of chloroplasts, and probably only one per cell becomes transformed, so the inserted DNA must carry a selectable marker such as the kanamycin resistance gene, and the embryos must be treated with the antibiotic for a considerable period to ensure that the transformed genomes propagate within the cell. Although this means that chloroplast transformation is a difficult method to carry out successfully, it could become an important adjunct to the more traditional methods for obtaining GM crops. As each cell has many chloroplasts, but only one nucleus, a gene inserted into the chloroplast genome is likely to be expressed at a higher level than one placed in the nucleus. This is particularly important when the engineered plants are to be used for production of pharmaceutical proteins (Chapter 13). So far the approach has been most successful with tobacco but chloroplast transformation has also been achieved with more useful crops such as soybean and cotton.

7.2.3 Attempts to use plant viruses as cloning vectors

Modified versions of λ and M13 bacteriophages are important cloning vectors for *E. coli* (Chapter 6). Most plants are subject to viral infection, so could viruses be used to clone genes in plants? If they could, then they would be much more convenient to use than other types of vector, because with many viruses transformation can be achieved simply by rubbing the virus DNA onto the surface of a leaf. The natural infection process then spreads the virus throughout the plant.

The potential of plant viruses as cloning vectors has been explored for several years but without great success. One problem is that the vast majority of plant viruses have genomes not of DNA but of RNA. RNA viruses are not so useful as potential cloning vectors because manipulations with RNA are more difficult to carry out. Only two classes of DNA virus are known to infect higher plants, the caulimoviruses and geminiviruses, and neither is ideally suited for gene cloning.

Caulimovirus vectors

Although one of the first successful plant genetic engineering experiments, back in 1984, used a caulimovirus vector to clone a new gene into turnip plants, two general difficulties with these viruses have limited their usefulness.

The first is that the total size of a caulimovirus genome is, like that of λ, constrained by the need to package it into its protein coat. Even after deletion of non-essential sections of the virus genome, the capacity for carrying inserted DNA is still very limited. Recent research has shown that it might be possible to circumvent this problem by adopting a helper virus strategy, similar to that used with phagemids (p. 96). In this strategy, the cloning vector is a cauliflower mosaic virus (CaMV) genome that lacks several of the essential genes, which means that it can carry a large DNA insert but cannot by itself direct infection. Plants are inoculated with the vector DNA along with a normal CaMV genome. The normal viral genome provides the genes needed for the cloning vector to be packaged into virus proteins and spread through the plant.

This approach has considerable potential, but does not solve the second problem, which is the extremely narrow host range of caulimoviruses. This restricts cloning experiments to just a few plants, mainly brassicas such as turnips, cabbages, and cauliflowers. Caulimoviruses have, however, been important in genetic engineering as the source of highly active promoters that work in all plants and that are used to obtain expression of genes introduced by Ti plasmid cloning or direct gene transfer.

Geminivirus vectors

What of the geminiviruses? These are particularly interesting because their natural hosts include plants such as maize and wheat, and they could therefore be potential vectors for these and other monocots. But geminiviruses have presented their own set of difficulties, one problem being that during the infection cycle the genomes of some geminiviruses undergo rearrangements and deletions, which would scramble up any additional DNA that has been inserted, an obvious disadvantage for a cloning vector. Research over the years has addressed these problems, and geminiviruses are beginning to find some specialist applications in plant gene cloning. One of these is in virus-induced gene silencing (VIGS), a technique used to investigate the functions of individual plant genes. This method exploits one of the natural defence mechanisms that plants use to protect themselves against viral attack. This method, called RNA silencing, results in degradation of viral mRNAs. If one of the viral RNAs is transcribed from a cloned gene contained within a geminivirus genome, then not only the viral transcripts but also the cellular mRNAs derived from the plant's copy of the gene are degraded (Figure 7.17). The plant gene therefore becomes silenced and the effect of its inactivation on the phenotype of the plant can be studied.

7.3 Cloning vectors for animals

Considerable effort has been put into the development of vector systems for cloning genes in animal cells. These vectors are needed in biotechnology for the synthesis of recombinant protein from genes that are not expressed correctly when cloned in *E. coli* or yeast (Chapter 13), and methods for cloning in humans are being sought by clinical molecular biologists attempting to devise techniques for gene therapy (p. 259), in which a disease is treated by introduction of a cloned gene into the patient.

The clinical aspect has meant that most attention has been directed at cloning systems for mammals, but important progress has also been made with insects. Cloning in insects

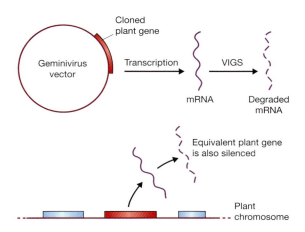

Figure 7.17

The use of a geminivirus vector to silence a plant gene via virus-induced gene silencing.

is interesting because it makes use of a novel type of vector that we have not met so far. We will therefore examine insect vectors before concluding the chapter with an overview of the cloning methods used with mammals.

7.3.1 Cloning vectors for insects

The fruit fly, *Drosophila melanogaster*, has been and still is one of the most important model organisms used by biologists. Its potential was first recognized by the famous geneticist Thomas Hunt Morgan, who in 1910 started to carry out genetic crosses between fruit flies with different eye colors, body shapes, and other inherited characteristics. These experiments led to the techniques still used today for gene mapping in insects and other animals. More recently, the discovery that the homeotic selector genes of *Drosophila*—the genes that control the overall body plan of the fly—are closely related to equivalent genes in mammals, has led to *D. melanogaster* being used as a model for the study of human developmental processes. The importance of the fruit fly in modern biology makes it imperative that vectors for cloning genes in this organism are available.

P elements as cloning vectors for Drosophila

The development of cloning vectors for *Drosophila* has taken a different route to that followed with bacteria, yeast, plants, and mammals. No plasmids are known in *Drosophila* and although fruit flies are, like all organisms, susceptible to infection with viruses, these have not been used as the basis for cloning vectors. Instead, cloning in *Drosophila* makes use of a **transposon** called the **P element**.

Transposons are common in all types of organism. They are short pieces of DNA (usually less than 10 kb in length) that can move from one position to another in the chromosomes of a cell. P elements, which are one of several types of transposon in *Drosophila*, are 2.9 kb in length and contain three genes flanked by short inverted repeat sequences at either end of the element (Figure 7.18a). The genes code for transposase, the enzyme that carries out the transposition process, and the inverted repeats form the recognition sequences that enable the enzyme to identify the two ends of the inserted transposon.

As well as moving from one site to another within a single chromosome, P elements can also jump between chromosomes, or between a plasmid carrying a P element and one of the fly's chromosomes (Figure 7.18b). The latter is the key to the use of P

Figure 7.18

Cloning in *Drosophila* with a P element vector.
(a) The structure of a P element. (b) Transposition
of a P element from a plasmid to a fly
chromosome. (c) The structure of a P element
cloning vector. The left-hand P element contains
a cloning site (R) that disrupts its transposase
gene. The right-hand P element has an intact
transposase gene but cannot itself transpose
because it is "wings-clipped"—it lacks terminal
inverted repeats.

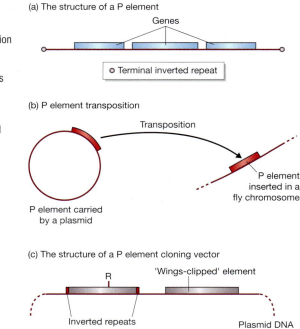

(a) The structure of a P element

Genes

o Terminal inverted repeat

(b) P element transposition

Transposition

P element carried
by a plasmid

P element
inserted in a
fly chromosome

(c) The structure of a P element cloning vector

R   'Wings-clipped' element

Inverted repeats   Plasmid DNA

elements as cloning vectors. The vector is a plasmid that carries two P elements, one of which contains the insertion site for the DNA that will be cloned. Insertion of the new DNA into this P element results in disruption of its transposase gene, so this element is inactive. The second P element carried by the plasmid is therefore one that has an intact version of the transposase gene. Ideally this second element should not itself be transferred to the *Drosophila* chromosomes, so it has its "wings clipped": its inverted repeats are removed so that the transposase does not recognize it as being a real P element (Figure 7.17c). Once the gene to be cloned has been inserted into the vector, the plasmid DNA is microinjected into fruit fly embryos. The transposase from the wings-clipped P element directs transfer of the engineered P element into one of the fruit fly chromosomes. If this happens within a germline nucleus, then the adult fly that develops from the embryo will carry copies of the cloned gene in all its cells. P element cloning was first developed in the 1980s and has made a number of important contributions to *Drosophila* genetics.

Cloning vectors based on insect viruses

Although virus vectors have not been developed for cloning genes in *Drosophila*, one type of virus, the baculovirus, has played an important role in gene cloning with other insects. The main use of baculovirus vectors is in the production of recombinant protein, and we will return to them when we consider this topic in Chapter 13.

7.3.2 Cloning in mammals

At present, gene cloning in mammals is carried out for one of three reasons:

- To achieve a gene knockout, which is an important technique used to help determine the function of an unidentified gene (p. 213). These experiments are usually carried out with rodents such as mice.

- For production of recombinant protein in a mammalian cell culture, and in the related technique of pharming, which involves genetic engineering of a farm animal so that it synthesizes an important protein such as a pharmaceutical, often in its milk (p. 241).
- In gene therapy, in which human cells are engineered in order to treat a disease (p. 259).

Viruses as cloning vectors for mammals

For many years it was thought that viruses would prove to be the key to cloning in mammals. This expectation has only partially been realized. The first cloning experiment involving mammalian cells was carried out in 1979 with a vector based on simian virus 40 (SV40). This virus is capable of infecting several mammalian species, following a lytic cycle in some hosts and a lysogenic cycle in others. The genome is 5.2 kb in size (Figure 7.19a) and contains two sets of genes, the "early" genes, expressed early in the infection cycle and coding for proteins involved in viral DNA replication, and the "late" genes, coding for viral capsid proteins. SV40 suffers from the same problem as λ and the plant caulimoviruses, in that packaging constraints limit the amount of new DNA that can be inserted into the genome. Cloning with SV40 therefore involves replacing one or more of the existing genes with the DNA to be cloned. In the original experiment a segment of the late gene region was replaced (Figure 7.19b), but early gene replacement is also an option.

In the years since 1979, a number of other types of virus have been used to clone genes in mammals. These include:

- **Adenoviruses**, which enable DNA fragments of up to 8 kb to be cloned, longer than is possible with an SV40 vector, though adenoviruses are more difficult to handle because their genomes are bigger.
- **Papillomaviruses**, which also have a relatively high capacity for inserted DNA. Bovine papillomavirus (BPV), which causes warts on cattle, is particularly attractive because it has an unusual infection cycle in mouse cells, taking the form of a multicopy plasmid with about 100 molecules present per cell. It does not

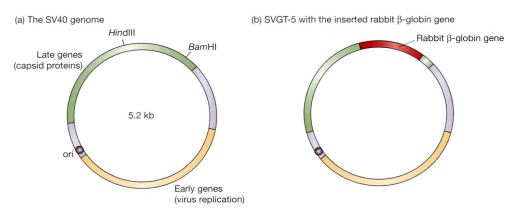

(a) The SV40 genome

HindIII
BamHI
Late genes
(capsid proteins)
5.2 kb
ori
Early genes
(virus replication)

(b) SVGT-5 with the inserted rabbit β-globin gene

Rabbit β-globin gene

Figure 7.19

SV40 and an example of its use as a cloning vector. To clone the rabbit β-globin gene the HindIII to BamHI restriction fragment was deleted (resulting in SVGT-5) and replaced with the rabbit gene.

cause the death of the mouse cell, and BPV molecules are passed to daughter cells on cell division, giving rise to a permanently transformed cell line. Shuttle vectors consisting of BPV and *E. coli* sequences, and capable of replication in both mouse and bacterial cells, have been used for the production of recombinant proteins in mouse cell lines.

- **Adeno-associated virus (AAV)**, which is unrelated to adenovirus but often found in the same infected tissues, because AAV makes use of some of the proteins synthesized by adenovirus in order to complete its replication cycle. In the absence of this helper virus, the AAV genome inserts into its host's DNA. With most integrative viruses this is a random event, but AAV has the unusual property of always inserting at the same position, within human chromosome 19. Knowing exactly where the cloned gene will be in the host genome is important if the outcome of the cloning experiment must be checked rigorously, as is the case in applications such as gene therapy. AAV vectors are therefore looked on as having major potential in this area.
- **Retroviruses**, which are the most commonly-used vectors for gene therapy. Although they insert at random positions, the resulting integrants are very stable, which means that the therapeutic effects of the cloned gene will persist for some time. We will return to gene therapy in Chapter 14.

Gene cloning without a vector

One of the reasons why virus vectors have not become widespread in mammalian gene cloning is because it was discovered in the early 1990s that the most effective way of transferring new genes into mammalian cells is by microinjection. Although a difficult procedure to carry out, microinjection of bacterial plasmids, or linear DNA copies of genes, into mammalian nuclei results in the DNA being inserted into the chromosomes, possibly as multiple copies in a tandem, head-to-tail arrangement (Figure 7.20). This procedure is generally looked on as more satisfactory than the use of a viral vector because it avoids the possibility that viral DNA will infect the cells and cause defects of one kind or another.

Microinjection of DNA is the basis to creation of a **transgenic animal**, one that contains a cloned gene in all of its cells. A transgenic mouse can be generated by micro-injection of a fertilized egg cell which is subsequently cultured *in vitro* for several cell divisions and then implanted into a foster mother. Alternatively, an **embryonic stem (ES) cell** can be used. These are obtained from within an early embryo and, unlike most mammalian cells, are **totipotent**, meaning that their developmental pattern is not pre-set and cells descended from them can form many different structures in the adult mouse. After microinjection, the ES cell is placed back in an embryo which is implanted into the foster mother. The resulting mouse is a **chimera**, comprising a mixture of engineered and non-engineered cells, because the embryo that receives the ES cell also contains a number of ordinary cells that contribute, along with the ES cell, to the make-up of the adult mouse. Non-chimeric mice, which contain the cloned gene in all their cells, are obtained by allowing the chimera to reproduce, as some of the offspring will be derived from egg cells that contain the cloned gene.

Figure 7.20

Multiple copies of cloned DNA molecules inserted as a tandem array in a chromosomal DNA molecule.

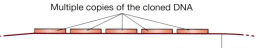

Multiple copies of the cloned DNA

Chromosomal DNA

Further reading

Brisson, N., Paszkowski, J., Penswick, J.R. et al. (1984) Expression of a bacterial gene in plants by using a viral vector. *Nature*, 310, 511–114. [The first cloning experiment with a caulimovirus.]

Broach, J.R. (1982) The yeast 2 μm circle. *Cell*, 28, 203–204.

Burke, D.T., Carle, G.F. & Olson, M.V. (1987) Cloning of large segments of exogenous DNA into yeast by means of artificial chromosome vectors. *Science*, 236, 806–812.

Carrillo-Tripp, J., Shimada-Beltrán, H. & Rivera-Bustamante, R. (2006) Use of geminiviral vectors for functional genomics. *Current Opinion in Plant Biology*, 9, 209–215. [Virus-induced gene silencing.]

Chilton, M.D. (1983) A vector for introducing new genes into plants. *Scientific American*, 248 (June), 50–59. [The Ti plasmid.]

Colosimo, A., Goncz, K.K., Holmes, A.R. et al. (2000) Transfer and expression of foreign genes in mammalian cells. *Biotechniques*, 29, 314–321.

Daniel, H., Kumar, S. & Dufourmantel, N. (2005) Breakthrough in chloroplast genetic engineering of agronomically important crops. *Trends in Biotechnology*, 23, 238–245.

Evans, M.J., Carlton, M.B.L. & Russ, A.P. (1997) Gene trapping and functional genomics. *Trends in Genetics*, 13, 370–374. [Includes a description of the use of ES cells.]

Graham, F.L. (1990) Adenoviruses as expression vectors and recombinant vaccines. *Trends in Biotechnology*, 8, 20–25.

Guillon, S., Trémouillaux-Guiller, J., Pati, P.K., Rideau, M. & Gantet, P. (2006) Hairy root research: recent scenario and exciting prospects. *Current Opinion in Plant Biology*, 9, 341–346. [Applications of the Ri plasmid.]

Hamer, D.H. & Leder, P. (1979) Expression of the chromosomal mouse β-maj-globin gene cloned in SV40. *Nature*, 281, 35–40.

Komori, T., Imayama, T., Kato, N., Ishida, Y., Ueki, J. & Komari, T. (2007) Current status of binary vectors and superbinary vectors. *Plant Physiology*, 145, 1155–1160. [Latest versions of Ti plasmid vectors.]

Maliga, P. (2004) Plastid transformation of higher plants. *Annual Reviews of Plant Biology*, 55, 289–313.

Parent, S.A. et al. (1985) Vector systems for the expression, analysis and cloning of DNA sequences in *S. cerevisiae*. *Yeast*, 1, 83–138. [Details of different yeast cloning vectors.]

Paszkowski, J., Shillito, R.D., Saul, M. et al. (1984) Direct gene transfer to plants. *EMBO Journal*, 3, 2717–2722.

Rubin, G.M. & Spradling, A.C. (1982) Genetic transformation of *Drosophila* with transposable element vectors. *Science*, 218, 348–353. [Cloning with P elements.]

Viaplana, R., Turner, D.S. & Covey, S.N. (2001) Transient expression of a GUS reporter gene from cauliflower mosaic virus replacement vectors in the presence and absence of helper virus. *Journal of General Virology*, 82, 59–65. [The helper virus approach to cloning with CaMV.]

Chapter 8

How to Obtain a Clone of a Specific Gene

In the preceding chapters we have examined the basic methodology used to clone genes, and surveyed the range of vector types that are used with bacteria, yeast, plants, and animals. Now we must look at the methods available for obtaining a clone of an individual, specified gene. This is the critical test of a gene cloning experiment, success or failure often depending on whether or not a strategy can be devised by which clones of the desired gene can be selected directly, or alternatively, distinguished from other recombinants. Once this problem has been resolved, and a clone has been obtained, the molecular biologist is able to make use of a wide variety of different techniques that will extract information about the gene. The most important of these will be described in Chapters 10 and 11.

8.1 The problem of selection

The problem faced by the molecular biologist wishing to obtain a clone of a single, specified gene was illustrated in Figure 1.4. Even the simplest organisms, such as *E. coli*, contain several thousand genes, and a restriction digest of total cell DNA produces not only the fragment carrying the desired gene, but also many other fragments carrying all the other genes (Figure 8.1a). During the ligation reaction there is no selection for an individual fragment: numerous different recombinant DNA molecules are produced, all containing different pieces of DNA (Figure 8.1b). Consequently, a variety of

Gene Cloning and DNA Analysis: An Introduction. 6th edition. By T.A. Brown. Published 2010 by Blackwell Publishing.

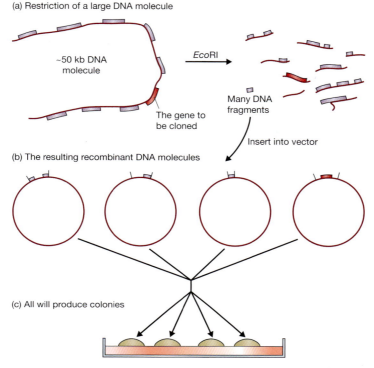

(a) Restriction of a large DNA molecule

~50 kb DNA molecule

EcoRI

Many DNA fragments

The gene to be cloned

Insert into vector

(b) The resulting recombinant DNA molecules

(c) All will produce colonies

Figure 8.1
The problem of selection.

recombinant clones are obtained after transformation and plating out (Figure 8.1c). Somehow the correct one must be identified.

8.1.1 There are two basic strategies for obtaining the clone you want

Although there are many different procedures by which the desired clone can be obtained, all are variations on two basic themes:

- **Direct selection for the desired gene** (Figure 8.2a), which means that the cloning experiment is designed in such a way that the only clones that are obtained are clones of the required gene. Almost invariably, selection occurs at the plating-out stage.

- **Identification of the clone from a gene library** (Figure 8.2b), which entails an initial "**shotgun**" cloning experiment, to produce a clone library representing all or most of the genes present in the cell, followed by analysis of the individual clones to identify the correct one.

In general terms, direct selection is the preferred method, as it is quick and usually unambiguous. However, as we shall see, it is not applicable to all genes, and techniques for clone identification are therefore very important.

Figure 8.2

The basic strategies that can be used to obtain a particular clone: (a) direct selection; (b) identification of the desired recombinant from a clone library.

(a) Direct selection

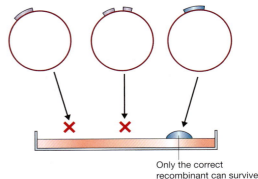

Only the correct
recombinant can survive

(b) Clone identification

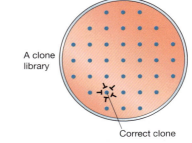

A clone
library

Correct clone

8.2 Direct selection

To be able to select for a cloned gene it is necessary to plate the transformants onto an agar medium on which only the desired recombinants, and no others, can grow. The only colonies that are obtained will therefore be ones that comprise cells containing the desired recombinant DNA molecule.

The simplest example of direct selection occurs when the desired gene specifies resistance to an antibiotic. As an example we will consider an experiment to clone the gene for kanamycin resistance from plasmid R6-5. This plasmid carries genes for resistances to four antibiotics: kanamycin, chloramphenicol, streptomycin, and sulphonamide. The kanamycin resistance gene lies within one of the 13 *Eco*RI fragments (Figure 8.3a).

To clone this gene, the *Eco*RI fragments of R6-5 could be inserted into the *Eco*RI site of a vector such as pBR322. The ligated mix will comprise many copies of 13 different recombinant DNA molecules, one set of which carries the gene for kanamycin resistance (Figure 8.3b).

Insertional inactivation cannot be used to select recombinants when the *Eco*RI site of pBR322 is used. This is because this site does not lie in either the ampicillin or the tetracycline resistance genes of this plasmid (see Figure 6.1). But this is immaterial for cloning the kanamycin resistance gene because in this case the cloned gene can be used as the selectable marker. Transformants are plated onto kanamycin agar, on which the only cells able to survive and produce colonies are those recombinants that contain the cloned kanamycin resistance gene (Figure 8.3c).

(a) Plasmid R6-5

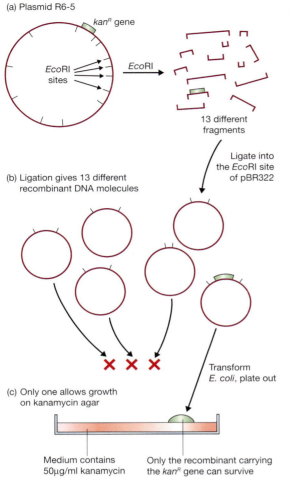

kan^R gene

EcoRI sites

EcoRI →

13 different fragments

(b) Ligation gives 13 different recombinant DNA molecules

Ligate into the *Eco*RI site of pBR322

(c) Only one allows growth on kanamycin agar

× × ×

Transform *E. coli*, plate out

Medium contains 50μg/ml kanamycin

Only the recombinant carrying the *kan^R* gene can survive

Figure 8.3

Direct selection for the cloned R6-5 kanamycin resistance (*kan^R*) gene.

8.2.1 Marker rescue extends the scope of direct selection

Direct selection would be very limited indeed if it could be used only for cloning anti-biotic resistance genes. Fortunately the technique can be extended by making use of mutant strains of *E. coli* as the hosts for transformation.

As an example, consider an experiment to clone the gene *trpA* from *E. coli*. This gene codes for the enzyme tryptophan synthase, which is involved in biosynthesis of the essential amino acid tryptophan. A mutant strain of *E. coli* that has a non-functional *trpA* gene is called *trpA⁻*, and is able to survive only if tryptophan is added to the growth medium. *E. coli trpA⁻* is therefore another example of an auxotroph (p. 106).

The *E. coli trpA⁻* auxotroph can be used to clone the correct version of the *trpA* gene. Total DNA is first purified from a normal (wild-type) strain of the bacterium. Digestion with a restriction endonuclease, followed by ligation into a vector, produces numerous recombinant DNA molecules, one of which may, with luck, carry an intact copy of the *trpA* gene (Figure 8.4a). This is, of course, the functional gene, as it has been obtained from the wild-type strain.

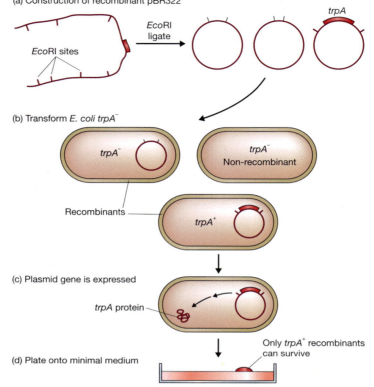

Figure 8.4

Direct selection for the *trpA* gene cloned in a *trpA*⁻ strain of *E. coli*.

The ligation mixture is now used to transform the auxotrophic *E. coli trpA⁻* cells (Figure 8.4b). The vast majority of the resulting transformants will be auxotrophic, but a few now have the plasmid-borne copy of the correct *trpA* gene. These recombinants are non-auxotrophic—they no longer require tryptophan as the cloned gene is able to direct production of tryptophan synthase (Figure 8.4c). Direct selection is therefore performed by plating transformants onto minimal medium, which lacks any added supplements, and in particular has no tryptophan (Figure 8.4d). Auxotrophs cannot grow on minimal medium, so the only colonies to appear are recombinants that contain the cloned *trpA* gene.

8.2.2 The scope and limitations of marker rescue

Although marker rescue can be used to obtain clones of many genes, the technique is subject to two limitations:

- A mutant strain must be available for the gene in question.
- A medium on which only the wild-type can survive is needed.

Marker rescue is applicable for most genes that code for biosynthetic enzymes, as clones of these genes can be selected on minimal medium in the manner described for *trpA*. The technique is not limited to *E. coli* or even bacteria. Auxotrophic strains of yeast and filamentous fungi are also available, and marker rescue has been used to select genes cloned into these organisms.

In addition, *E. coli* auxotrophs can be used as hosts for the selection of some genes from other organisms. Often there is sufficient similarity between equivalent enzymes from different bacteria, or even from yeast, for the foreign enzyme to function in *E. coli*, so that the cloned gene is able to transform the host to wild type.

8.3 Identification of a clone from a gene library

Although marker rescue is a powerful technique, it is not all-embracing and there are many important genes that cannot be selected by this method. Many bacterial mutants are not auxotrophs, so the mutant and wild-type strains cannot be distinguished by plating onto minimal or any other special medium. In addition, neither marker rescue nor any other direct selection method is of much use in providing bacterial clones of genes from animals or plants, as in these cases the differences are usually so great that the foreign enzymes do not function in the bacterial cell.

The alternative strategy must therefore be considered. This is where a large number of different clones are obtained and the desired one identified in some way.

8.3.1 Gene libraries

Before looking at the methods used to identify individual clones, the library itself must be considered. A genomic library (p. 102) is a collection of clones sufficient in number to be likely to contain every single gene present in a particular organism. Genomic libraries are prepared by purifying total cell DNA, and then making a partial restriction digest, resulting in fragments that can be cloned into a suitable vector (Figure 8.5), usually a λ replacement vector, a cosmid, or possibly a yeast artificial chromosome (YAC), bacterial artificial chromosome (BAC) or P1 vector.

For bacteria, yeast, and fungi, the number of clones needed for a complete genomic library is not so large as to be unmanageable (see Table 6.1). For plants and animals though, a complete library contains so many different clones that identification of the desired one may prove a mammoth task. With these organisms a second type of library, specific not to the whole organism but to a particular cell type, may be more useful.

8.3.2 Not all genes are expressed at the same time

A characteristic of most multicellular organisms is specialization of individual cells. A human being, for example, is made up of a large number of different cell types—brain cells, blood cells, liver cells, etc. Each cell contains the same complement of genes, but in different cell types different sets of genes are switched on, while others are silent (Figure 8.6).

The fact that only relatively few genes are expressed in any one type of cell can be utilized in preparation of a library if the material that is cloned is not DNA but messenger RNA (mRNA). Only those genes that are being expressed are transcribed into mRNA, so if mRNA is used as the starting material then the resulting clones comprise only a selection of the total number of genes in the cell.

A cloning method that uses mRNA would be particularly useful if the desired gene is expressed at a high rate in an individual cell type. For example, the gene for gliadin, one of the nutritionally important proteins present in wheat, is expressed at a very high

Figure 8.5

Preparation of a gene library in a cosmid vector.

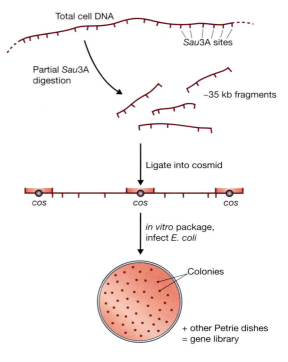

Figure 8.6

Different genes are expressed in different types of cell.

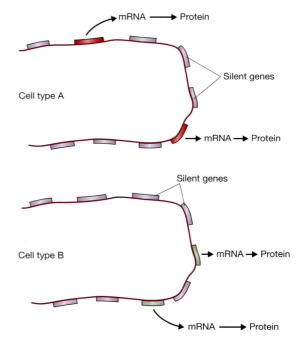

level in the cells of developing wheat seeds. In these cells over 30% of the total mRNA specifies gliadin. Clearly, if we could clone the mRNA from wheat seeds we would obtain a large number of clones specific for gliadin.

8.3.3 mRNA can be cloned as complementary DNA

Messenger RNA cannot itself be ligated into a cloning vector. However, mRNA can be converted into DNA by complementary DNA (cDNA) synthesis.

The key to this method is the enzyme reverse transcriptase (p. 49), which synthesizes a DNA polynucleotide complementary to an existing RNA strand (Figure 8.7a). Once the cDNA strand has been synthesized the RNA member of the hybrid molecule can be partially degraded by treating with ribonuclease (RNase) HI (Figure 8.7b). The remaining RNA fragments then serve as primers (p. 48) for DNA polymerase I, which synthesizes the second cDNA strand (Figure 8.7c), resulting in a double-stranded DNA fragment that can be ligated into a vector and cloned (Figure 8.7d).

The resulting cDNA clones are representative of the mRNA present in the original preparation. In the case of mRNA prepared from wheat seeds, the cDNA library would contain a large proportion of clones representing gliadin mRNA (Figure 8.7e). Other clones will also be present, but locating the cloned gliadin cDNA is a much easier process than identifying the equivalent gene from a complete wheat genomic library.

8.4 Methods for clone identification

Once a suitable library has been prepared, a number of procedures can be employed to attempt identification of the desired clone. Although a few of these procedures are based on detection of the translation product of the cloned gene, it is usually easier to identify directly the correct recombinant DNA molecule. This can be achieved by the important technique of hybridization probing.

8.4.1 Complementary nucleic acid strands hybridize to each other

Any two single-stranded nucleic acid molecules have the potential to form base pairs with one another. With most pairs of molecules the resulting hybrid structures are unstable, as only a small number of individual interstrand bonds are formed (Figure 8.8a). However, if the polynucleotides are complementary, extensive base pairing can occur to form a stable double-stranded molecule (Figure 8.8b). Not only can this occur between single-stranded DNA molecules to form the DNA double helix, but also between a pair of single-stranded RNA molecules or between combinations of one DNA strand and one RNA strand (Figure 8.8c).

Nucleic acid hybridization can be used to identify a particular recombinant clone if a DNA or RNA probe, complementary to the desired gene, is available. The exact nature of the probe will be discussed later in this chapter. First we must consider the technique itself.

8.4.2 Colony and plaque hybridization probing

Hybridization probing can be used to identify recombinant DNA molecules contained in either bacterial colonies or bacteriophage plaques. First the colonies or plaques are

Figure 8.7

One possible scheme for cDNA cloning. Poly(A) = polyadenosine, oligo(dT) = oligodeoxythymidine.

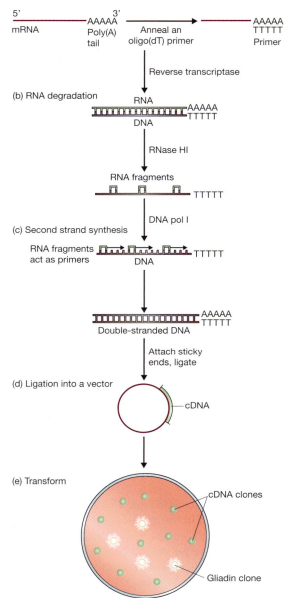

(a) First strand synthesis

5'   3'
mRNA ————————AAAAA   Anneal an   ————————AAAAA
            Poly(A)   oligo(dT) primer   TTTTT
            tail   Primer

Reverse transcriptase

(b) RNA degradation

RNA
□□□□□□□□□□□□AAAAA
            TTTTT
DNA

RNase HI

RNA fragments
▯   ▯   ▯ TTTTT

DNA pol I

(c) Second strand synthesis

RNA fragments ▯▭▯▭▯▭ TTTTT
act as primers   DNA

□□□□□□□□□□□□AAAAA
            TTTTT
Double-stranded DNA

Attach sticky
ends, ligate

(d) Ligation into a vector

cDNA

(e) Transform

cDNA clones

Gliadin clone

transferred to a nitrocellulose or nylon membrane (Figure 8.9a) and then treated to remove all contaminating material, leaving just DNA (Figure 8.9b). Usually this treatment also results in denaturation of the DNA molecules, so that the hydrogen bonds between individual strands in the double helix are broken. These single-stranded molecules can then be bound tightly to the membrane by a short period at 80°C if a nitrocellulose membrane is being used, or with a nylon membrane by ultraviolet irradiation. The molecules become attached to the membrane through their sugar–phosphate backbones, so the bases are free to pair with complementary nucleic acid molecules.

(a) An unstable hybrid

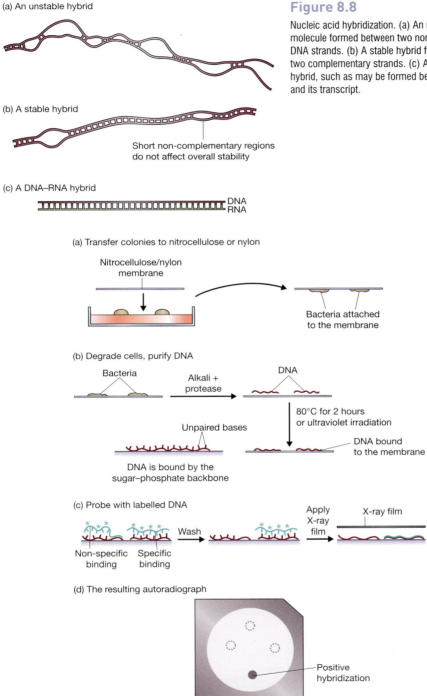

Figure 8.8

Nucleic acid hybridization. (a) An unstable hybrid molecule formed between two non-homologous DNA strands. (b) A stable hybrid formed between two complementary strands. (c) A DNA–RNA hybrid, such as may be formed between a gene and its transcript.

(b) A stable hybrid

Short non-complementary regions do not affect overall stability

(c) A DNA–RNA hybrid

DNA
RNA

(a) Transfer colonies to nitrocellulose or nylon

Nitrocellulose/nylon membrane

Bacteria attached to the membrane

(b) Degrade cells, purify DNA

Bacteria

Alkali + protease

DNA

80°C for 2 hours or ultraviolet irradiation

DNA bound to the membrane

Unpaired bases

DNA is bound by the sugar–phosphate backbone

(c) Probe with labelled DNA

Wash

Apply X-ray film

X-ray film

Non-specific binding

Specific binding

(d) The resulting autoradiograph

Positive hybridization

Figure 8.9

Colony hybridization probing. In this example, the probe is labelled with a radioactive marker and hybridization detected by autoradiography, but other types of label and detection system can also be used.

The probe must now be labeled with a radioactive or other type of marker, denatured by heating, and applied to the membrane in a solution of chemicals that promote nucleic acid hybridization (Figure 8.9c). After a period to allow hybridization to take place, the filter is washed to remove unbound probe, dried, and the label detected in order to identify the colonies or plaques to which the probe has become bound (Figure 8.9d).

Labeling with a radioactive marker

A DNA molecule is usually labeled by incorporating nucleotides that carry a radioactive isotope of phosphorus, ^{32}P (Figure 8.10). Several methods are available:

- **Nick translation**. Most purified samples of DNA contain some nicked molecules, however carefully the preparation has been carried out, which means that DNA polymerase I is able to attach to the DNA and catalyze a strand replacement reaction (Figure 8.11a). This reaction requires a supply of nucleotides: if one of these is radioactively labeled, the DNA molecule will itself become labeled. Nick translation can be used to label any DNA molecule but might under some circumstances also cause DNA cleavage.

- **End filling** is a gentler method than nick translation and rarely causes breakage of the DNA, but unfortunately can only be used to label DNA molecules that have sticky ends. The enzyme used is the Klenow fragment (p. 49), which "fills in" a sticky end by synthesizing the complementary strand (Figure 8.11b)). As with nick translation, if the end filling reaction is carried out in the presence of labeled nucleotides, the DNA becomes labeled.

Figure 8.10

The structure of α-^{32}P-deoxyadenosine triphosphate ([α-^{32}P]dATP).

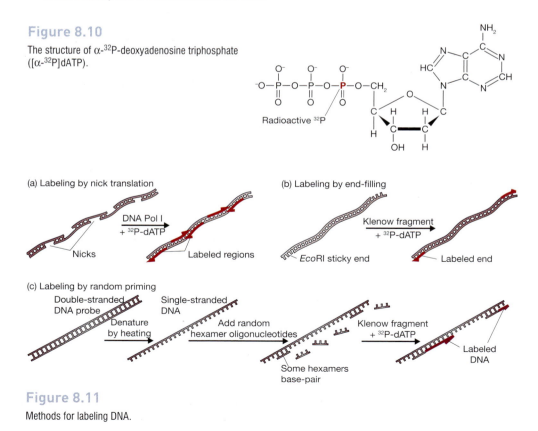

Figure 8.11

Methods for labeling DNA.

- **Random priming** results in a probe with higher activity and therefore able to detect smaller amounts of membrane-bound DNA. The denatured DNA is mixed with a set of hexameric oligonucleotides of random sequence. By chance, these random hexamers will contain a few molecules that will base pair with the probe and prime new DNA synthesis. The Klenow fragment is used as this enzyme lacks the nuclease activity of DNA polymerase I (p. 48) and so only fills in the gaps between adjacent primers (Figure 8.11c). Labeled nucleotides are incorporated into the new DNA that is synthesized.

After hybridization, the location of the bound probe is detected by autoradiography. A sheet of X-ray-sensitive photographic film is placed over the membrane. The radioactive DNA exposes the film, which is developed to reveal the positions of the colonies or plaques to which the probe has hybridized (see Figure 8.9d).

Non-radioactive labeling

Radioactive labeling methods are starting to fall out of favor, partly because of the hazard to the researcher and partly because of the problems associated with disposal of radioactive waste. As an alternative, the hybridization probe can be labeled in a non-radioactive manner. A number of methods have been developed, two of which are illustrated in Figure 8.12. The first makes use of deoxyuridine triphosphate (dUTP) nucleotides modified by reaction with biotin, an organic molecule that has a high affinity for a protein called avidin. After hybridization the positions of the bound biotinylated probe can be determined by washing with avidin coupled to a fluorescent marker (Figure 8.12a). This method is as sensitive as radioactive probing and is becoming increasingly popular.

The same is true for a second procedure for non-radioactive hybridization probing, in which the probe DNA is complexed with the enzyme horseradish peroxidase, and is detected through the enzyme's ability to degrade luminol with the emission of chemiluminescence (Figure 8.12b). The signal can be recorded on normal photographic film in a manner analogous to autoradiography.

8.4.3 Examples of the practical use of hybridization probing

Clearly, the success of colony or plaque hybridization as a means of identifying a particular recombinant clone depends on the availability of a DNA molecule that can be used as a probe. This probe must share at least a part of the sequence of the cloned gene. If the gene itself is not available (which presumably is the case if the aim of the experiment is to provide a clone of it), then what can be used as the probe?

In practice, the nature of the probe is determined by the information available about the desired gene. We will consider three possibilities:

- Where the desired gene is expressed at a high level in a cell type from which a cDNA clone library has been prepared;
- Where the amino acid sequence of the protein coded by the gene is completely or partially known;
- Where the gene is a member of a family of related genes.

Abundancy probing to analyze a cDNA library

As described earlier in this chapter, a cDNA library is often prepared in order to obtain a clone of a gene expressed at a relatively high level in a particular cell type. In the

Figure 8.12

Two methods for the non-radioactive labelling of DNA probes.

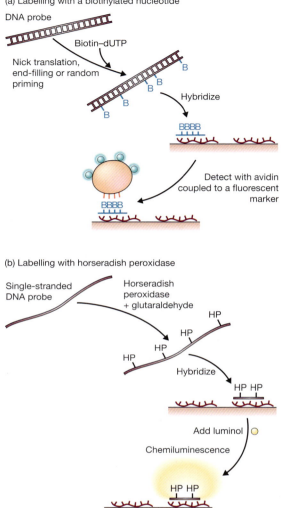

(a) Labelling with a biotinylated nucleotide

DNA probe

Biotin–dUTP

Nick translation, end-filling or random priming

Hybridize

Detect with avidin coupled to a fluorescent marker

(b) Labelling with horseradish peroxidase

Single-stranded DNA probe

Horseradish peroxidase + glutaraldehyde

Hybridize

Add luminol

Chemiluminescence

example of a cDNA library from developing wheat seeds, a large proportion of the clones are copies of the mRNA transcripts of the gliadin gene (see Figure 8.7e).

Identification of the gliadin clones is simply a case of using individual cDNA clones from the library to probe all the other members of the library (Figure 8.13). A clone is selected at random and the recombinant DNA molecule purified, labeled, and used to probe the remaining clones. This is repeated with different clones as probes until one that hybridizes to a large proportion of the library is obtained. This abundant cDNA is considered a possible gliadin clone and analyzed in greater detail (e.g., by DNA sequencing and isolation of the translation product) to confirm the identification.

Oligonucleotide probes for genes whose translation products have been characterized
Often the gene to be cloned codes for a protein that has already been studied in some detail. In particular, the amino acid sequence of the protein might have been determined, using sequencing techniques that have been available for over 50 years. If the

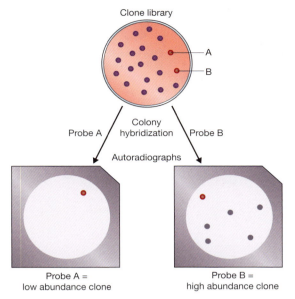

Figure 8.13

Probing within a library to identify an abundant clone.

Clone library

Colony hybridization

Probe A   Probe B

Autoradiographs

Probe A = low abundance clone

Probe B = high abundance clone

amino acid sequence is known, then it is possible to use the genetic code to predict the nucleotide sequence of the relevant gene. This prediction is always an approximation, as only methionine and tryptophan can be assigned unambiguously to triplet codons, all other amino acids being coded by at least two codons each. Nevertheless, in most cases, the different codons for an individual amino acid are related. Alanine, for example, is coded by GCA, GCC, GCG, and GCT, so two out of the three nucleotides of the triplet coding for alanine can be predicted with certainty.

As an example to clarify how these predictions are made, consider cytochrome c, a protein that plays an important role in the respiratory chain of all aerobic organisms. The cytochrome c protein from yeast was sequenced in 1963, with the result shown in Figure 8.14. This sequence contains a segment, starting at amino acid 59, that runs Trp–Asp–Glu– Asn–Asn–Met. The genetic code states that this hexapeptide is coded by TGG–GA$_C^T$–GA$_G^A$–AA$_C^T$–AA$_C^T$–ATG. Although this represents a total of 16 different possible sequences, 14 of the 18 nucleotides can be predicted with certainty.

Oligonucleotides of up to about 150 nucleotides in length can easily be synthesized in the laboratory (Figure 8.15). An oligonucleotide probe could therefore be constructed according to the predicted nucleotide sequence, and this probe might be able to

GLY–SER–ALA–LYS–LYS–GLY–ALA–THR–LEU–PHE–LYS–THR–ARG–CYS–GLU– 15
LEU–CYS–HIS–THR–VAL–GLU–LYS–GLY–GLY–PRO–HIS–LYS–VAL–GLY–PRO– 30
ASN–LEU–HIS–GLY–ILE–PHE–GLY–ARG–HIS–SER–GLY–GLN–ALA–GLN–GLY– 45
TYR–SER–TYR–THR–ASP–ALA–ASN–ILE–LYS–LYS–ASN–VAL–LEU–TRP–ASP– 60
GLU–ASN–ASN–MET–SER–GLU–TYR–LEU–THR–ASN–PRO–LYS–LYS–TYR–ILE– 75
PRO–GLY–THR–LYS–MET–ALA–PHE–GLY–GLY–LEU–LYS–LYS–GLU–LYS–ASP– 90
ARG–ASN–ASP–LEU–ILE–THR–TYR–LEU–LYS–LYS–ALA–CYS–GLU 103

Figure 8.14

The amino acid sequence of yeast cytochrome c. The hexapeptide that is highlighted red is the one used to illustrate how a nucleotide sequence can be predicted from an amino acid sequence.

Figure 8.15

A simplified scheme for oligonucleotide synthesis. Each nucleotide is modified by attachment of an activating group to the 3′ carbon and a protecting group to the 5′ carbon. The activating group enables the normally inefficient process of nucleotide joining to proceed much more rapidly. The protecting group ensures that individual nucleotides cannot attach to one another, and instead react only with the terminal 5′ group of the growing oligonucleotide, this 5′ group being deprotected by chemical treatment at the appropriate point in each synthesis cycle.

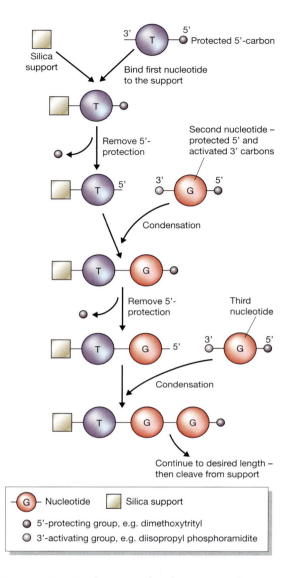

identify the gene coding for the protein in question. In the example of yeast cytochrome c, the 16 possible oligonucleotides that can code for Trp–Asp–Glu–Asn–Asn–Met would be synthesized, either separately or as a pool, and then used to probe a yeast genomic or cDNA library (Figure 8.16). One of the oligonucleotides in the probe will have the correct sequence for this region of the cytochrome c gene, and its hybridization signal will indicate which clones carry this gene.

The result can be checked by carrying out a second probing with a mixture of oligonucleotides whose sequences are predicted from a different segment of the cytochrome c protein (Figure 8.16). However, the segment of the protein used for nucleotide sequence prediction must be chosen with care: the hexapeptide Ser–Glu–Tyr–Leu–Thr–Asn, which immediately follows our first choice, could be coded by several thousand different 18-nucleotide sequences, clearly an unsuitable choice for a synthetic probe.

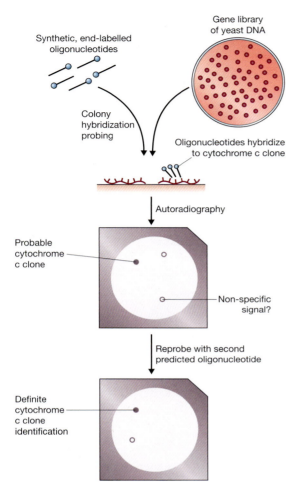

Synthetic, end-labelled oligonucleotides

Gene library of yeast DNA

Colony hybridization probing

Oligonucleotides hybridize to cytochrome c clone

Autoradiography

Probable cytochrome c clone

Non-specific signal?

Reprobe with second predicted oligonucleotide

Definite cytochrome c clone identification

Figure 8.16

The use of a synthetic, end-labelled oligonucleotide to identify a clone of the yeast cytochrome c gene.

Heterologous probing allows related genes to be identified

Often a substantial amount of nucleotide similarity is seen when two genes for the same protein, but from different organisms, are compared, a reflection of the conservation of gene structure during evolution. Frequently, two genes from related organisms are sufficiently similar for a single-stranded probe prepared from one gene to form a stable hybrid with the second gene. Although the two molecules are not entirely complementary, enough base pairs are formed to produce a stable structure (Figure 8.17a).

Heterologous probing makes use of hybridization between related sequences for clone identification. For example, the yeast cytochrome c gene, identified in the previous section by oligonucleotide probing, could itself be used as a hybridization probe to identify cytochrome c genes in clone libraries of other organisms. A probe prepared from the yeast gene would not be entirely complementary to the gene from, say, the fungus *Neurospora crassa*, but sufficient base pairing should occur for a hybrid to be formed (Figure 8.17b).

Heterologous probing can also identify related genes in the *same* organism. If the wheat gliadin cDNA clone, identified earlier in the chapter by abundancy probing, is used to probe a genomic library, it will hybridize not only to its own gene but to a

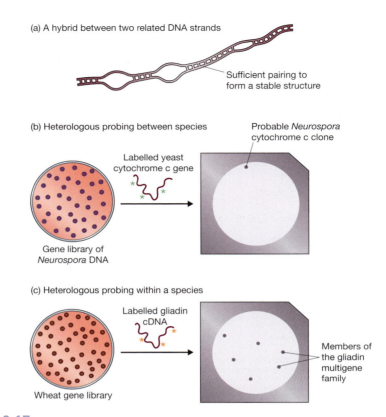

(a) A hybrid between two related DNA strands

Sufficient pairing to form a stable structure

(b) Heterologous probing between species

Labelled yeast cytochrome c gene

Probable *Neurospora* cytochrome c clone

Gene library of *Neurospora* DNA

(c) Heterologous probing within a species

Labelled gliadin cDNA

Members of the gliadin multigene family

Wheat gene library

Figure 8.17

Heterologous probing.

variety of other genes as well (Figure 8.17c). These are all related to the gliadin cDNA, but have slightly different nucleotide sequences. This is because the wheat gliadins form a complex group of related proteins that are coded by the members of a **multigene family**. Once one gene in the family has been cloned, then all the other members can be isolated by heterologous probing.

Southern hybridization enables a specific restriction fragment containing a gene to be identified

As well as colony and plaque hybridization analysis, there are also occasions when it is necessary to use hybridization probing to identify which of a series of restriction fragments contains a gene of interest. As an example, we will return to the genomic clone of the yeast cytochrome c gene, which we identified by oligonucleotide hybridization probing. Let us imagine that this particular genomic library was prepared by partial restriction of yeast DNA with *Bam*HI followed by cloning in the cosmid vector pJB8 (p. 101). The cloned fragment containing the cytochrome c gene will therefore be approximately 40 kb in length, and will probably contain about ten *Bam*HI fragments, remembering that the hexanucleotide recognition site for this enzyme will be present, on average, once every $4^6 = 4096$ bp.

The cytochrome c gene, on the other hand, is predicted to be just 309 bp in length (we know the protein has 103 amino acids; see Figure 8.14). The gene therefore makes up less than 1% of the cloned DNA fragment, and it is quite possible that other genes

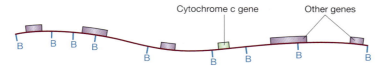

Cytochrome c gene   Other genes

B   B B B   B   B   B   B   B

Figure 8.18

A long cloned DNA fragment may contain several genes in addition to the one in which we are interested. B = *Bam*HI restriction site.

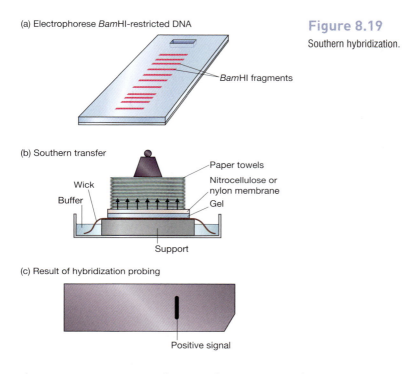

(a) Electrophorese *Bam*HI-restricted DNA

*Bam*HI fragments

Figure 8.19

Southern hybridization.

(b) Southern transfer

Paper towels

Wick

Nitrocellulose or nylon membrane

Buffer

Gel

Support

(c) Result of hybridization probing

Positive signal

that we are not interested in are also present in this insert (Figure 8.18). The method called **Southern hybridization** enables the individual restriction fragment containing the cytochrome c gene to be identified.

The first step in using Southern hybridization for this purpose would be to digest the clone with *Bam*HI and then separate the restriction fragments by electrophoresis in an agarose gel (Figure 8.19a). The aim is to use the oligonucleotide probe for cytochrome c to identify the fragment that contains the gene. This can be attempted while the restriction fragments are still contained in the electrophoresis gel, but the results are usually not very good, as the gel matrix causes a lot of spurious background hybridization that obscures the specific hybridization signal. Instead, the DNA bands in the agarose gel are transferred to a nitrocellulose or nylon membrane, providing a much "cleaner" environment for the hybridization experiment.

Transfer of DNA bands from an agarose gel to a membrane makes use of the technique perfected in 1975 by Professor E.M. Southern and referred to as **Southern transfer**. The membrane is placed on the gel, and buffer allowed to soak through, carrying the DNA from the gel to the membrane where the DNA is bound. Sophisticated pieces of apparatus can be purchased to assist this process, but many molecular biologists prefer a homemade set-up incorporating a lot of paper towels and considerable balancing skills (Figure 8.19b). The same method can also be used for the transfer of RNA molecules ("northern" transfer) or proteins ("western" transfer). So far no one has come up with "eastern" transfers!

Southern transfer results in a membrane that carries a replica of the DNA bands from the agarose gel. If the labeled probe is now applied, hybridization occurs and autoradiography (or the equivalent detection system for a non-radioactive probe) reveals which restriction fragment contains the cloned gene (Figure 8.19c).

8.4.4 Identification methods based on detection of the translation product of the cloned gene

Hybridization probing is usually the preferred method for identification of a particular recombinant from a clone library. The technique is easy to perform and, with modifications introduced in recent years, can be used to check up to 10,000 recombinants per experiment, allowing large genomic libraries to be screened in a reasonably short time. Nevertheless, the requirement for a probe that is at least partly complementary to the desired gene sometimes makes it impossible to use hybridization in clone identification. On these occasions a different strategy is needed.

The main alternative to hybridization probing is immunological screening. The distinction is that, whereas with hybridization probing the cloned DNA fragment is itself directly identified, an immunological method detects the protein coded by the cloned gene. Immunological techniques therefore presuppose that the cloned gene is being expressed, so that the protein is being made, and that this protein is not normally present in the host cells.

Antibodies are required for immunological detection methods

If a purified sample of a protein is injected into the bloodstream of a rabbit, the immune system of the animal responds by synthesizing antibodies that bind to and help degrade the foreign molecule (Figure 8.20a). This is a version of the natural defense

Figure 8.20

Antibodies. (a) Antibodies in the bloodstream bind to foreign molecules and help degrade them. (b) Purified antibodies can be obtained from a small volume of blood taken from a rabbit injected with the foreign protein.

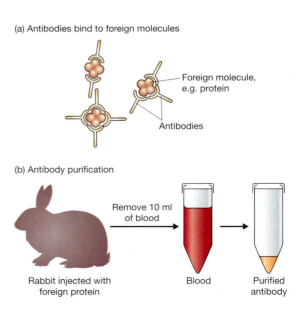

(a) Antibodies bind to foreign molecules

Foreign molecule, e.g. protein

Antibodies

(b) Antibody purification

Remove 10 ml of blood

Rabbit injected with foreign protein

Blood

Purified antibody

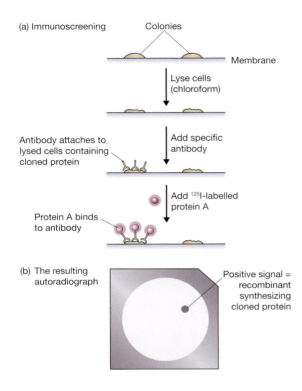

(a) Immunoscreening

Colonies

Membrane

Lyse cells
(chloroform)

Antibody attaches to
lysed cells containing
cloned protein

Add specific
antibody

Add ^{125}I-labelled
protein A

Protein A binds
to antibody

(b) The resulting
autoradiograph

Positive signal =
recombinant
synthesizing
cloned protein

Figure 8.21

Using a purified antibody to detect protein
in recombinant colonies. Instead of labelled
protein A, the antibody itself can be labelled, or
alternatively a second labelled antibody which
binds specifically to the primary antibody can
be used.

mechanism that the animal uses to deal with invasion by bacteria, viruses, and other
infective agents.

Once a rabbit is challenged with a protein, the levels of antibody present in its bloodstream remain high enough over the next few days for substantial quantities to be
purified. It is not necessary to kill the rabbit, because as little as 10 ml of blood provides
a considerable amount of antibody (Figure 8.20b). This purified antibody binds only to
the protein with which the animal was originally challenged.

Using a purified antibody to detect protein in recombinant colonies

There are several versions of immunological screening, but the most useful method is a
direct counterpart of colony hybridization probing. Recombinant colonies are transferred to a polyvinyl or nitrocellulose membrane, the cells are lysed, and a solution containing the specific antibody is added (Figure 8.21a). In the original methods, either the
antibody itself was labeled, or the membrane was subsequently washed with a solution
of labeled protein A, a bacterial protein that specifically binds to the immunoglobulins
that antibodies are made of (as shown in Figure 8.21a). In the more modern methods,
the bound antibody—the primary antibody—is detected by washing the membrane with
a labeled secondary antibody, which binds specifically to the primary antibody. Several
secondary antibody molecules can bind to a single primary antibody molecule, increasing
the amount of signal that is produced and enabling a clearer detection of each positive
colony. In all three methods, the label can be a radioactive one, in which case the
colonies that bind the label are detected by autoradiography (Figure 8.21b), or non-
radioactive labels resulting in a fluorescent or chemiluminescent signal can be used.

The problem of gene expression

Immunological screening depends on the cloned gene being expressed so that the protein translation product is present in the recombinant cells. However, as will be discussed in greater detail in Chapter 13, a gene from one organism is often not expressed in a different organism. In particular, it is very unlikely that a cloned animal or plant gene (with the exception of chloroplast genes) will be expressed in *E. coli* cells. This problem can be circumvented by using a special type of vector, called an **expression vector** (p. 227), designed specifically to promote expression of the cloned gene in a bacterial host. Immunological screening of recombinant *E. coli* colonies carrying animal genes cloned into expression vectors has been very useful in identifying genes for several important hormones.

Further reading

Benton, W.D. & Davis, R.W. (1977) Screening λgt recombinant clones by hybridization to single plaques in situ. *Science*, 196, 180–182.

Feinberg, A.P. & Vogelstein, B. (1983) A technique for labelling DNA restriction fragments to high specific activity. *Analytical Biochemistry*, 132, 6–13. [Random priming labeling.]

Grunstein, M. & Hogness, D.S. (1975) Colony hybridization: a method for the isolation of cloned cDNAs that contain a specific gene. *Proceedings of the National Academy of Sciences of the USA*, 72, 3961–3965.

Gubler, U. & Hoffman, B.J. (1983) A simple and very efficient method for generating cDNA libraries. *Gene*, 25, 263–269.

Southern, E.M. (2000) Blotting at 25. *Trends in Biochemical Science*, 25, 585–588. [The origins of Southern hybridization.]

Thorpe, G.H.G., Kricka, L.J., Moseley, S.B. & Whitehead, T.P. (1985) Phenols as enhancers of the chemiluminescent horseradish peroxidase–luminol–hydrogen peroxide reaction: application in luminescence-monitored enzyme immunoassays. *Clinical Chemistry*, 31, 1335–1341. [Describes the basis to a non-radioactive labeling method.]

Young, R.A. & Davis, R.W. (1983) Efficient isolation of genes by using antibody probes. *Proceedings of the National Academy of Sciences of the USA*, 80, 1194–1198.

Chapter 9

The Polymerase Chain Reaction

As a result of the last seven chapters we have become familiar not only with the basic principles of gene cloning, but also with fundamental molecular biology techniques such as restriction analysis, gel electrophoresis, DNA labeling, and DNA–DNA hybridization. To complete our basic education in DNA analysis we must now return to the second major technique for studying genes, the polymerase chain reaction (PCR). PCR is a very uncomplicated technique: all that happens is that a short region of a DNA molecule, a single gene for instance, is copied many times by a DNA polymerase enzyme (see Figure 1.2). This might seem a rather trivial exercise, but it has a multitude of applications in genetics research and in broader areas of biology.

We begin this chapter with an outline of the polymerase chain reaction in order to understand exactly what it achieves. Then we will look at the key issues that determine whether or not an individual PCR experiment is successful, before examining some of the methods that have been devised for studying the amplified DNA fragments that are obtained.

9.1 The polymerase chain reaction in outline

The polymerase chain reaction results in the selective amplification of a chosen region of a DNA molecule. Any region of any DNA molecule can be chosen, so long as the sequences at the borders of the region are known. The border sequences must be known because in order to carry out a PCR, two short oligonucleotides must hybridize to the

Gene Cloning and DNA Analysis: An Introduction. 6th edition. By T.A. Brown. Published 2010 by Blackwell Publishing.

Figure 9.1

Hybridization of the oligonucleotide primers to the template DNA at the beginning of a PCR.

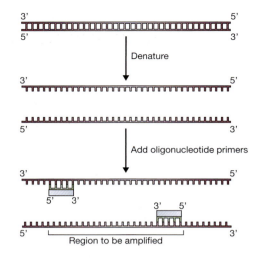

DNA molecule, one to each strand of the double helix (Figure 9.1). These oligonucleotides, which act as primers for the DNA synthesis reactions, delimit the region that will be amplified.

Amplification is usually carried out by the DNA polymerase I enzyme from *Thermus aquaticus*. As mentioned on p. 49, this organism lives in hot springs, and many of its enzymes, including *Taq* polymerase, are thermostable, meaning that they are resistant to denaturation by heat treatment. As will be apparent in a moment, the thermostability of *Taq* polymerase is an essential requirement in PCR methodology.

To carry out a PCR experiment, the target DNA is mixed with *Taq* polymerase, the two oligonucleotide primers, and a supply of nucleotides. The amount of target DNA can be very small because PCR is extremely sensitive and will work with just a single starting molecule. The reaction is started by heating the mixture to 94°C. At this temperature the hydrogen bonds that hold together the two polynucleotides of the double helix are broken, so the target DNA becomes denatured into single-stranded molecules (Figure 9.2). The temperature is then reduced to 50–60°C, which results in some rejoining of the single strands of the target DNA, but also allows the primers to attach to their annealing positions. DNA synthesis can now begin, so the temperature is raised to 74°C, just below the optimum for *Taq* polymerase. In this first stage of the PCR, a set of "long products" is synthesized from each strand of the target DNA. These polynucleotides have identical 5′ ends but random 3′ ends, the latter representing positions where DNA synthesis terminates by chance.

The cycle of denaturation–annealing–synthesis is now repeated (Figure 9.3). The long products denature and the four resulting strands are copied during the DNA synthesis stage. This gives four double-stranded molecules, two of which are identical to the long products from the first cycle and two of which are made entirely of new DNA. During the third cycle, the latter give rise to "short products", the 5′ and 3′ ends of which are both set by the primer annealing positions. In subsequent cycles, the number of short products accumulates in an exponential fashion (doubling during each cycle) until one of the components of the reaction becomes depleted. This means that after 30 cycles, there will be over 130 million short products derived from each starting molecule. In real terms, this equates to several micrograms of PCR product from a few nanograms or less of target DNA.

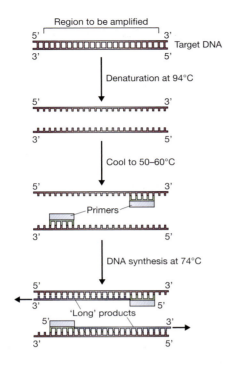

Figure 9.2

The first stage of a PCR, resulting in synthesis of the long products.

At the end of a PCR a sample of the reaction mixture is usually analyzed by agarose gel electrophoresis, sufficient DNA having been produced for the amplified fragment to be visible as a discrete band after staining with ethidium bromide. This may by itself provide useful information about the DNA region that has been amplified, or alternatively the PCR product can be examined by techniques such as DNA sequencing.

9.2 PCR in more detail

Although PCR experiments are very easy to set up, they must be planned carefully if the results are to be of any value. The sequences of the primers are critical to the success of the experiment, as are the precise temperatures used in the heating and cooling stages of the reaction cycle.

9.2.1 Designing the oligonucleotide primers for a PCR

The primers are the key to the success or failure of a PCR experiment. If the primers are designed correctly the experiment results in amplification of a single DNA fragment, corresponding to the target region of the template molecule. If the primers are incorrectly designed the experiment will fail, possibly because no amplification occurs, or possibly because the wrong fragment, or more than one fragment, is amplified (Figure 9.4). Clearly a great deal of thought must be put into the design of the primers.

Working out appropriate sequences for the primers is not a problem: they must correspond with the sequences flanking the target region on the template molecule. Each primer must, of course, be complementary (not identical) to its template strand in

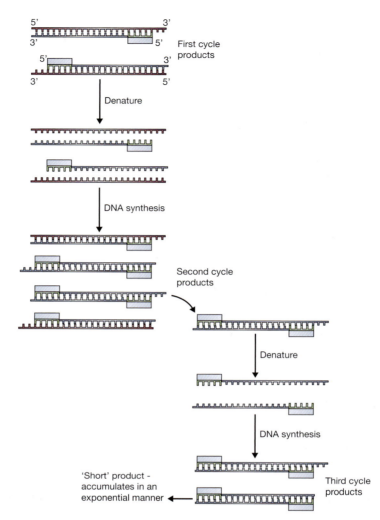

Figure 9.3

The second and third cycles of a PCR, during which the first short products are synthesized.

Figure 9.4

The results of PCRs with well designed and poorly designed primers. Lane 1 shows a single amplified fragment of the expected size, the result of a well designed experiment. In lane 2 there is no amplification product, suggesting that one or both of the primers were unable to hybridize to the template DNA. Lanes 3 and 4 show, respectively, an amplification product of the wrong size, and a mixture of products (the correct product plus two wrong ones); both results are due to hybridization of one or both of the primers to non-target sites on the template DNA molecule.

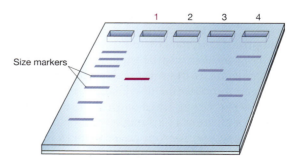

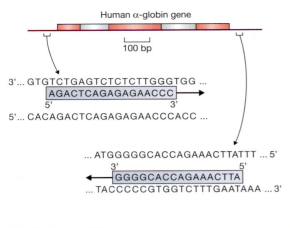

Human α-globin gene

100 bp

3'... GTGTCTGAGTCTCTCTTGGGTGG ...
AGACTCAGAGAGAACCC
5'   3'
5'... CACAGACTCAGAGAGAACCCACC ...

... ATGGGGGCACCAGAAACTTATTT ... 5'
3'   5'
GGGGCACCAGAAACTTA
... TACCCCCGTGGTCTTTGAATAAA ... 3'

Figure 9.5

A pair of primers designed to amplify the human α₁-globin gene. The exons of the gene are shown as closed boxes, the introns as open boxes.

(a) PCR of human DNA with 8-mer primers

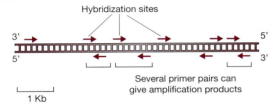

Hybridization sites

3'   5'
5'   3'

1 Kb

Several primer pairs can give amplification products

Figure 9.6

The lengths of the primers are critical for the specificity of the PCR.

(b) PCR of human DNA with 17-mer primers

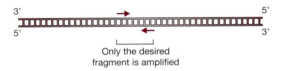

3'   5'
5'   3'

Only the desired fragment is amplified

order for hybridization to occur, and the 3' ends of the hybridized primers should point toward one another (Figure 9.5). The DNA fragment to be amplified should not be greater than about 3 kb in length and ideally less than 1 kb. Fragments up to 10 kb can be amplified by standard PCR techniques, but the longer the fragment the less efficient the amplification and the more difficult it is to obtain consistent results. Amplification of very long fragments—up to 40 kb—is possible, but requires special methods.

The first important issue to address is the length of the primers. If the primers are too short they might hybridize to non-target sites and give undesired amplification products. To illustrate this point, imagine that total human DNA is used in a PCR experiment with a pair of primers eight nucleotides in length (in PCR jargon, these are called "8-mers"). The likely result is that a number of different fragments will be amplified. This is because attachment sites for these primers are expected to occur, on average, once every $4^8 = 65,536$ bp, giving approximately 49,000 possible sites in the 3,200,000 kb of nucleotide sequence that makes up the human genome. This means that it would be very unlikely that a pair of 8-mer primers would give a single, specific amplification product with human DNA (Figure 9.6a).

What if the 17-mer primers shown in Figure 9.5 are used? The expected frequency of a 17-mer sequence is once every $4^{17} = 17,179,869,184$ bp. This figure is over five

Figure 9.7

A typical temperature profile for a PCR.

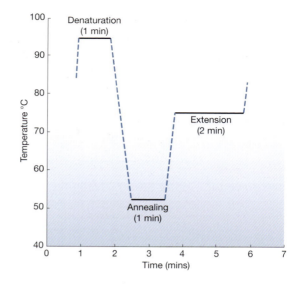

times greater than the length of the human genome, so a 17-mer primer would be expected to have just one hybridization site in total human DNA. A pair of 17-mer primers should therefore give a single, specific amplification product (Figure 9.6b).

Why not simply make the primers as long as possible? The length of the primer influences the rate at which it hybridizes to the template DNA, long primers hybridizing at a slower rate. The efficiency of the PCR, measured by the number of amplified molecules produced during the experiment, is therefore reduced if the primers are too long, as complete hybridization to the template molecules cannot occur in the time allowed during the reaction cycle. In practice, primers longer than 30-mer are rarely used.

9.2.2 Working out the correct temperatures to use

During each cycle of a PCR, the reaction mixture is transferred between three temperatures (Figure 9.7):

- The denaturation temperature, usually 94°C, which breaks the base pairs and releases single-stranded DNA to act as templates in the next round of DNA synthesis;
- The hybridization or annealing temperature, at which the primers attach to the templates;
- The extension temperature, at which DNA synthesis occurs. This is usually set at 74°C, just below the optimum for *Taq* polymerase.

The annealing temperature is the important one because, again, this can affect the specificity of the reaction. DNA–DNA hybridization is a temperature-dependent phenomenon. If the temperature is too high no hybridization takes place; instead the primers and templates remain dissociated (Figure 9.8a). However, if the temperature is too low, mismatched hybrids—ones in which not all the correct base pairs have formed—are stable (Figure 9.8b). If this occurs the earlier calculations regarding the appropriate lengths for the primers become irrelevant, as these calculations assumed that only perfect primer–template hybrids are able to form. If mismatches are tolerated,

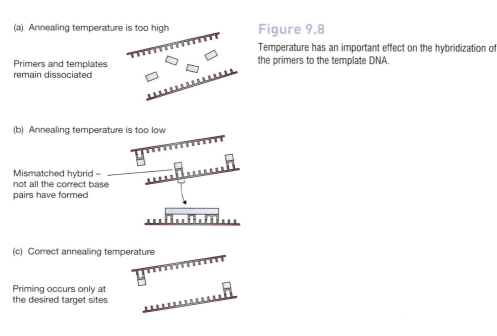

(a) Annealing temperature is too high

Primers and templates remain dissociated

(b) Annealing temperature is too low

Mismatched hybrid – not all the correct base pairs have formed

(c) Correct annealing temperature

Priming occurs only at the desired target sites

Figure 9.8

Temperature has an important effect on the hybridization of the primers to the template DNA.

the number of potential hybridization sites for each primer is greatly increased, and amplification is more likely to occur at non-target sites in the template molecule.

The ideal annealing temperature must be low enough to enable hybridization between primer and template, but high enough to prevent mismatched hybrids from forming (Figure 9.8c). This temperature can be estimated by determining the **melting temperature** or T_m of the primer–template hybrid. The T_m is the temperature at which the correctly base-paired hybrid dissociates ("melts"). A temperature 1–2°C below this should be low enough to allow the correct primer–template hybrid to form, but too high for a hybrid with a single mismatch to be stable. The T_m can be determined experimentally but is more usually calculated from the simple formula (Figure 9.9):

$$T_m = (4 \times [G + C]) + (2 \times [A + T])°C$$

in which $[G + C]$ is the number of G and C nucleotides in the primer sequence, and $[A + T]$ is the number of A and T nucleotides.

The annealing temperature for a PCR experiment is therefore determined by calculating the T_m for each primer and using a temperature of 1–2°C below this figure. Note that this means the two primers should be designed so that they have identical T_ms. If this is not the case, the appropriate annealing temperature for one primer may be too high or too low for the other member of the pair.

9.3 After the PCR: studying PCR products

PCR is often the starting point for a longer series of experiments in which the amplification product is studied in various ways in order to gain information about the DNA molecule that acted as the original template. We will encounter many studies of this type in Parts II and III, when we examine the applications of gene cloning and PCR

Figure 9.9

Calculating the T_m of a primer.

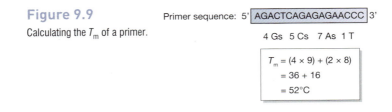

Primer sequence: 5' AGACTCAGAGAGAACCC 3'

4 Gs 5 Cs 7 As 1 T

$$T_m = (4 \times 9) + (2 \times 8)$$
$$= 36 + 16$$
$$= 52°C$$

in research and biotechnology. Although a wide range of procedures have been devised for studying PCR products, three techniques are particularly important:

- Gel electrophoresis of PCR products
- Cloning of PCR products
- Sequencing of PCR products.

The first two of these techniques are dealt with in this chapter. The third technique is deferred until Chapter 10, when all aspects of DNA sequencing will be covered.

9.3.1 Gel electrophoresis of PCR products

The results of most PCR experiments are checked by running a portion of the amplified reaction mixture in an agarose gel. A band representing the amplified DNA may be visible after staining, or if the DNA yield is low the product can be detected by Southern hybridization (p. 142). If the expected band is absent, or if additional bands are present, something has gone wrong and the experiment must be repeated.

In some cases, agarose gel electrophoresis is used not only to determine if a PCR experiment has worked, but also to obtain additional information. For example, the presence of restriction sites in the amplified region of the template DNA can be determined by treating the PCR product with a restriction endonuclease before running the sample in the agarose gel (Figure 9.10). This is a type of restriction fragment length polymorphism (RFLP) analysis and is important both in the construction of genome maps (p. 180) and in studying genetic diseases (p. 257).

Alternatively, the exact size of the PCR product can be used to establish if the template DNA contains an insertion or deletion mutation in the amplified region (Figure 9.10). Length mutations of this type form the basis of DNA profiling, a central technique in forensic science (Chapter 16).

In some experiments, the mere presence or absence of the PCR product is the diagnostic feature. An example is when PCR is used as the screening procedure to identify a desired gene from a genomic or cDNA library. Carrying out PCRs with every clone in a genomic library might seem to be a tedious task, but one of the advantages of PCR is that individual experiments are quick to set up and many PCRs can be performed in parallel. The workload can also be reduced by combinatorial screening, an example of which is shown in Figure 9.11.

9.3.2 Cloning PCR products

Some applications require that after a PCR the resulting products are ligated into a vector and examined by any of the standard methods used for studying cloned DNA. This may sound easy, but there are complications.

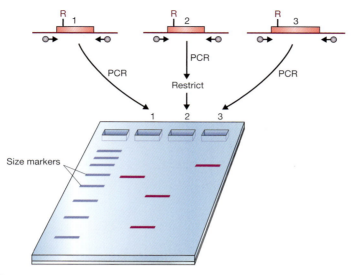

Figure 9.10

Gel electrophoresis of the PCR product can provide information on the template DNA molecule. Lanes 1 and 2 show, respectively, an unrestricted PCR product and a product restricted with the enzyme that cuts at site R. Lane 3 shows the result obtained when the template DNA contains an insertion in the amplified region.

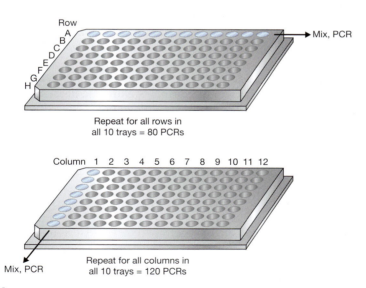

Figure 9.11

Combinatorial screening of clones in microtiter trays. A library of 960 clones is screened by a series of PCRs, each with a combination of clones. The clone combinations that give positive results enable the well(s) containing positive clone(s) to be identified. For example, if positive PCRs are given with row A of tray 2, row D of tray 6, column 7 of tray 2, and column 9 of tray 6, then it can be deduced that there are positive clones in well A7 of tray 2 and well D9 of tray 6. Although there are 960 clones, unambiguous identification of the positive clones is therefore achieved after just 200 PCRs.

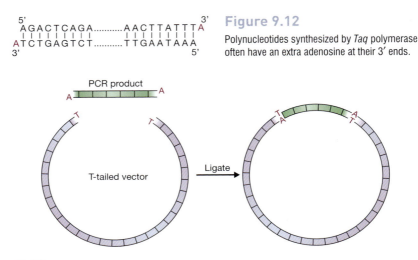

Figure 9.12

Polynucleotides synthesized by *Taq* polymerase often have an extra adenosine at their 3′ ends.

Figure 9.13

Using a special T-tailed vector to clone a PCR product.

The first problem concerns the ends of the PCR products. From an examination of Figure 9.3 it might be imagined that the short products resulting from PCR amplification are blunt-ended. If this was the case they could be inserted into a cloning vector by blunt-end ligation, or alternatively the PCR products could be provided with sticky ends by the attachment of linkers or adaptors (p. 64). Unfortunately, the situation is not so straightforward. *Taq* polymerase tends to add an additional nucleotide, usually an adenosine, to the end of each strand that it synthesizes. This means that a double-stranded PCR product is not blunt-ended, and instead most 3′ termini have a single nucleotide overhang (Figure 9.12). The overhangs could be removed by treatment with an exonuclease enzyme, resulting in PCR products with true blunt ends, but this is not a popular approach as it is difficult to prevent the exonuclease from becoming over-active and causing further damage to the ends of the molecules.

One solution is to use a special cloning vector which carries thymidine (T) overhangs and which can therefore be ligated to a PCR product (Figure 9.13). These vectors are usually prepared by restricting a standard vector at a blunt-end site, and then treating with *Taq* polymerase in the presence of just 2′-deoxythymidine 5′-triphosphate (dTTP). No primer is present so all the polymerase can do is add a T nucleotide to the 3′ ends of the blunt-ended vector molecule, resulting in the T-tailed vector into which the PCR products can be inserted. Special vectors of this type have also been designed for use with the topoisomerase ligation method described on p. 69, and this is currently the most popular way of cloning PCR products.

A second solution is to design primers that contain restriction sites. After PCR the products are treated with the restriction endonuclease, which cuts each molecule within the primer sequence, leaving sticky-ended fragments that can be ligated efficiently into a standard cloning vector (Figure 9.14). The approach is not limited to those instances where the primers span restriction sites that are present in the template DNA. Instead, the restriction site can be included within a short extension at the 5′ end of each primer (Figure 9.15). These extensions cannot hybridize to the template molecule, but they are copied during the PCR, resulting in PCR products that carry terminal restriction sites.

Primer sequence

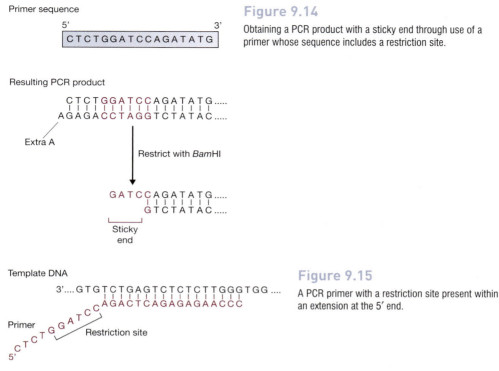

5'   3'

C T C T G G A T C C A G A T A T G

Figure 9.14

Obtaining a PCR product with a sticky end through use of a primer whose sequence includes a restriction site.

Resulting PCR product

C T C T G G A T C C A G A T A T G .....
| | | | | | | | | | | | | | | | | |
A G A G A C C T A G G T C T A T A C .....

Extra A

Restrict with *Bam*HI

G A T C C A G A T A T G .....
| | | | | | | |
G T C T A T A C .....

Sticky end

Template DNA

3' .... G T G T C T G A G T C T C T C T T G G G T G G ....
| | | | | | | | | | | | | | | | | | |
A G A C T C A G A G A G A A C C C

Primer   G A T C C
            G G
         C T G
      C T
   C
5'

Restriction site

Figure 9.15

A PCR primer with a restriction site present within an extension at the 5' end.

9.3.3 *Problems with the error rate of* Taq *polymerase*

All DNA polymerases make mistakes during DNA synthesis, occasionally inserting an incorrect nucleotide into the growing DNA strand. Most polymerases, however, are able to rectify these errors by reversing over the mistake and resynthesizing the correct sequence. This property is referred to as the "proofreading" function and depends on the polymerase possessing a 3' to 5' exonuclease activity (p. 168).

Taq polymerase lacks a proofreading activity and as a result is unable to correct its errors. This means that the DNA synthesized by *Taq* polymerase is not always an accurate copy of the template molecule. The error rate has been estimated at one mistake for every 9000 nucleotides of DNA that is synthesized, which might appear to be almost insignificant but which translates to one error in every 300 bp for the PCR products obtained after 30 cycles. This is because PCR involves copies being made of copies of copies, so the polymerase-induced errors gradually accumulate, the fragments produced at the end of a PCR containing copies of earlier errors together with any new errors introduced during the final round of synthesis.

For many applications this high error rate does not present a problem. In particular, sequencing of a PCR product provides the correct sequence of the template, even though the PCR products contain the errors introduced by *Taq* polymerase. This is because the errors are distributed randomly, so for every molecule that has an error at a particular nucleotide position, there will be many molecules with the correct sequence. In this context the error rate is indeed insignificant.

This is not the case if the PCR products are cloned. Each resulting clone contains multiple copies of a single amplified fragment, so the cloned DNA does not necessarily

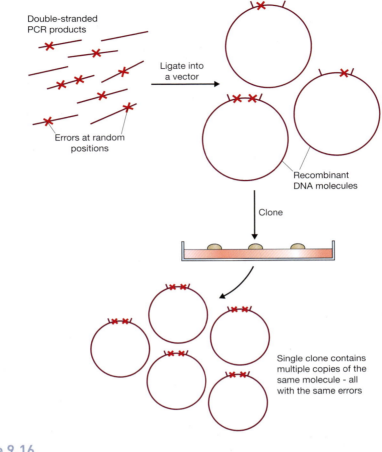

Double-stranded PCR products

Ligate into a vector

Errors at random positions

Recombinant DNA molecules

Clone

Single clone contains multiple copies of the same molecule - all with the same errors

Figure 9.16

The high error rate of *Taq* polymerase becomes a factor when PCR products are cloned.

have the same sequence as the original template molecule used in the PCR (Figure 9.16). This possibility lends an uncertainty to all experiments carried out with cloned PCR products and dictates that, whenever possible, the amplified DNA should be studied directly rather than being cloned.

9.4 Real-time PCR enables the amount of starting material to be quantified

The amount of product that is synthesized during a set number of cycles of a PCR depends on the number of DNA molecules that are present in the starting mixture (Table 9.1). If there are only a few DNA molecules at the beginning of the PCR then relatively little product will be made, but if there are many starting molecules then the product yield will be higher. This relationship enables PCR to be used to quantify the number of DNA molecules present in an extract.

Table 9.1

Number of short products synthesized after 25 cycles of PCR with different numbers of starting molecule.

NUMBER OF STARTING MOLECULES	NUMBER OF SHORT PRODUCTS
1	4,194,304
2	8,388,608
5	20,971,520
10	41,943,040
25	104,857,600
50	209,715,200
100	419,430,400

Note: The numbers assume that amplification is 100% efficient, none of the reactants becoming limiting during the course of the PCR.

9.4.1 Carrying out a quantitative PCR experiment

In **quantitative PCR (qPCR)** the amount of product synthesized during a test PCR is compared with the amounts synthesized during PCRs with known quantities of starting DNA. In the early procedures, agarose gel electrophoresis was used to make these comparisons. After staining the gel, the band intensities were examined to identify the control PCR whose product was most similar to that of the test (Figure 9.17). Although easy to perform, this type of qPCR is imprecise, because large differences in the amount of starting DNA give relatively small differences in the band intensities of the resulting PCR products.

Today, quantification is carried out by **real-time PCR**, a modification of the standard PCR technique in which synthesis of the product is measured over time, as the PCR proceeds through its series of cycles. There are two ways of following product synthesis in real time:

- A dye that gives a fluorescent signal when it binds to double-stranded DNA can be included in the PCR mixture. This method measures the total amount of double-stranded DNA in the PCR at any particular time, which may over-estimate the actual amount of the product because sometimes the primers anneal to one another in various non-specific ways, increasing the amount of double-stranded DNA that is present.
- A short oligonucleotide called a **reporter probe**, which gives a fluorescent signal when it hybridizes to the PCR product, can be used. Because the probe only hybridizes to the PCR product, this method is less prone to inaccuracies caused by primer-primer annealing. Each probe molecule has pair of labels. A fluorescent dye

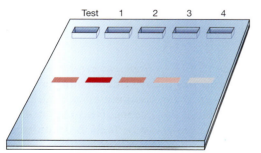

Test 1 2 3 4

Figure 9.17

Using agarose gel electrophoresis to quantify the amount of DNA in a test PCR. Lanes 1 to 4 are control PCRs carried out with decreasing amounts of template DNA. The intensity of staining for the test band suggests that this PCR contained approximately the same amount of DNA as the control run in lane 2.

Figure 9.18

Hybridization of a reporter probe to its target DNA.

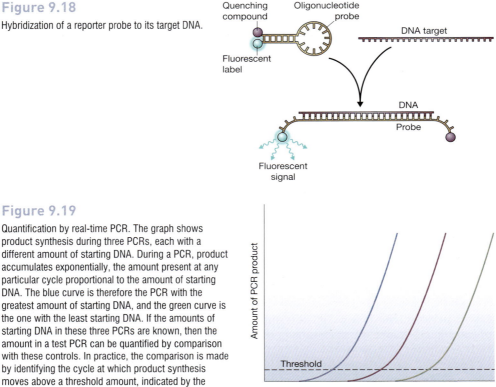

Figure 9.19

Quantification by real-time PCR. The graph shows product synthesis during three PCRs, each with a different amount of starting DNA. During a PCR, product accumulates exponentially, the amount present at any particular cycle proportional to the amount of starting DNA. The blue curve is therefore the PCR with the greatest amount of starting DNA, and the green curve is the one with the least starting DNA. If the amounts of starting DNA in these three PCRs are known, then the amount in a test PCR can be quantified by comparison with these controls. In practice, the comparison is made by identifying the cycle at which product synthesis moves above a threshold amount, indicated by the horizontal line on the graph.

is attached to one end of the oligonucleotide, and a quenching compound, which inhibits the fluorescent signal, is attached to the other end (Figure 9.18). Normally there is no fluorescence because the oligonucleotide is designed in such a way that its two ends base pair to one another, placing the quencher next to the dye. Hybridization between the oligonucleotide and the PCR product disrupts this base pairing, moving the quencher away from the dye and enabling the fluorescent signal to be generated.

Both systems enable synthesis of the PCR product to be followed by measuring the fluorescent signal. Quantification again requires comparison between test and control PCRs, usually by identifying the stage in the PCR at which the amount of fluorescent signal reaches a pre-set threshold (Figure 9.19). The more rapidly the threshold is reached, the greater the amount of DNA in the starting mixture.

9.4.2 Real-time PCR can also quantify RNA

Real-time PCR is often used to quantify the amount of DNA in an extract, for example to follow the progression of a viral infection by measuring the amount of pathogen DNA that is present in a tissue. More frequently, however, the method is used as a means of measuring RNA amounts, in particular to determine the extent of expression of a particular gene by quantifying its mRNA. The gene under study might be one that is switched on in cancerous cells, in which case quantifying its mRNA will enable the

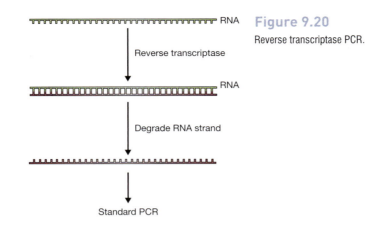

Figure 9.20

Reverse transcriptase PCR.

development of the cancer to be monitored and the effects of subsequent treatment to be assessed.

How do we carry out PCR if RNA is the starting material? The answer is to use reverse transcriptase PCR. The first step in this procedure is to convert the RNA molecules into single-stranded complementary DNA (cDNA) (Figure 9.20). Once this preliminary step has been carried out, the PCR primers and *Taq* polymerase are added and the experiment proceeds exactly as in the standard technique. Some thermostable polymerases are able to make DNA copies of both RNA and DNA molecules (i.e., they have both reverse transcriptase and DNA-dependent DNA polymerase activities) and so can carry out all the steps of this type of PCR in a single reaction.

Further reading

Higuchi, R., Dollinger, G., Walsh, P.S. & Griffith, R. (1992) Simultaneous amplification and detection of specific DNA sequences. *Biotechnology*, 10, 413–417. [The first description of real-time PCR].

Marchuk, D., Drumm, M., Saulino, A. & Collins, F.S. (1991) Construction of T-vectors, a rapid and general system for direct cloning of unmodified PCR products. *Nucleic Acids Research*, 19, 1154.

Rychlik, W., Spencer, W.J. & Rhoads, R.E. (1990) Optimization of the annealing temperature for DNA amplification *in vitro*. *Nucleic Acids Research*, 18, 6409–6412.

Saiki, R.K., Gelfand, D.H., Stoffel, S. et al. (1988) Primer-directed enzymatic amplification of DNA with a thermostable DNA polymerase. *Science*, 239, 487–491. [The first description of PCR with *Taq* polymerase.]

Tindall, K.R. & Kunkel, T.A. (1988) Fidelity of DNA synthesis by the Thermus aquaticus DNA polymerase. *Biochemistry*, 27, 6008–6013. [Describes the error rate of *Taq* polymerase.]

VanGuilder, H.D., Vrana, K.E. & Freeman, W.M. (2008) Twenty-five years of quantitative PCR for gene expression analysis. *Biotechniques*, 44, 619–624.

PART II

The Applications of Gene Cloning and DNA Analysis in Research

Sequencing Genes and Genomes

Part I of this book has shown how a skilfully performed cloning or PCR experiment can provide a pure sample of an individual gene, or any other DNA sequence, separated from all the other genes and DNA sequences in the cell. Now we can turn our attention to the ways in which cloning, PCR, and other DNA analysis techniques are used to study genes and genomes. We will consider three aspects of molecular biology research:

- The techniques used to obtain the nucleotide sequence of individual genes and entire genomes (this Chapter);
- The methods used to study the expression and function of individual genes (Chapter 11);
- The techniques that are used to study entire genomes (Chapter 12).

Probably the most important technique available to the molecular biologist is DNA sequencing, by which the precise order of nucleotides in a piece of DNA can be determined. DNA sequencing methods have been around for 40 years, and since the mid-1970s rapid and efficient sequencing has been possible. Initially these techniques were applied to individual genes, but since the early 1990s an increasing number of entire genome sequences have been obtained. In this chapter we will study the methodology used in DNA sequencing and then examine how these techniques are used in genome projects.

10.1 The methodology for DNA sequencing

There are several procedures for DNA sequencing, the most popular being the chain termination method first devised by Fred Sanger and colleagues in the mid-1970s. Chain

Gene Cloning and DNA Analysis: An Introduction. 6th edition. By T.A. Brown. Published 2010 by Blackwell Publishing.

termination sequencing has gained pre-eminence for several reasons, not least being the relative ease with which the technique can be automated. As we will see later in this chapter, in order to sequence an entire genome a huge number of individual sequencing experiments must be carried out, and it would take many years to perform all of these by hand. Automated sequencing techniques are therefore essential if a genome project is to be completed in a reasonable timespan.

Part of the automation strategy is to design systems that enable many individual sequencing experiments to be carried out at once. With the chain termination method, up to 96 sequences can be obtained simultaneously in a single run of a sequencing machine. This is still not enough to fully satisfy the demands of genome sequencing, and during the last few years an alternative method called **pyrosequencing** has become popular. Pyrosequencing, which was invented in 1998, forms the basis to a **massively parallel** strategy that enables hundreds of thousands of short sequences to be generated at the same time.

10.1.1 Chain termination DNA sequencing

Chain termination DNA sequencing is based on the principle that single-stranded DNA molecules that differ in length by just a single nucleotide can be separated from one another by polyacrylamide gel electrophoresis. This means that it is possible to resolve a family of molecules, representing all lengths from 10 to 1500 nucleotides, into a series of bands in a slab or capillary gel (Figure 10.1).

Chain termination sequencing in outline

The starting material for a chain termination sequencing experiment is a preparation of identical single-stranded DNA molecules. The first step is to anneal a short oligonucleotide to the same position on each molecule, this oligonucleotide subsequently acting as the primer for synthesis of a new DNA strand that is complementary to the template (Figure 10.2a).

The strand synthesis reaction, which is catalyzed by a DNA polymerase enzyme and requires the four deoxyribonucleotide triphosphates (dNTPs—dATP, dCTP, dGTP, and dTTP) as substrates, would normally continue until several thousand nucleotides had been polymerized. This does not occur in a chain termination sequencing experiment

Figure 10.1

Polyacrylamide gel electrophoresis can resolve single-stranded DNA molecules that differ in length by just one nucleotide. The banding pattern shown here is produced after separation of single-stranded DNA molecules by denaturing polyacrylamide gel electrophoresis. The molecules have been labeled with a radioactive marker and the bands visualized by autoradiography.

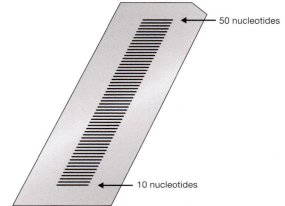

50 nucleotides

10 nucleotides

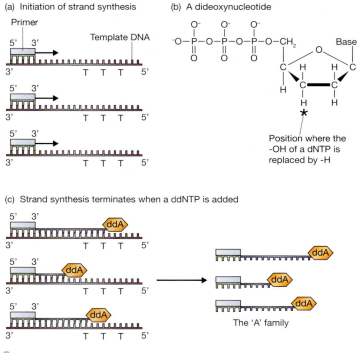

Figure 10.2

Chain termination DNA sequencing.

because, as well as the four deoxynucleotides, a small amount of each of four dideoxynucleotides (ddNTPs—ddATP, ddCTP, ddGTP, and ddTTP) is added to the reaction. Each of these dideoxynucleotides is labeled with a different fluorescent marker.

The polymerase enzyme does not discriminate between deoxy- and dideoxynucleotides, but once incorporated a dideoxynucleotide blocks further elongation because it lacks the 3′-hydroxyl group needed to form a connection with the next nucleotide (Figure 10.2b). Because the normal deoxynucleotides are also present, in larger amounts than the dideoxynucleotides, the strand synthesis does not always terminate close to the primer: in fact, several hundred nucleotides may be polymerized before a dideoxynucleotide is eventually incorporated. The result is a set of new molecules, all of different lengths, and each ending in a dideoxynucleotide whose identity indicates the nucleotide—A, C, G, or T—that is present at the equivalent position in the template DNA (Figure 10.2c).

To work out the DNA sequence, all that we have to do is identify the dideoxynucleotide at the end of each chain-terminated molecule. This is where the polyacrylamide gel comes into play. The mixture is loaded into a well of a polyacrylamide slab gel, or into a tube of a capillary gel system, and electrophoresis carried out to separate the molecules according to their lengths. After separation, the molecules are run past a fluorescent detector capable of discriminating the labels attached to the dideoxynucleotides (Figure 10.3a). The detector therefore determines if each molecule ends in an A, C, G, or T. The sequence can be printed out for examination by the operator (Figure 10.3b), or entered directly into a storage device for future analysis.

Figure 10.3

Reading the sequence generated by a chain termination experiment. (a) Each dideoxynucleotide is labeled with a different fluorochrome, so the chain-terminated polynucleotides are distinguished as they pass by the detector. (b) An example of a sequence print out.

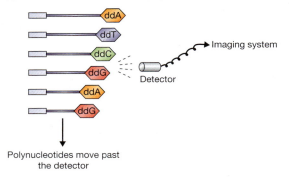

(a) Detection of chain-terminated polynucleotides

Imaging system

Detector

Polynucleotides move past the detector

(b) The print out from an automated sequencer

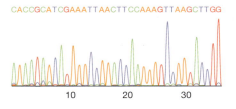

CACCGCATCGAAATTAACTTCCAAAGTTAAGCTTGG

10   20   30

Not all DNA polymerases can be used for sequencing

Any DNA polymerase is capable of extending a primer that has been annealed to a single-stranded DNA molecule, but not all polymerases can be used for DNA sequencing. This is because many DNA polymerases have a mixed enzymatic activity, being able to degrade as well as synthesize DNA (p. 48). Degradation can occur in either the 5'→3' or 3'→5' direction (Figure 10.4), and both activities are detrimental to accurate chain termination sequencing. The 5'→3' exonuclease activity enables the polymerase to remove nucleotides from the 5' ends of the newly-synthesized strands, changing the lengths of these strands so that they no longer run through the polyacrylamide gel in the appropriate order. The 3'→5' activity could have the same effect, but more importantly

(a) 5' ⟶ 3' exonuclease activity

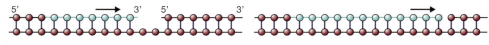

(b) 3' ⟶ 5' exonuclease activity

Figure 10.4

The exonuclease activities of DNA polymerases. (a) The 5'→3' activity has an important role in DNA repair in the cell, as it enables the polymerase to replace a damaged DNA strand. In DNA sequencing this activity can result in the 5' ends of newly-synthesized strands becoming shortened. (b) The 3'→5' activity also has an important role in the cell, as it allows the polymerase to correct its own mistakes, by reversing and replacing a nucleotide that has been added in error (e.g., a T instead of a G). This is called proofreading. During DNA sequencing, this activity can result in removal of a dideoxynucleotide that has just been added to the newly-synthesized strand, so that chain termination does not occur.

will remove a dideoxynucleotide that has just been added at the 3′ end, preventing chain termination from occurring.

In the original method for chain termination sequencing, the Klenow polymerase was used as the sequencing enzyme. As described on p. 49, this is a modified version of the DNA polymerase I enzyme from *E. coli*, the modification removing the 5′→3′ exonuclease activity of the standard enzyme. However, the Klenow polymerase has low **processivity**, meaning that it can only synthesize a relatively short DNA strand before dissociating from the template due to natural causes. This limits the length of sequence that can be obtained from a single experiment to about 250 bp. To avoid this problem, most sequencing today makes use of a more specialized enzyme, such as **Sequenase**, a modified version of the DNA polymerase encoded by bacteriophage T7. Sequenase has high processivity and no exonuclease activity and so is ideal for chain termination sequencing, enabling sequences of up to 750 bp to be obtained in a single experiment.

Chain termination sequencing requires a single-stranded DNA template

The template for a chain termination experiment is a single-stranded version of the DNA molecule to be sequenced. One way of obtaining single-stranded DNA is to use an M13 vector, but the M13 system, although designed specifically to provide DNA for chain termination sequencing, is not ideal for this purpose. The problem is that cloned DNA fragments that are longer than about 3 kb are unstable in an M13 vector and can undergo deletions and rearrangements. This means that M13 cloning can only be used with short pieces of DNA.

Plasmid vectors, which do not suffer instability problems, are therefore more popular, but some means is needed of converting the double-stranded plasmid into a single-stranded form. There are two possibilities:

- Double-stranded plasmid DNA can be converted into single-stranded DNA by denaturation with alkali or by boiling. This is a common method for obtaining template DNA for DNA sequencing, but a shortcoming is that it can be difficult to prepare plasmid DNA that is not contaminated with small quantities of bacterial DNA and RNA, which can act as spurious templates or primers in the DNA sequencing experiment.
- The DNA can be cloned in a phagemid, a plasmid vector that contains an M13 origin of replication and which can therefore be obtained as both double- and single-stranded DNA versions (p. 96). Phagemids avoid the instabilities of M13 cloning and can be used with fragments up to 10 kb or more.

The need for single-stranded DNA can also be sidestepped by using a thermostable DNA polymerase as the sequencing enzyme. This method, called **thermal cycle sequencing**, is carried out in a similar way to PCR, but just one primer is used and the reaction mixture includes the four dideoxynucleotides (Figure 10.5). Because there is only one primer, only one of the strands of the starting molecule is copied, and the product accumulates in a linear fashion, not exponentially as is the case in a real PCR. The presence of the dideoxynucleotides in the reaction mixture causes chain termination, as in the standard methodology, and the family of resulting strands can be analyzed and the sequence read in the usual way. Thermal cycle sequencing can therefore be used with DNA cloned in any type of vector.

The primer determines the region of the template DNA that will be sequenced

In the first stage of a chain termination sequencing experiment, an oligonucleotide primer is annealed onto the template DNA (see Figure 10.2a). The main function of the

Figure 10.5

The basis to thermal cycle sequencing. A PCR is set up with just one primer and one of the dideoxynucleotides. One of the template strands is copied into a family of chain-terminated polynucleotides. ddA = dideoxyATP.

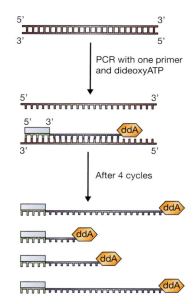

Figure 10.6

Different types of primer for chain termination sequencing.

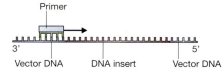

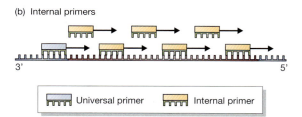

primer is to provide the short double-stranded region that is needed in order for the DNA polymerase to initiate DNA synthesis. The primer also plays a second critical role in determining the region of the template molecule that will be sequenced.

For most sequencing experiments a **universal primer** is used, this being one that is complementary to the part of the vector DNA immediately adjacent to the point into which new DNA is ligated (Figure 10.6a). The 3′ end of the primer points toward the inserted DNA, so the sequence that is obtained starts with a short stretch of the vector and then progresses into the cloned DNA fragment. If the DNA is cloned in a plasmid vector, then both forward and reverse universal primers can be used, enabling sequences to be obtained from both ends of the insert. This is an advantage if the cloned DNA is more than 750 bp and hence too long to be sequenced completely in one experiment. Alternatively, it is possible to extend the sequence in one direction by synthesizing a non-universal primer, designed to anneal at a position within the insert DNA (Figure 10.6b).

An experiment with this primer will provide a second short sequence that overlaps the previous one.

10.1.2 Pyrosequencing

Pyrosequencing is the second important type of DNA sequencing methodology that is in use today. Pyrosequencing does not require electrophoresis or any other fragment separation procedure and so is more rapid than chain termination sequencing. It is only able to generate up to 150 bp in a single experiment, and at first glance might appear to be less useful than the chain termination method, especially if the objective is to sequence a genome. The advantage with pyrosequencing is that it can be automated in a massively parallel manner that enables hundreds of thousands of sequences to be obtained at once, perhaps as much as 1000 Mb in a single run. Sequence is therefore produced much more quickly than is possible by the chain termination method, which explains why pyrosequencing is gradually taking over as the method of choice for genome projects.

Pyrosequencing involves detection of pulses of chemiluminescence

Pyrosequencing, like the chain termination method, requires a preparation of identical single-stranded DNA molecules as the starting material. These are obtained by alkali denaturation of PCR products or, more rarely, recombinant plasmid molecules. After attachment of the primer, the template is copied by a DNA polymerase in a straight-forward manner without added dideoxynucleotides. As the new strand is being made, the order in which the deoxynucleotides are incorporated is detected, so the sequence can be "read" as the reaction proceeds.

The addition of a deoxynucleotide to the end of the growing strand is detectable because it is accompanied by release of a molecule of pyrophosphate, which can be converted by the enzyme sulfurylase into a flash of chemiluminescence. Of course, if all four deoxynucleotides were added at once, then flashes of light would be seen all the time and no useful sequence information would be obtained. Each deoxynucleotide is therefore added separately, one after the other, with a nucleotidase enzyme also present in the reaction mixture so that if a deoxynucleotide is not incorporated into the poly-nucleotide then it is rapidly degraded before the next one is added (Figure 10.7). This procedure makes it possible to follow the order in which the deoxynucleotides are incor-porated into the growing strand. The technique sounds complicated, but it simply requires that a repetitive series of additions be made to the reaction mixture, precisely the type of procedure that is easily automated.

Massively parallel pyrosequencing

The high throughput version of pyrosequencing usually begins with genomic DNA, rather than PCR products or clones. The DNA is broken into fragments between 300 and 500 bp in length (Figure 10.8a), and each fragment is ligated to a pair of adaptors (p. 65), one adaptor to either end (Figure 10.8b). These adaptors play two important roles. First, they enable the DNA fragments to be attached to small metallic beads. This is because one of the adaptors has a biotin label attached to its 5′ end, and the beads are coated with streptavidin, to which biotin binds with great affinity (p. 137). DNA fragments therefore become attached to the beads via biotin-streptavidin linkages (Figure 10.8c). The ratio of DNA fragments to beads is set so that, on average, just one fragment becomes attached to each bead.

Figure 10.7

Pyrosequencing.

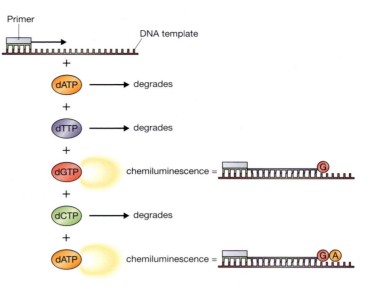

Figure 10.8

One method for massively parallel DNA sequencing. B = biotin, S = streptavidin.

(a) Break genomic DNA into fragments

Genomic DNA

(b) Ligate adaptors

(c) Separate strands and attach to beads

Bead

Streptavidin–biotin attachment

(d) PCRs in an oil emulsion

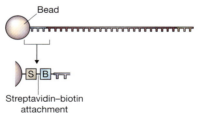

Water droplets

Oil emulsion

PCR products

Each DNA fragment will now be amplified by PCR so that enough copies are made for sequencing. The adaptors now play their second role as they provide the annealing sites for the primers for this PCR. The same pair of primers can therefore be used for all the fragments, even though the fragments themselves have many different sequences. If the PCR is carried out immediately then all we will obtain is a mixture of all the products, which will not enable us to obtain the individual sequences of each one. To solve this problem, PCR is carried out in an oil emulsion, each bead residing in its own aqueous droplet within the emulsion (Figure 10.8d). Each droplet contains all the reagents needed for PCR, and is physically separated from all the other droplets by the barrier provided by the oil component of emulsion. After PCR, the aqueous droplets are transferred into wells on a plastic strip so there is one droplet and hence once PCR product per well, and the pyrosequencing reactions are carried out in each well.

10.2 How to sequence a genome

The first DNA molecule to be completely sequenced was the 5386 nucleotide genome of bacteriophage φX174, which was completed in 1975. This was quickly followed by sequences for SV40 virus (5243 bp) in 1977 and pBR322 (4363 bp) in 1978. Gradually sequencing was applied to larger molecules. Professor Sanger's group published the sequence of the human mitochondrial genome (16.6 kb) in 1981 and of bacteriophage λ (49 kb) in 1982. Nowadays sequences of 100–200 kb are routine and most research laboratories have the necessary expertise to generate this amount of information.

The pioneering projects today are the massive genome initiatives, each aimed at obtaining the nucleotide sequence of the entire genome of a particular organism. The first chromosome sequence, for chromosome III of the yeast *Saccharomyces cerevisiae*, was published in 1992, and the entire yeast genome was completed in 1996. There are now complete genome sequences for the worm *Caenorhabditis elegans*, the fly *Drosophila melanogaster*, the plant *Arabidopsis thaliana*, the human *Homo sapiens*, and over 1000 other species. It is even possible to obtain genome sequences for extinct species such as mammoths and Neanderthals.

A single chain termination sequencing experiment produces about 750 bp of sequence, and a single pyrosequence yields up to 150 bp. But the total size of a fairly typical bacterial genome is 4,000,000 bp and the human genome is 3,200,000,000 bp (Table 10.1). Clearly a large number of sequencing experiments must be carried out in

Table 10.1

Sizes of representative genomes.

SPECIES	TYPE OF ORGANISM	GENOME SIZE (Mb)
Mycoplasma genitalium	Bacterium	0.58
Haemophilus influenzae	Bacterium	1.83
Escherichia coli	Bacterium	4.64
Saccharomyces cerevisiae	Yeast	12.10
Caenorhabditis elegans	Nematode worm	97.00
Drosophila melanogaster	Insect	180.00
Arabidopsis thaliana	Plant	125.00
Homo sapiens	Mammal	3200.00
Triticum aestivum	Plant (wheat)	16,000.00

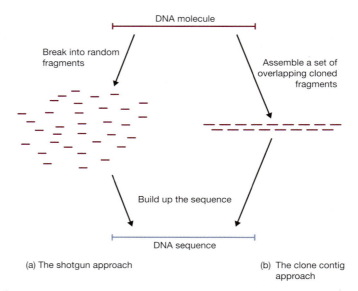

Figure 10.9

Strategies for assembly of a contiguous genome sequence: (a) the shotgun approach; (b) the clone contig approach.

order to determine the sequence of an entire genome. In practice, thanks to automated systems, the generation of sufficient sequence data is one of the more routine aspects of a genome project. The first real problem that arises is the need to assemble the thousands or perhaps millions of individual sequences into a contiguous genome sequence. Two different strategies have been developed for sequence assembly (Figure 10.9):

- The shotgun approach, in which the genome is randomly broken into short fragments. The resulting sequences are examined for overlaps and these are used to build up the contiguous genome sequence.
- The clone contig approach, which involves a pre-sequencing phase during which a series of overlapping clones is identified. This contiguous series is called a contig. Each piece of cloned DNA is then sequenced, and this sequence placed at its appropriate position on the contig map in order to gradually build up the overlapping genome sequence.

10.2.1 The shotgun approach to genome sequencing

The key requirement of the shotgun approach is that it must be possible to identify overlaps between all the individual sequences that are generated, and this identification process must be accurate and unambiguous so that the correct genome sequence is obtained. An error in identifying a pair of overlapping sequences could lead to the genome sequence becoming scrambled, or parts being missed out entirely. The probability of making mistakes increases with larger genome sizes, so the shotgun approach has been used mainly with the smaller bacterial genomes.

The Haemophilus influenzae *genome sequencing project*
The shotgun approach was first used successfully with the bacterium *Haemophilus. influenzae*, which was the first free-living organism whose genome was entirely

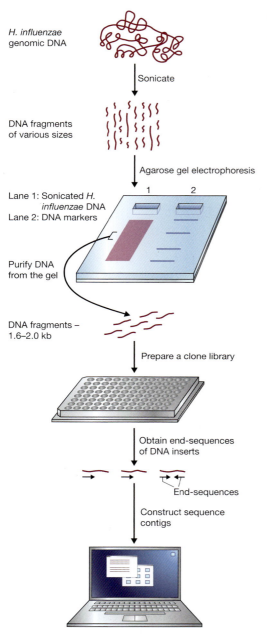

Figure 10.10
A schematic of the key steps in the *H. influenzae* genome sequencing project.

sequenced, the results being published in 1995. The first step was to break the 1830 kb genome of the bacterium into short fragments, which would provide the templates for the sequencing experiments (Figure 10.10). A restriction endonuclease could have been used but **sonication** was chosen because this technique cleaves DNA in a more random fashion and hence reduces the possibility of gaps appearing in the genome sequence.

It was decided to concentrate on fragments of 1.6–2.0 kb because these could yield two DNA sequences, one from each end, reducing the amount of cloning and DNA

Figure 10.11

Using oligonucleotide hybridization to close gaps in the *H. influenzae* genome sequence. Oligonucleotides 2 and 5 both hybridize to the same λ clone, indicating that contigs I and III are adjacent. The gap between them can be closed by sequencing the appropriate part of the λ clone.

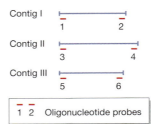

(a) Prepare oligonucleotide probes

Contig I

    1   2

Contig II

    3   4

Contig III

    5   6

1 2   Oligonucleotide probes

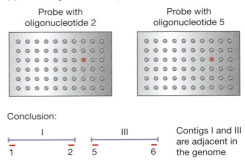

(b) Probe a genomic library

Probe with oligonucleotide 2   Probe with oligonucleotide 5

Conclusion:

I   III

1   2   5   6

Contigs I and III are adjacent in the genome

preparation that was required. The sonicated DNA was therefore fractionated by agarose gel electrophoresis and fragments of the desired size purified from the gel. After cloning, 28,643 chain termination sequencing experiments were carried out with 19,687 of the clones. A few of these sequences—4339 in all—were rejected because they were less than 400 bp in length. The remaining 24,304 sequences were entered into a computer, which spent 30 hours analyzing the data. The result was 140 contiguous sequences, each a different segment of the *H. influenzae* genome.

It might have been possible to continue sequencing more of the sonicated fragments in order eventually to close the gaps between the individual segments. However, 11,631,485 bp of sequence had already been generated—six times the length of the genome—suggesting that a large amount of additional work would be needed before the correct fragments were, by chance, sequenced. At this stage of the project the most time-effective approach was to use a more directed strategy in order to close each of the gaps individually. Several approaches were used for gap closure, the most successful of these involving hybridization analysis of a clone library prepared in a λ vector (Figure 10.11). The library was probed in turn with a series of oligonucleotides whose sequences corresponded with the ends of each of the 140 segments. In some cases, two oligonucleotides hybridized to the same λ clone, indicating that the two segment ends represented by those oligonucleotides lay adjacent to one another in the genome. The gap between these two ends could then be closed by sequencing the appropriate part of the λ clone.

Problems with shotgun sequencing

Shotgun sequencing has been successful with many bacterial genomes. Not only are these genomes small, so the computational requirements for finding sequence overlaps are not too great, but they contain little or no repetitive DNA sequences. These are

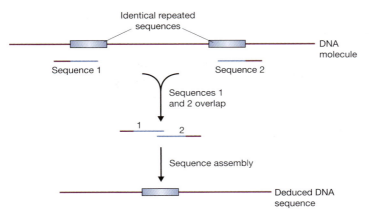

Figure 10.12

One problem with the shotgun approach. An incorrect overlap is made between two sequences that both terminate within a repeated element. The result is that a segment of the DNA molecule is absent from the DNA sequence.

sequences, from a few base pairs to several kilobases, which are repeated at two or more places in a genome. They cause problems for the shotgun approach because when sequences are assembled those that lie partly or wholly within one repeat element might accidentally be assigned an overlap with the identical sequence present in a different repeat element (Figure 10.12). This could lead to a part of the genome sequence being placed at the incorrect position or left out entirely. For this reason, it has generally been thought that shotgun sequencing is inappropriate for eukaryotic genomes, as these have many repeat elements. Later in this chapter (p. 183) we will see how this limitation can be circumvented by using a genome map to direct assembly of sequences obtained by the shotgun approach.

10.2.2 The clone contig approach

The clone contig approach does not suffer from the limitations of shotgun sequencing and so can provide an accurate sequence of a large genome that contains repetitive DNA. Its drawback is that it involves much more work and so takes longer and costs more money. The additional time and effort is needed to construct the overlapping series of cloned DNA fragments. Once this has been done, each cloned fragment is sequenced by the shotgun method and the genome sequence built up step by step (see Figure 10.9).

The cloned fragments should be as long as possible in order to minimize the total number needed to cover the entire genome. A high capacity vector is therefore used. The first eukaryotic chromosome to be sequenced—chromosome III of *Saccharomyces cerevisiae* – was initially cloned in a cosmid vector (p. 101) with the resulting contig comprising 29 cloned fragments. Chromosome III is relatively short, however, and the average size of the cloned fragments was just 10.8 kb. Sequencing of the much longer human genome required 300,000 bacterial artificial chromosome (BAC) clones (p. 103). Assembling all of these into chromosome-specific contigs was a massive task.

Clone contig assembly by chromosome walking

One technique that can be used to assemble a clone contig is **chromosome walking**. To begin a chromosome walk a clone is selected at random from the library, labeled, and

Figure 10.13

Chromosome walking.

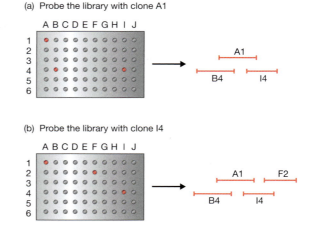

used as a hybridization probe against all the other clones in the library (Figure 10.13a). Those clones that give hybridization signals are ones that overlap with the probe. One of these overlapping clones is now labeled and a second round of probing carried out. More hybridization signals are seen, some of these indicating additional overlaps (Figure 10.13b). Gradually the clone contig is built up in a step-by-step fashion. But this is a laborious process and is only attempted when the contig is for a short chromosome and so involves relatively few clones, or when the aim is to close one or more small gaps between contigs that have been built up by more rapid methods.

Rapid methods for clone contig assembly

The weakness of chromosome walking is that it begins at a fixed starting point and builds up the clone contig step by step, and hence slowly, from that fixed point. The more rapid techniques for clone contig assembly do not use a fixed starting point and instead aim to identify pairs of overlapping clones: when enough overlapping pairs have been identified the contig is revealed (Figure 10.14). The various techniques that can be used to identify overlaps are collectively known as **clone fingerprinting**.

Clone fingerprinting is based on the identification of sequence features that are shared by a pair of clones. The simplest approach is to digest each clone with one or more restriction endonucleases and to look for pairs of clones that share restriction fragments of the same size, excluding those fragments that derive from the vector rather than

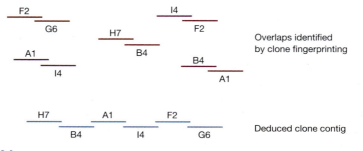

Figure 10.14

Building up a clone contig by a clone fingerprinting technique.

(a) The basis to IRE–PCR

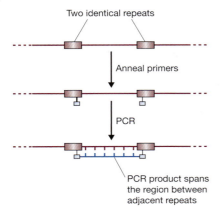

Two identical repeats

Anneal primers

PCR

PCR product spans
the region between
adjacent repeats

Figure 10.15

Interspersed repeat element PCR (IRE–PCR).

(b) Interpreting the results

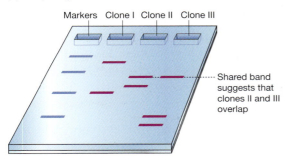

Markers   Clone I   Clone II   Clone III

Shared band
suggests that
clones II and III
overlap

the inserted DNA. This technique might appear to be easy to carry out, but in practice it takes a great deal of time to scan the resulting agarose gels for shared fragments. There is also a relatively high possibility that two clones that do not overlap will, by chance, share restriction fragments whose sizes are indistinguishable by agarose gel electrophoresis.

More accurate results can be obtained by repetitive DNA PCR, also known as interspersed repeat element PCR (IRE–PCR). This type of PCR uses primers that are designed to anneal within repetitive DNA sequences and direct amplification of the DNA between adjacent repeats (Figure 10.15). Repeats of a particular type are distributed fairly randomly in a eukaryotic genome, with varying distances between them, so a variety of product sizes are obtained when these primers are used with clones of eukaryotic DNA. If a pair of clones gives PCR products of the same size, they must contain repeats that are identically spaced, possibly because the cloned DNA fragments overlap.

Clone contig assembly by sequence tagged site content analysis

A third way to assemble a clone contig is to search for pairs of clones that contain a specific DNA sequence that occurs at just one position in the genome under study. If two clones contain this feature, then clearly they must overlap (Figure 10.16). A sequence of this type is called a sequence tagged site (STS). Often an STS is a gene that has been sequenced in an earlier project. As the sequence is known, a pair of PCR primers can be designed that are specific for that gene and then used to identify which members of a clone library contain the gene. The STS does not have to be a gene and can be any

Figure 10.16

The basis to STS content mapping.

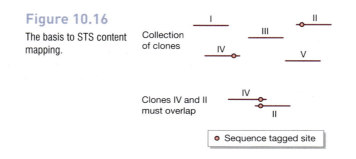

short piece of DNA sequence, the only requirement being that it occurs just once in the genome.

10.2.3 Using a map to aid sequence assembly

Sequence tagged site content mapping is a particularly important method for clone contig assembly, because often the positions of STSs within the genome will have been determined by **genetic mapping** or **physical mapping**. This means that the STS positions can be used to anchor the clone contig onto a genome map, enabling the position of the contig within a chromosome to be determined. We will now look at how these maps are obtained.

Genetic maps

A genetic map is one that is obtained by genetic studies using Mendelian principles and involving directed breeding programmes for experimental organisms or **pedigree analysis** for humans. In many cases the loci that are studied are genes, whose inheritance patterns are followed by monitoring the phenotypes of the offspring produced after a cross between parents with contrasting characteristics (e.g., tall and short for the pea plants studied by Mendel). The inheritance patterns reveal the extent of genetic linkage between genes present on the same chromosome, enabling the relative positions of those genes to be deduced and a genetic map to be built up.

More recently, techniques have been devised for genetic mapping of DNA sequences that are not genes but which still display variability in the human population. The most important of these **DNA markers** are:

- **Single nucleotide polymorphisms (SNPs)**, which are positions in a genome where either of two different nucleotides can occur (Figure 10.17). Some members of the species have one version of the SNP and some have the other version. SNPs are usually typed with short oligonucleotide probes that hybridize to the alternative forms and hence distinguish which is present.

- **Restriction fragment length polymorphisms (RFLPs)** are special types of SNPs, ones which result in a restriction site being changed. When digested with a restriction endonuclease the loss of the site is revealed because two fragments remain joined together. Originally, RFLPs were typed by Southern hybridization

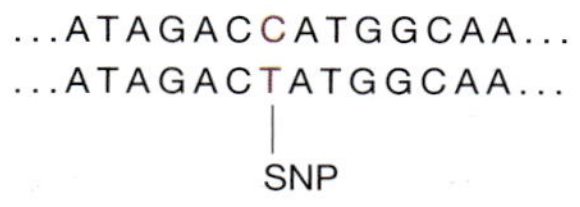

Figure 10.17

Two versions of an SNP.

```
...ATAGACCATGGCAA...
...ATAGACTATGGCAA...
              |
             SNP
```

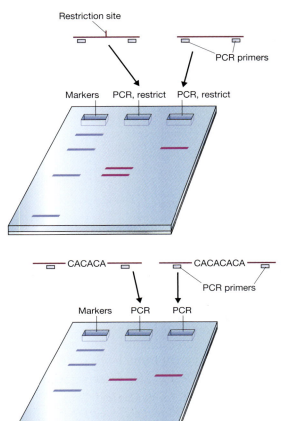

Figure 10.18

Typing a restriction site polymorphism by PCR. In the middle lane the PCR product gives two bands because it is cut by treatment with the restriction enzyme. In the right-hand lane there is just one band because the template DNA lacks the restriction site.

Figure 10.19

Typing an STR by PCR. The PCR product in the right-hand lane is slightly longer than that in the middle lane, because the template DNA from which it is generated contains an additional CA unit.

of restricted genomic DNA, but this is a time-consuming process, so nowadays the presence or absence of the restriction site is usually determined by PCR (Figure 10.18).

- **Short tandem repeats (STRs)**, also called **microsatellites**, are made up of short repetitive sequences of 1–13 nucleotides in length, linked head to tail. The number of repeats present in a particular STR varies, usually between 5 and 20. The number can be determined by carrying out a PCR using primers that anneal either side of the STR, and then examining the size of the resulting product by agarose or polyacrylamide gel electrophoresis (Figure 10.19).

All of these DNA markers are variable and so exist in two or more allelic forms. Their inheritance patterns can be determined by analysis of DNA prepared from the parents and offspring from a genetic cross, and the data used to place the DNA markers on a genetic map, in exactly the same way as genes are mapped.

Physical maps

A physical map is generated by methods that directly locate the positions of specific sequences on a chromosomal DNA molecule. As in genetic mapping, the loci that are studied can be genes or DNA markers. The latter might include **expressed sequence tags**

Figure 10.20

Fluorescent *in situ* hybridization.

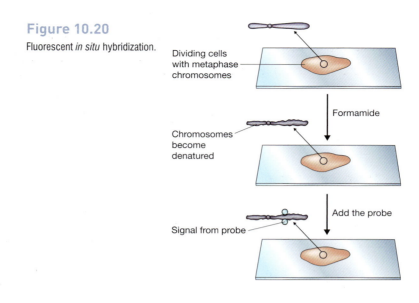

(ESTs), which are short sequences obtained from the ends of complementary DNAs (cDNAs) (p. 133). Expressed sequence tags are therefore partial gene sequences, and when used in map construction they provide a quick way of locating the positions of genes, even though the identity of the gene might not be apparent from the EST sequence.

Two types of technique are used in physical mapping:

- Direct examination of chromosomal DNA molecules, for example by fluorescence *in situ* hybridization (FISH). In this technique, a cloned DNA fragment is labeled with a fluorescent marker and then hybridized to a preparation of chromosomes immobilized on a glass slide. The physical position of the DNA fragment within the chromosome is then revealed simply by examining the preparation with a microscope (Figure 10.20). If FISH is carried out simultaneously with two DNA probes, each labeled with a different fluorochrome, the relevant positions on the chromosome of the two markers represented by the probes can be visualized. Special techniques for working with extended chromosomes, whose DNA molecules are stretched out rather than tightly coiled as in normal chromosomes, enable markers to be positioned with a high degree of accuracy.
- Physical mapping with a mapping reagent, which is a collection of overlapping DNA fragments spanning the chromosome or genome that is being studied. Pairs of markers that lie within a single fragment must be located close to each other on the chromosome: how close can be determined by measuring the frequency with which the pair occurs together in different fragments in the mapping reagent (Figure 10.21). The mapping reagent could be a clone library, possibly one that is also being assembled into a contig prior to DNA sequencing. Radiation hybrids are a second type of mapping reagent and were particularly important in the Human Genome Project. These are hamster cell lines that contain fragments of human chromosomes, prepared by a treatment involving irradiation (hence their name). Mapping is carried out by hybridization of marker probes to a panel of cell lines, each one containing a different part of the human genome.

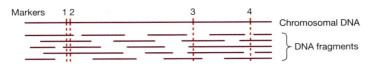

Figure 10.21

The principle behind the use of a mapping reagent. It can be deduced that markers 1 and 2 are relatively close because they are present together on four DNA fragments. In contrast, markers 3 and 4 must be relatively far apart because they occur together on just one fragment.

The importance of a map in sequence assembly

It is possible to obtain a genome sequence without the use of a genetic or physical map. This is illustrated by the *H. influenzae* project that we followed on p. 174, and many other bacterial genomes have been sequenced without the aid of a map. But a map is very important when a larger genome is being sequenced because it provides a guide that can be used to check that the genome sequence is being assembled correctly from the many short sequences that emerge from the automated sequencer. If a marker that has been mapped by genetic and/or physical means appears in the genome sequence at an unexpected position, then an error in sequence assembly is suspected.

Detailed genetic and/or physical maps have been important in the Human Genome Project, as well as those for yeast, fruit fly, *C. elegans*, and *A. thaliana*, all of which were based on the clone contig approach. Maps are also being used to direct sequence assembly in projects that use the shotgun approach. As described on p. 176, the major problem when applying shotgun sequencing to a large genome is the presence of repeated sequences and the possibility that the assembled sequence "jumps" between two repeats, so part of the genome is misplaced or left out (see Figure 10.12). These errors can be avoided if sequence assembly makes constant reference to a genome map. Because it avoids the laborious construction of clone contigs, this **directed shotgun approach** is becoming the method of choice for sequencing large genomes.

Further reading

Adams, M.D., Celnicker, S.E., Holt, R.A. et al. (2000) The genome sequence of *Drosophila melanogaster. Science*, 287, 2185–2195. [A clear description of this genome project.]

Brown, T.A. (2007) *Genomes*, 3rd edn. Garland Science, Abingdon. [Gives details of techniques for studying genomes, including genetic and physical mapping.]

Fleischmann, R.D., Adams, M.D., White, O. et al. (1995) Whole genome random sequencing and assembly of *Haemophilus influenzae* Rd. *Science*, 269, 496–512. [The first complete bacterial genome sequence to be published.]

Heiskanen, M., Peltonen, L. & Palotie, A. (1996) Visual mapping by high resolution FISH. *Trends in Genetics*, 12, 379–382.

Margulies, M., Egholm, M., Altman, W.E. et al. (2005) Genome sequencing in micro-fabricated high-density picolitre reactors. *Nature*, 437, 376–380. [Massively parallel pyrosequencing.]

Prober, J.M., Trainor, G.L., Dam, R.J. et al. (1987) A system for rapid DNA sequencing with fluorescent chain-terminating dideoxynucleotides. *Science*, 238, 336–341. [The chain termination method as used today.]

Ronaghi, M., Ehleen, M. & Nyrn, P. (1998) A sequencing method based on real-time pyrophosphate. *Science*, 281, 363–365. [Pyrosequencing.]

Sanger, F., Nicklen, S. & Coulson, A.R. (1977) DNA sequencing with chain-terminating inhibitors. *Proceedings of the National Academy of Sciences of the USA*, 74, 5463–5467. [The first description of chain termination sequencing.]

Sears, L.E., Moran, L.S., Kisinger, C. et al. (1992) CircumVent thermal cycle sequencing and alternative manual and automated DNA sequencing protocols using the highly thermostable Vent (exo⁻) DNA polymerase. *Biotechniques*, 13, 626–633. [Thermal cycle sequencing.]

Walter, M.A., Spillett, D.J., Thomas, P., et al. (1994) A method for constructing radiation hybrid maps of whole genomes. *Nature Genetics*, 7, 22–28.

Chapter 11

Studying Gene Expression and Function

All genes have to be expressed in order to function. The first step in expression is transcription of the gene into a complementary RNA strand (Figure 11.1a). For some genes—for example, those coding for transfer RNA (tRNA) and ribosomal RNA (rRNA) molecules—the transcript itself is the functionally important molecule. For other genes, the transcript is translated into a protein molecule.

To understand how a gene is expressed, the RNA transcript must be studied. In particular, the molecular biologist will want to know whether the transcript is a faithful copy of the gene, or whether segments of the gene are missing from the transcript (Figure 11.1b). These missing pieces are called introns and considerable interest centers on their structure and possible function. In addition to introns, the exact locations of the start and end points of transcription are important. Most transcripts are copies not only of the gene itself, but also of the nucleotide regions either side of it (Figure 11.1c). The signals that determine the start and finish of the transcription process are only partly understood, and their positions must be located if the expression of a gene is to be studied.

In this chapter we will begin by looking at the methods used for transcript analysis. These methods can be used to map the positions of the start and end points for transcription and also to determine if a gene contains introns. Then we will briefly consider a few of the numerous techniques developed in recent years for examining how expression of a gene is regulated. These techniques are important as aberrations in gene regulation underlie many clinical disorders. Finally, we will tackle the difficult problem of how to identify the translation product of a gene.

Gene Cloning and DNA Analysis: An Introduction. 6th edition. By T.A. Brown. Published 2010 by Blackwell Publishing.

Figure 11.1

Some fundamentals of gene expression.
mRNA = messenger RNA, tRNA = transfer RNA,
rRNA = ribosomal RNA.

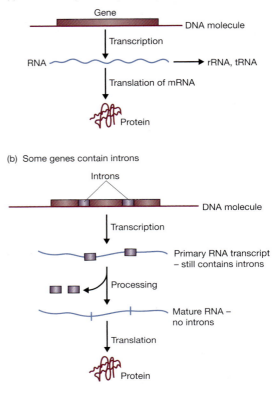

(a) Genes are expressed by transcription and translation

(b) Some genes contain introns

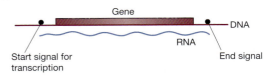

(c) RNA transcripts include regions either side of the gene

11.1 Studying the RNA transcript of a gene

Over the years a variety of techniques have been devised for studying RNA transcripts. Some of these techniques merely detect the presence of a transcript and give some indication of its length, others enable the start and end of the transcript to be mapped and the positions of introns to be located.

11.1.1 Detecting the presence of a transcript and determining its nucleotide sequence

Before studying the more sophisticated techniques for RNA analysis, we must consider the methods used to obtain basic information about a transcript. The first of these methods is **northern hybridization**, the RNA equivalent of Southern hybridization

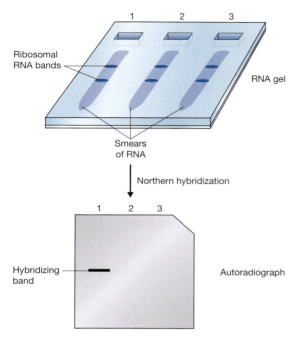

1 2 3

Ribosomal
RNA bands

RNA gel

Smears
of RNA

Northern hybridization

1 2 3

Hybridizing
band

Autoradiograph

Figure 11.2

Northern hybridization. Three RNA extracts from different tissues have been electrophoresed in an agarose gel. The extracts are made up of many RNAs of different lengths so each gives a smear of RNA, but two distinct bands are seen, one for each of the abundant ribosomal RNAs. The sizes of these rRNAs are known (e.g. 4718 and 1874 nucleotides in mammals), so they can be used as internal size markers. The gel is transferred to a membrane, probed with a cloned gene, and the results visualized, for example by autoradiography if the probe has been radioactively labeled. Only lane 1 gives a band, showing that the cloned gene is expressed only in the tissue from which this RNA extract was obtained.

(p. 142), which is used to measure the length of a transcript. An RNA extract is electrophoresed in an agarose gel, using a denaturing electrophoresis buffer (e.g., one containing formaldehyde) to ensure that the RNAs do not form inter- or intramolecular base pairs, as base pairing would affect the rate at which the molecules migrate through the gel. After electrophoresis, the gel is blotted onto a nylon or nitrocellulose membrane, and hybridized with a labeled probe (Figure 11.2). If the probe is a cloned gene, the band that appears in the autoradiograph is the transcript of that gene. The size of the transcript can be determined from its position within the gel, and if RNA from different tissues is run in different lanes of the gel, then the possibility that the gene is differentially expressed can be examined.

Once a transcript has been identified, cDNA synthesis (p. 133) can be used to convert it into a double-stranded DNA copy, which can be cloned and sequenced. Comparison between the sequence of the cDNA and the sequence of its gene will reveal the positions of introns and possibly the start and end points of the transcript. For this to be possible, the cDNA must be a full length copy of the mRNA from which it is derived. The 3′ end of the transcript will usually be represented in the cDNA, because most methods for cDNA synthesis begin with a primer that anneals to the poly(A) tail of the mRNA, which means that the 3′ end is the first part to be copied (see Figure 8.7). The cDNA synthesis might not, however, continue all the way to the 5′ end of the transcript, especially if the RNA is more than a few hundred nucleotides in length. Premature termination of cDNA synthesis will result in a cDNA that is not a full length copy of its transcript, and whose 3′ end does not correspond to the true 5′ end of the mRNA (Figure 11.3). If this is the case, then the start point for transcription cannot be identified from the sequence of the cDNA. It will also be impossible to map the positions of any introns that are located near the start of the gene.

Figure 11.3

The 5' end of an mRNA cannot be identified from the sequence of an incomplete cDNA.

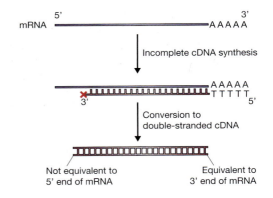

11.1.2 Transcript mapping by hybridization between gene and RNA

The problems with incomplete cDNA synthesis mean that other, more direct methods are needed to identify the start point of a transcript. The most informative of these methods are based on examination of the hybridization product formed between a cloned gene and its RNA.

Nucleic acid hybridization occurs just as readily between complementary DNA and RNA strands as between single-stranded DNA molecules. If a hybrid is formed between a DNA strand, containing a gene, and its mRNA, then the boundaries between the double- and single-stranded regions will mark the start and end points of the mRNA (Figure 11.4a). Introns, which are present in the DNA but not in the mRNA, will "loop out" as additional single-stranded regions.

Now consider the result of treating the DNA–RNA hybrid with a single-strand-specific nuclease such as S1 (p. 46). S1 nuclease degrades single-stranded DNA or RNA polynucleotides, including single-stranded regions at the ends of predominantly double-stranded molecules, but has no effect on double-stranded DNA or on DNA–RNA hybrids. S1 nuclease will therefore digest the non-hybridized single-stranded DNA

Figure 11.4

A DNA-mRNA hybrid and the effect of treating this hybrid with a single-strand-specific nuclease such as S1.

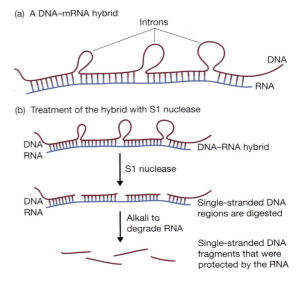

regions at each end of the DNA–RNA hybrid, along with any looped-out introns (Figure 11.4b). The single-stranded DNA fragments protected from S1 nuclease digestion can be recovered if the RNA strand is degraded by treatment with alkali.

Unfortunately, the manipulations shown in Figure 11.4 are not very informative. The sizes of the protected DNA fragments could be measured by gel electrophoresis, but this does not allow their order or relative positions in the DNA sequence to be determined. However, a few subtle modifications to the technique allow the precise start and end points of the transcript and of any introns it contains to be mapped onto the DNA sequence.

An example of the way in which S1 nuclease mapping is used to locate the start point of a transcript is shown in Figure 11.5. Here, a *Sau*3A fragment that contains 100 bp of coding region, along with 300 bp of the leader sequence preceding the gene, has been cloned into an M13 vector and obtained as a single-stranded molecule. A sample of the RNA transcript is added and allowed to anneal to the DNA molecule. The DNA molecule is still primarily single-stranded but now has a small region protected by the RNA transcript. All but this protected region is digested by S1 nuclease and the RNA is degraded with alkali, leaving a short single-stranded DNA fragment. If these

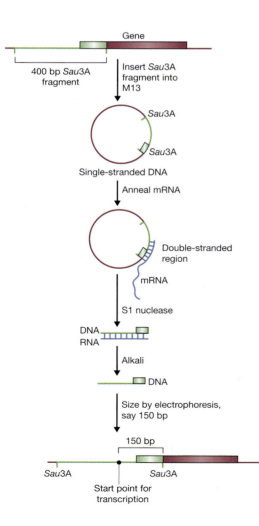

Figure 11.5

Locating a transcription start point by S1 nuclease mapping.

manipulations are examined closely it will become clear that the size of this single-stranded fragment corresponds to the distance between the transcription start point and the right-hand *Sau*3A site. The size of the single-stranded fragment is therefore determined by gel electrophoresis and this information is used to locate the transcription start point on the DNA sequence. Exactly the same strategy could locate the end point of transcription and the junction points between introns and exons. The only difference would be the position of the restriction site chosen to delimit one end of the protected single-stranded DNA fragment.

11.1.3 Transcript analysis by primer extension

S1 nuclease analysis is a powerful technique that allows both the 5′ and 3′ termini of a transcript, as well as the positions of intron–exon boundaries, to be identified. The second method of transcript analysis that we will consider—**primer extension**—is less adaptable, because it can only identify the 5′ end of an RNA. It is, nonetheless, an important technique that is frequently used to check the results of S1 analyses.

Primer extension can only be used if at least part of the sequence of the transcript is known. This is because a short oligonucleotide primer must be annealed to the RNA at a known position, ideally within 100–200 nucleotides of the 5′ end of the transcript. Once annealed, the primer is extended by reverse transcriptase (p. 49). This is a cDNA synthesis reaction, but one that is very likely to proceed to completion as only a short segment of RNA has to be copied (Figure 11.6). The 3′ end of the newly synthesized strand of DNA will therefore correspond with the 5′ terminus of the transcript. Locating

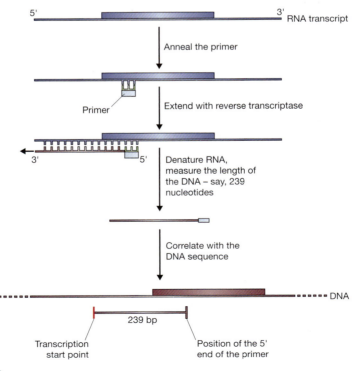

Figure 11.6
Locating a transcription start point by primer extension.

the position of this terminus on the DNA sequence is achieved simply by determining the length of the single-stranded DNA molecule and correlating this information with the annealing position of the primer.

11.1.4 Transcript analysis by PCR

We have already studied the way in which reverse transcriptase PCR is used to quantify the amount of a particular RNA in an extract (p. 160). The standard reverse transcriptase PCR procedure provides a copy of the internal region of an RNA molecule but does not give any information about the ends of the molecule. A modified version called rapid amplification of cDNA ends (RACE) can be used to identify the 5' and 3' termini of RNA molecules and so, like S1 analysis, can be used to map the ends of transcripts.

There are several variations to the RACE method. Here we will consider how the 5' end of an RNA molecule can be mapped (Figure 11.7). This procedure uses a primer

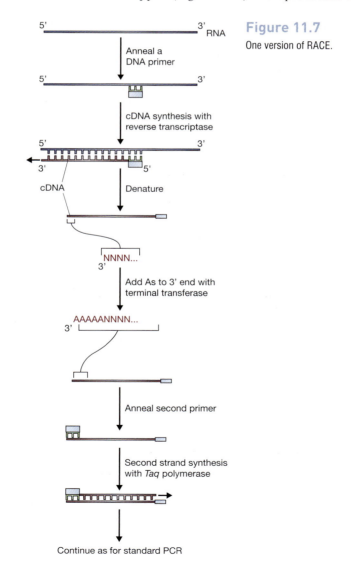

Figure 11.7

One version of RACE.

that is specific for an internal region of the RNA molecule. The primer attaches to the RNA and directs the first, reverse transcriptase catalyzed, stage of the process, during which a single-stranded cDNA is made. As in the primer extension method, the 3′ end of the cDNA corresponds with the 5′ end of the RNA. Terminal deoxynucleotidyl transferase (p. 50) is now used to attach a series of A nucleotides to the 3′ end of the cDNA, forming the priming site for a second PCR primer, which is made up entirely of Ts and hence anneals to the poly(A) tail created by terminal transferase. Now the standard PCR begins, first converting the single-stranded cDNA into a double-stranded molecule, and then amplifying this molecule as the PCR proceeds. The PCR product is then sequenced to reveal the precise position of the start of the transcript.

11.2 Studying the regulation of gene expression

Few genes are expressed all the time. Most are subject to regulation and are switched on only when their gene product is required by the cell. The simplest gene regulation systems are found in bacteria such as *E. coli*, which can regulate expression of genes for biosynthetic and metabolic processes, so that gene products that are not needed are not synthesized. For instance, the genes coding for the enzymes involved in tryptophan biosynthesis can be switched off when there are abundant amounts of tryptophan in the cell, and switched on again when tryptophan levels drop. Similarly, genes for the utilization of sugars such as lactose are activated only when the relevant sugar is there to be metabolized. In higher organisms, gene regulation is more complex because there are many more genes to control. Differentiation of cells involves wholesale changes in gene expression patterns, and the process of development from fertilized egg cell to adult requires coordination between different cells as well as time-dependent changes in gene expression.

Many of the problems in gene regulation require a classical genetic approach: genetics enables genes that control regulation to be distinguished, allows the biochemical signals that influence gene expression to be identified, and can explore the interactions between different genes and gene families. It is for this reason that most of the breakthroughs in understanding development in higher organisms have started with studies of the fruit fly *Drosophila melanogaster*. Gene cloning and DNA analysis complement classical genetics as they provide much more detailed information on the molecular events involved in regulating the expression of a single gene.

We now know that a gene subject to regulation has one or more control sequences in its upstream region (Figure 11.8) and that the gene is switched on and off by the attachment of regulatory proteins to these sequences. A regulatory protein may repress gene expression, in which case the gene is switched off when the protein is bound to the control sequence, or alternatively the protein may have a positive or enhancing role, switching on or increasing expression of the target gene. In this section we will examine methods for locating control sequences and determining their roles in regulating gene expression.

Figure 11.8

Possible positions for control sequences in the region upstream of a gene.

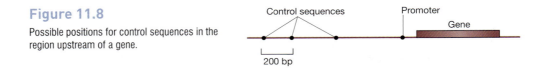

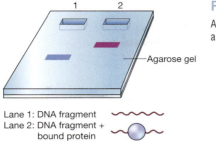

Figure 11.9

A bound protein decreases the mobility of a DNA fragment during gel electrophoresis.

Lane 1: DNA fragment
Lane 2: DNA fragment + bound protein

11.2.1 Identifying protein binding sites on a DNA molecule

A control sequence is a region of DNA that can bind a regulatory protein. It should therefore be possible to identify control sequences upstream of a gene by searching the relevant region for protein binding sites. There are three different ways of doing this.

Gel retardation of DNA–protein complexes

Proteins are quite substantial structures and a protein attached to a DNA molecule results in a large increase in overall molecular mass. If this increase can be detected, a DNA fragment containing a protein binding site will have been identified. In practice a DNA fragment carrying a bound protein is identified by gel electrophoresis, as it has a lower mobility than the uncomplexed DNA molecule (Figure 11.9). The procedure is referred to as gel retardation.

In a gel retardation experiment (Figure 11.10), the region of DNA upstream of the gene being studied is digested with a restriction endonuclease and then mixed with the regulatory protein or, if the protein has not yet been purified, with an unfractionated extract of nuclear protein (remember that gene regulation occurs in the nucleus). The

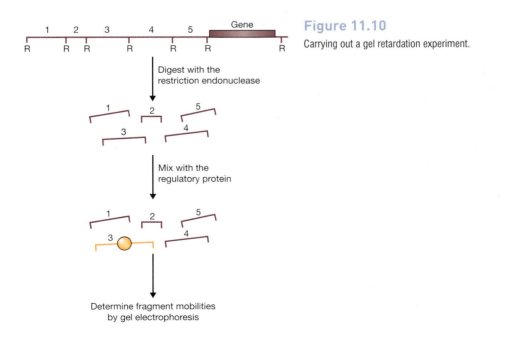

Figure 11.10

Carrying out a gel retardation experiment.

Digest with the restriction endonuclease

Mix with the regulatory protein

Determine fragment mobilities by gel electrophoresis

Figure 11.11

A bound protein protects a region of a DNA molecule from degradation by a nuclease such as DNase I.

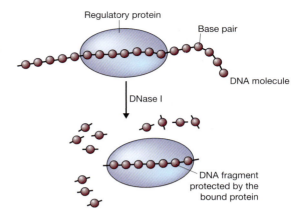

restriction fragment containing the control sequence forms a complex with the regulatory protein: all the other fragments remain as "naked" DNA. The location of the control sequence is then determined by finding the position on the restriction map of the fragment that is retarded during gel electrophoresis. The precision with which the control sequence can be located depends on how detailed the restriction map is and how conveniently placed the restriction sites are. A single control sequence may be less than 10 bp in size, so gel retardation is rarely able to pinpoint it exactly. More precise techniques are therefore needed to delineate the position of the control sequence within the fragment identified by gel retardation.

Footprinting with DNase I

The procedure generally called footprinting enables a control region to be positioned within a restriction fragment that has been identified by gel retardation. Footprinting works on the basis that the interaction with a regulatory protein protects the DNA in the region of a control sequence from the degradative action of an endonuclease such as deoxyribonuclease (DNase) I (Figure 11.11). This phenomenon can be used to locate the protein binding site on the DNA molecule.

The DNA fragment being studied is first labeled at one end, and then complexed with the regulatory protein (Figure 11.12a). Then DNase I is added, but the amount used is limited so that complete degradation of the DNA fragment does not occur. Instead the aim is to cut each molecule at just a single phosphodiester bond (Figure 11.12b). If the DNA fragment has no protein attached to it, the result of this treatment is a family of labeled fragments, differing in size by just one nucleotide each.

After removal of the bound protein and separation on a polyacrylamide gel, the family of labeled fragments appears as a ladder of bands (Figure 11.12c). However, the bound protein protected certain phosphodiester bonds from being cut by DNase I, meaning that in this case the family of fragments is not complete, as the fragments resulting from cleavage within the control sequence are absent. Their absence shows up as a "footprint", clearly seen in Figure 11.12c. The region of the DNA molecule containing the control sequence can now be worked out from the sizes of the fragments on either side of the footprint.

Modification interference assays

Gel retardation analysis and footprinting enable control sequences to be located, but do not give information about the interaction between the binding protein and the DNA

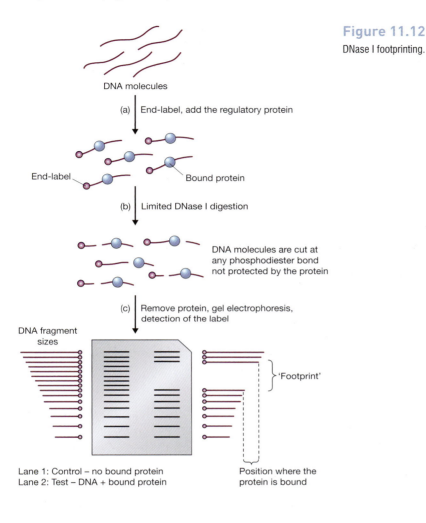

Figure 11.12
DNase I footprinting.

DNA molecules

(a) End-label, add the regulatory protein

End-label

Bound protein

(b) Limited DNase I digestion

DNA molecules are cut at
any phosphodiester bond
not protected by the protein

(c) Remove protein, gel electrophoresis,
detection of the label

DNA fragment
sizes

'Footprint'

Lane 1: Control – no bound protein
Lane 2: Test – DNA + bound protein

Position where the
protein is bound

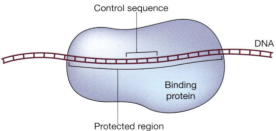

Control sequence

DNA

Binding
protein

Protected region

Figure 11.13
A bound protein can protect a region of DNA that
is much longer than the control sequence.

molecule. The more precise of these two techniques—footprinting—only reveals the region of DNA that is protected by the bound protein. Proteins are relatively large compared with a DNA double helix, and can protect several tens of base pairs when bound to a control sequence that is just a few base pairs in length (Figure 11.13). Footprinting therefore does not delineate the control region itself, only the region within which it is located.

Figure 11.14

A modification interference assay.

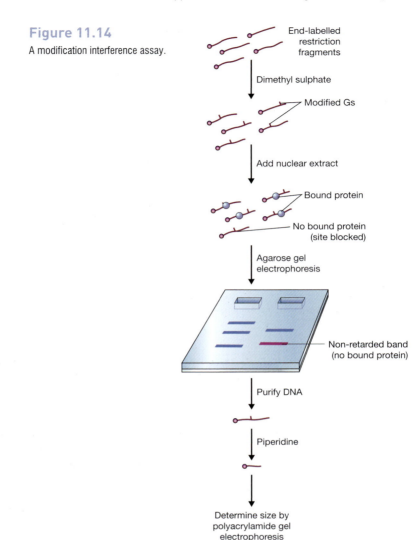

End-labelled restriction fragments

Dimethyl sulphate

Modified Gs

Add nuclear extract

Bound protein

No bound protein (site blocked)

Agarose gel electrophoresis

Non-retarded band (no bound protein)

Purify DNA

Piperidine

Determine size by polyacrylamide gel electrophoresis

Nucleotides that actually form attachments with a bound protein can be identified by the modification interference assay. As in footprinting, the DNA fragments must first be labeled at one end. Then they are treated with a chemical that modifies specific nucleotides, an example being dimethyl sulphate, which attaches methyl groups to guanine nucleotides (Figure 11.14). This modification is carried out under limiting conditions so an average of just one nucleotide per DNA fragment is modified. Now the DNA is mixed with the protein extract. The key to the assay is that the binding protein will probably not attach to the DNA if one of the guanines within the control region is modified, because methylation of a nucleotide interferes with the specific chemical reaction that enables it to form an attachment with a protein.

How is the absence of protein binding monitored? If the DNA and protein mixture is examined by agarose gel electrophoresis two bands will be seen, one comprising the DNA–protein complex and one containing DNA with no bound protein – in essence, this part of the procedure is a gel retardation assay (Figure 11.14). The band made up of unbound DNA is purified from the gel and treated with piperidine, a chemical which

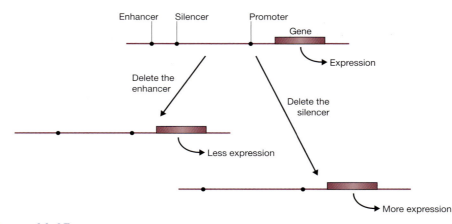

Figure 11.15
The principle behind deletion analysis.

cleaves DNA molecules at methylated nucleotides. The products of piperidine treatment are now separated in a polyacrylamide gel and the labeled bands visualized. The sizes of the bands that are seen indicate the position in the DNA fragment of guanines whose methylation prevented protein binding. These guanines lie within the control sequence. The modification assay can now be repeated with chemicals that target A, T, or C nucleotides to determine the precise position of the control sequence.

11.2.2 Identifying control sequences by deletion analysis

Gel retardation, footprinting, and modification interference assays are able to locate possible control sequences upstream of a gene, but they provide no information on the function of the individual sequences. Deletion analysis is a totally different approach that not only can locate control sequences (though only with the precision of gel retardation), but importantly also can indicate the function of each sequence.

The technique depends on the assumption that deletion of the control sequence will result in a change in the way in which expression of a cloned gene is regulated (Figure 11.15). For instance, deletion of a sequence that represses expression of a gene should result in that gene being expressed at a higher level. Similarly, tissue-specific control sequences can be identified as their deletion results in the target gene being expressed in tissues other than the correct one.

Reporter genes
Before carrying out a deletion analysis, a way must be found to assay the effect of a deletion on expression of the cloned gene. The effect will probably only be observed when the gene is cloned into the species from which it was originally obtained: it will be no good assaying for light regulation of a plant gene if the gene is cloned in a bacterium.

Cloning vectors have now been developed for most organisms (Chapter 7), so cloning the gene under study back into its host should not cause a problem. The difficulty is that in most cases the host already possesses a copy of the gene within its chromosomes. How can changes in the expression pattern of the cloned gene be distinguished from the normal pattern of expression displayed by the chromosomal copy of the gene? The answer is to use a reporter gene. This is a test gene that is fused to the upstream region

Figure 11.16

A reporter gene.

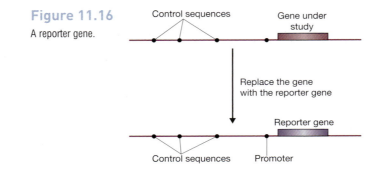

Table 11.1

A few examples of reporter genes used in studies of gene regulation in higher organisms.

GENE*	GENE PRODUCT	ASSAY
lacZ	β-Galactosidase	Histochemical test
neo	Neomycin phosphotransferase	Kanamycin resistance
cat	Chloramphenicol acetyltransferase	Chloramphenicol resistance
dhfr	Dihydrofolate reductase	Methotrexate resistance
aphIV	Hygromycin phosphotransferase	Hygromycin resistance
lux	Luciferase	Bioluminescence
GFP	Green fluorescent protein	Fluoresence
uidA	β-Glucuronidase	Histochemical test

*All of these genes are obtained from *E. coli*, except for: *lux* which has three sources: the luminescent bacteria *Vibrio harveyii* and *V. fischeri*, and the firefly *Photinus pyralis*; and GFP, which is obtained from the jellyfish *Aequorea victoria*.

of the cloned gene (Figure 11.16), replacing this gene. When cloned into the host organism the expression pattern of the reporter gene should exactly mimic that of the original gene, as the reporter gene is under the influence of exactly the same control sequences as the original gene.

The reporter gene must be chosen with care. The first criterion is that the reporter gene must code for a phenotype not already displayed by the host organism. The phenotype of the reporter gene must be relatively easy to detect after it has been cloned into the host, and ideally it should be possible to assay the phenotype quantitatively. These criteria have not proved difficult to meet and a variety of different reporter genes have been used in studies of gene regulation. A few examples are listed in Table 11.1.

Carrying out a deletion analysis

Once a reporter gene has been chosen and the necessary construction made, carrying out a deletion analysis is fairly straightforward. Deletions can be made in the upstream region of the construct by any one of several strategies, a simple example being shown in Figure 11.17. The effect of the deletion is then assessed by cloning the deleted construct into the host organism and determining the pattern and extent of expression of the reporter gene. An increase in expression will imply that a repressing or silencing sequence has been removed, a decrease will indicate removal of an activator or enhancer, and a change in tissue specificity (as shown in Figure 11.17) will pinpoint a tissue-responsive control sequence.

The results of a deletion analysis project have to be interpreted very carefully. Complications may arise if a single deletion removes two closely linked control

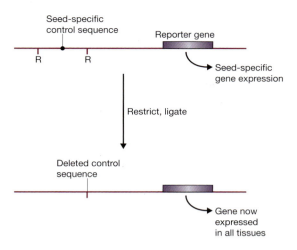

Figure 11.17

Deletion analysis. A reporter gene has been attached to the upstream region of a seed-specific gene from a plant. Removal of the restriction fragment bounded by the sites R deletes the control sequence that mediates seed-specific gene expression, so that the reporter gene is now expressed in all tissues of the plant.

sequences or, as is fairly common, two distinct control sequences cooperate to produce a single response. Despite these potential difficulties, deletion analyses, in combination with studies of protein binding sites, have provided important information about how the expression of individual genes is regulated, and have supplemented and extended the more broadly based genetic analyses of differentiation and development.

11.3 Identifying and studying the translation product of a cloned gene

Over the years gene cloning has become increasingly useful in the study not only of genes themselves but also of the proteins coded by cloned genes. Investigations into protein structure and function have benefited greatly from the development of techniques that allow mutations to be introduced at specific points in a cloned gene, resulting in directed changes in the structure of the encoded protein.

Before considering these procedures we should first look at the more mundane problem of how to isolate the protein coded by a cloned gene. In many cases this analysis will not be necessary, as the protein will have been characterized long before the gene cloning experiment is performed and pure samples of the protein will already be available. On the other hand, there are occasions when the translation product of a cloned gene has not been identified. A method for isolating the protein is then needed.

11.3.1 HRT and HART can identify the translation product of a cloned gene

Two related techniques, hybrid-release translation (HRT) and hybrid-arrest translation (HART), are used to identify the translation product encoded by a cloned gene. Both depend on the ability of purified mRNA to direct synthesis of proteins in cell-free translation systems. These are cell extracts, usually prepared from germinating wheat seeds or from rabbit reticulocyte cells (both of which are exceptionally active in protein synthesis) and containing ribosomes, tRNAs, and all the other molecules needed for protein synthesis. The mRNA sample is added to the cell-free translation system, along with a mixture of the 20 amino acids found in proteins, one of which is labeled (often

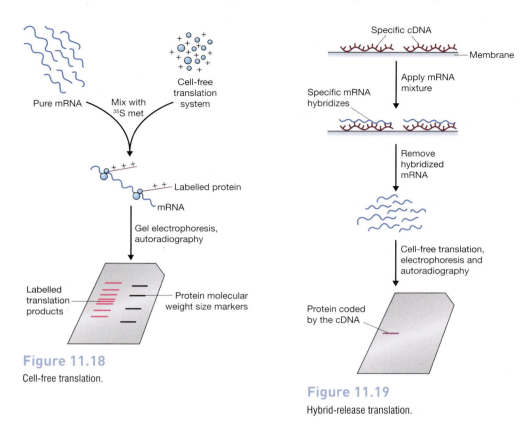

Figure 11.18

Cell-free translation.

Figure 11.19

Hybrid-release translation.

^{35}S-methionine is used). The mRNA molecules are translated into a mixture of radio-active proteins (Figure 11.18), which can be separated by gel electrophoresis and visualized by autoradiography. Each band represents a single protein coded by one of the mRNA molecules present in the sample.

Both HRT and HART work best when the gene being studied has been obtained as a cDNA clone. For HRT the cDNA is denatured, immobilized on a nitrocellulose or nylon membrane, and incubated with the mRNA sample (Figure 11.19). The specific mRNA counterpart of the cDNA hybridizes and remains attached to the membrane. After discarding the unbound molecules, the hybridized mRNA is recovered and trans-lated in a cell-free system. This provides a pure sample of the protein coded by the cDNA.

Hybrid-arrest translation is slightly different in that the denatured cDNA is added directly to the mRNA sample (Figure 11.20). Hybridization again occurs between the cDNA and its mRNA counterpart, but in this case the unbound mRNA is not discarded. Instead the entire sample is translated in the cell-free system. The hybridized mRNA is unable to direct translation, so all the proteins except the one coded by the cloned gene are synthesized. The cloned gene's translation product is therefore identified as the protein that is absent from the autoradiograph.

11.3.2 Analysis of proteins by *in vitro mutagenesis*

Although HRT and HART can identify the translation product of a cloned gene, these techniques tell us little about the protein itself. Some of the major questions asked by the molecular biologist today center on the relationship between the structure of a

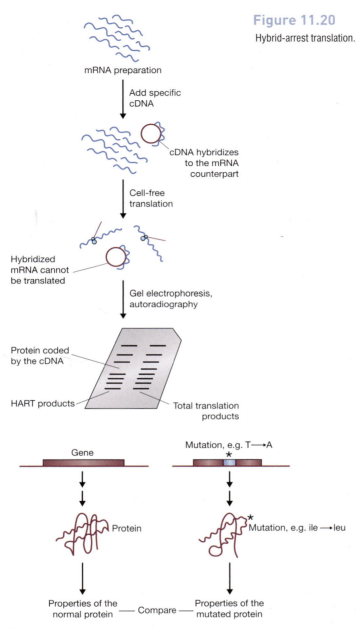

Figure 11.20

Hybrid-arrest translation.

mRNA preparation

Add specific cDNA

cDNA hybridizes to the mRNA counterpart

Cell-free translation

Hybridized mRNA cannot be translated

Gel electrophoresis, autoradiography

Protein coded by the cDNA

HART products

Total translation products

Gene

Mutation, e.g. T⟶A
*

Protein

Mutation, e.g. ile ⟶ leu

Properties of the normal protein — Compare — Properties of the mutated protein

Figure 11.21

A mutation may change the amino acid sequence of a protein, possibly affecting its properties.

protein and its mode of activity. The best way of tackling these problems is to induce a mutation in the gene coding for the protein and then to determine what effect the change in amino acid sequence has on the properties of the translation product (Figure 11.21). Under normal circumstances mutations occur randomly and a large number may have to be screened before one that gives useful information is found. A solution to this problem is provided by *in vitro* mutagenesis, a technique that enables a directed mutation to be made at a specific point in a cloned gene.

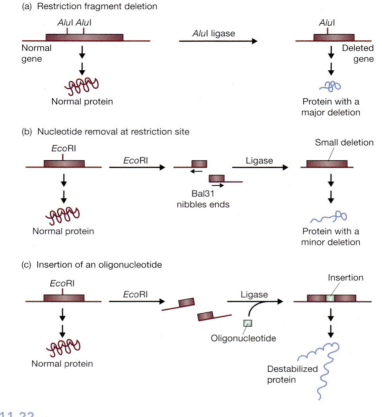

Figure 11.22
Various *in vitro* mutagenesis techniques.

Different types of in vitro mutagenesis techniques

An almost unlimited variety of DNA manipulations can be used to introduce mutations into cloned genes. The following are the simplest:

- A restriction fragment can be deleted (Figure 11.22a).
- The gene can be opened at a unique restriction site, a few nucleotides removed with a double-strand-specific endonuclease such as Bal31 (p. 46), and the gene religated (Figure 11.22b).
- A short, double-stranded oligonucleotide can be inserted at a restriction site (Figure 11.22c). The sequence of the oligonucleotide can be such that the additional stretch of amino acids inserted into the protein produces, for example, a new structure such as an α-helix, or destabilizes an existing structure.

Although potentially useful, these manipulations depend on the fortuitous occurrence of a restriction site at the area of interest in the cloned gene. **Oligonucleotide-directed mutagenesis** is a more versatile technique that can create a mutation at any point in the gene.

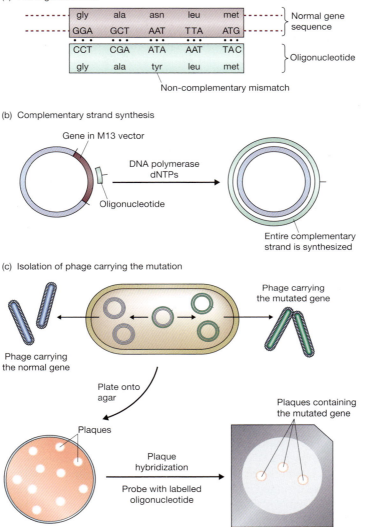

(a) The oligonucleotide

| gly | ala | asn | leu | met | Normal gene |
| GGA | GCT | AAT | TTA | ATG | sequence |

| CCT | CGA | ATA | AAT | TAC | Oligonucleotide |
| gly | ala | tyr | leu | met | |

Non-complementary mismatch

(b) Complementary strand synthesis

Gene in M13 vector

DNA polymerase
dNTPs

Oligonucleotide

Entire complementary
strand is synthesized

(c) Isolation of phage carrying the mutation

Phage carrying
the mutated gene

Phage carrying
the normal gene

Plate onto
agar

Plaques

Plaques containing
the mutated gene

Plaque
hybridization

Probe with labelled
oligonucleotide

Figure 11.23

One method for oligonucleotide-directed mutagenesis.

Using an oligonucleotide to create a point mutation in a cloned gene

There are a number of different ways of carrying out oligonucleotide-directed mutagenesis; we will consider one of the simplest of these methods. The gene to be mutated must be obtained in a single-stranded form and so is generally cloned into an M13 vector. The single-stranded DNA is purified and the region to be mutated identified by DNA sequencing. A short oligonucleotide is then synthesized, complementary to the relevant region, but containing the desired nucleotide alteration (Figure 11.23a). Despite this mismatch the oligonucleotide will anneal to the single-stranded DNA and act as a primer for complementary strand synthesis by a DNA polymerase (Figure 11.23b). This strand synthesis reaction is continued until an entire new strand is made and the recombinant molecule is completely double-stranded.

After introduction, by transfection, into competent *E. coli* cells, DNA replication produces numerous copies of the recombinant DNA molecule. The semi-conservative nature of DNA replication means that half the double-stranded molecules that are produced are unmutated in both strands, whereas half are mutated in both strands. Similarly, half the resulting phage progeny carry copies of the unmutated molecule and half carry the mutation. The phages produced by the transfected cells are plated onto solid agar so that plaques are produced. Half the plaques should contain the original recombinant molecule, and half the mutated version. Which are which is determined by plaque hybridization, using the oligonucleotide as the probe, and employing very strict conditions so that only the completely base-paired hybrid is stable.

Cells infected with M13 vectors do not lyse, but instead continue to divide (p. 19). The mutated gene can therefore be expressed in the host *E. coli* cells, resulting in production of recombinant protein. The protein coded by the mutated gene can be purified from the recombinant cells and its properties studied. The effect of a single base pair mutation on the activity of the protein can therefore be assessed.

Other methods of creating a point mutation in a cloned gene

The oligonucleotide-directed procedure illustrated in Figure 11.23 is an effective way of creating a single point mutation in a cloned gene. But what if the objective is to make a number of changes to the sequence of the gene? The oligonucleotide procedure could, of course, be repeated several times, introducing a new mutation at each stage, but this would be a very lengthy process.

An alternative for short genes (up to about 1 kb) would be to construct the gene in the test tube, placing mutations at all the desired positions. This approach is feasible now that oligonucleotides of 150 units and longer can be made by chemical synthesis (p. 140). The gene is constructed by synthesizing a series of partially overlapping oligonucleotides, the sequences of these oligonucleotides making up the desired sequence for the gene. The overlaps can be partial, rather than complete, because the gaps can be filled in with a DNA polymerase, and the final phosphodiester bonds synthesized with DNA ligase, to create the completed, double-stranded gene (Figure 11.24). If restriction sites are included in the end-sequences of the gene, then treatment with the appropriate enzyme will produce sticky ends that allow the gene to be ligated into a cloning vector. This procedure is called artificial gene synthesis.

PCR can also be used to create mutations in cloned genes, though like oligonucleotide-directed mutagenesis, only one mutation can be created per experiment. Various procedures have been devised, one of which is shown in Figure 11.25. In this example, the starting DNA molecule is amplified by two PCRs. In each of these, one primer is normal and forms a fully base-paired hybrid with the template DNA, but the second is mutagenic, as it contains a single base-pair mismatch corresponding to the mutation that we wish to introduce into the DNA sequence. This mutation is therefore present in the two PCR products, each of which represents one half of the starting DNA molecule. The two PCR products are mixed together and a final PCR cycle carried out. In this cycle, complementary strands from the two products anneal to one another and are then extended by the polymerase, producing the full-length, mutated DNA molecule. This technique, and related ones using PCR, is very quick and easy to carry out but a major problem is caused by the high error rate of the *Taq* polymerase used in PCR (p. 157). This error rate makes it likely that not only the desired mutation, but also

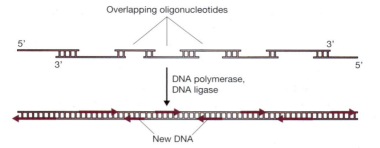

Figure 11.24

Artificial gene synthesis.

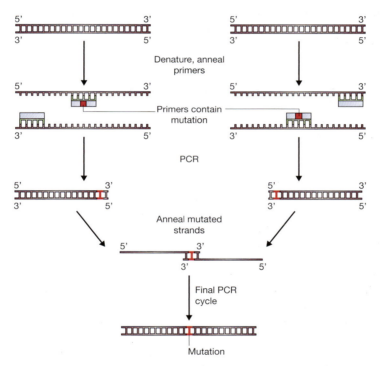

Figure 11.25

One method for using PCR to create a directed mutation.

random ones, will be present at the end of the experiment. The PCR product therefore has to be cloned and the sequences of individual clones checked to find one that has just the desired mutation.

The potential of in vitro *mutagenesis*

In vitro mutagenesis has remarkable potential, both for pure research and for applied biotechnology. For example, the biochemist can now ask very specific questions about the way that protein structure affects the action of an enzyme. In the past, it has been possible through biochemical analysis to gain some idea of the identity of the amino acids that provide the substrate binding and catalytic functions of an enzyme molecule. Mutagenesis techniques provide a much more detailed picture by enabling the role of each individual amino acid to be assessed by replacing it with an alternative residue and determining the effect this has on the enzymatic activity.

The ability to manipulate enzymes in this way has resulted in dramatic advances in our understanding of biological catalysis and has led to the new field of protein engineering, in which mutagenesis techniques are used to develop new enzymes for biotechnological purposes. For example, careful alterations to the amino acid sequence of subtilisin, an enzyme used in biological washing powders, have resulted in engineered versions with greater resistances to the thermal and bleaching (oxidative) stresses encountered in washing machines.

Further reading

Berk, A.J. (1989) Characterization of RNA molecules by S1 nuclease analysis. *Methods in Enzymology*, 180, 334–347.

Fried, M. & Crothers, D.M. (1981) Equilibria and kinetics of *lac* repressor–operator interactions by polyacrylamide gel electrophoresis. *Nucleic Acids Research*, 9, 6505–6525. [Gel retardation.]

Frohman, M.A., Dush, D.K. & Martin, G.R. (1988) Rapid production of full length cDNAs from rare transcripts: amplification using a single gene-specific oligonucleotide primer. *Proceedings of the National Academy of Sciences of the USA*, 85, 8998–9002. [An example of RACE.]

Galas, D.J. & Schmitz, A. (1978) DNase footprinting: a simple method for the detection of protein–DNA binding specificity. *Nucleic Acids Research*, 5, 3157–3170.

Garner, M.M. & Rezvin, A. (1981) A gel electrophoretic method for quantifying the binding of proteins to specific DNA regions: application to components of the *Escherichia coli* lactose operon regulatory system. *Nucleic Acids Research*, 9, 3047–3060. [Gel retardation.]

Hendrickson, W. & Schleif, R. (1985) A dimer of AraC protein contacts three adjacent major groove regions at the Ara I DNA site. *Proceedings of the National Academy of Sciences of the USA*, 82, 3129–3133. [An example of the modification interference assay.]

Kunkel, T.A. (1985) Rapid and efficient site-specific mutagenesis without phenotypic selection. *Proceedings of the National Academy of Sciences of the USA*, 82, 488–492. [Oligonucleotide-directed mutagenesis.]

Matzke, A.J.M., Stöger, E.M., Schernthaner, J.P. & Matzke, M.A. (1990) Deletion analysis of a zein gene promoter in transgenic tobacco plants. *Plant Molecular Biology*, 14, 323–332.

Paterson, B.M., Roberts, B.E. & Kuff, E.L. (1977) Structural gene identification and mapping by DNA.mRNA hybrid-arrested cell-free translation. *Proceedings of the National Academy of Sciences of the USA*, 74, 4370–4374.

Pellé, R. & Murphy, N.B. (1993) Northern hybridization: rapid and simple electrophoretic conditions. *Nucleic Acids Research*, 21, 2783–2784.

Tsien, R. (1998) The green fluorescent protein. *Annual Reviews of Biochemistry*, 67, 509–544. [A reporter gene system.]

Chapter 12

Studying Genomes

At the start of the 21st century the emphasis in molecular biology shifted from the study of individual genes to the study of entire genomes. This change in emphasis was prompted by the development during the 1990s of methods for sequencing large genomes. Genome sequencing predates the 1990s—we saw in Chapter 10 how the first genome, that of the phage φX174, was completed in 1975—but it was not until 20 years later, in 1995, that the first genome of a free-living organism, the bacterium *Haemophilus influenzae*, was completely sequenced. The next five years were a watershed with the genome sequences of almost 50 other bacteria published, along with complete sequences for the much larger genomes of yeast, fruit fly, *Caenorhabditis elegans* (a nematode worm), *Arabidopsis thaliana* (a plant), and humans. Today, the sequencing of bacterial genomes has become routine, with over 900 completed, and almost 100 eukaryotic genomes have also been sequenced.

Genome sequencing has led to the development of a new area of DNA research, loosely called post-genomics or functional genomics. Post-genomics includes the use of computer systems in genome annotation, the process by which the genes, control sequences, and other interesting features are identified in a genome sequence, as well as computer-based and experimental techniques aimed at determining the functions of any unknown genes that are discovered. Post-genomics also encompasses techniques designed to identify which genes are expressed in a particular type of cell or tissue, and how this pattern of genome expression changes over time.

12.1 Genome annotation

Once a genome sequence has been completed, the next step is to locate all the genes and determine their functions. It is in this area that bioinformatics, sometimes referred to

Gene Cloning and DNA Analysis: An Introduction. 6th edition. By T.A. Brown. Published 2010 by Blackwell Publishing.

as molecular biology *in silico*, is proving of major value as an adjunct to conventional experiments.

Genome annotation is a far from trivial process, even with genomes that have been extensively studied by genetic analysis and gene cloning techniques prior to complete sequencing. For example, the sequence of the yeast *Saccharomyces cerevisiae*, one of the best studied of all organisms, revealed that this genome contains about 6000 genes. Of these, some 3600 could be assigned a function either on the basis of previous studies that had been carried out with yeast or because the yeast gene had a similar sequence to a gene that had been studied in another organism. This left 2400 genes whose functions were not known. Despite a massive amount of work since the yeast genome was completed in 1996, the functions of many of these orphans have still not been determined.

12.1.1 Identifying the genes in a genome sequence

Locating a gene in a genome sequence is easy if the amino acid sequence of the protein product is known, allowing the nucleotide sequence of the gene to be predicted, or if the corresponding cDNA has been previously sequenced. But for many genes there is no prior information that enables the correct DNA sequence to be recognized. How can these genes be located in a genome sequence?

Searching for open reading frames

The DNA sequence of a gene is an open reading frame (ORF), a series of nucleotide triplets beginning with an initiation codon (usually but not always ATG) and ending in a termination codon (TAA, TAG, or TGA in most genomes). Searching a genome sequence for ORFs, by eye or more usually by computer, is therefore the first step in gene location. When carrying out the search it is important to remember that each DNA sequence has six reading frames, three in one direction and three in the reverse direction on the complementary strand (Figure 12.1).

The key to the success of ORF scanning is the frequency with which termination codons appear in the DNA sequence. If the DNA has a random sequence and a GC content of 50%, then each of the three termination codons will appear, on average, once every $4^3 = 64$ bp. This means that there should not be many ORFs longer than 30–40 codons in random DNA, and not all of these ORFs will start with ATG. Most genes are much longer than this: the average lengths are 317 codons for *Escherichia coli*, 483 codons for *S. cerevisiae*, and approximately 450 codons for humans. ORF scanning, in its simplest form, therefore takes a figure of 100 codons as the shortest length of a putative gene and records positive hits for all ORFs longer than this.

With bacterial genomes, simple ORF scanning is an effective way of locating most of the genes in a DNA sequence. Most bacterial genes are much longer than 100 codons

Figure 12.1

A double-stranded DNA molecule has six reading frames.

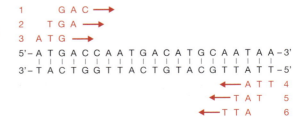

Figure 12.2

The typical result of a search for ORFs in a bacterial genome. The arrows indicate the directions in which the genes and spurious ORFs run.

in length and so are easily recognized (Figure 12.2). With bacteria the analysis is further simplified by the fact that most genes are closely spaced with very little intergenic DNA between them. If we assume that the real genes do not overlap, which is true for most bacterial genomes, then it is only in these short intergenic regions that there is a possibility of mistaking a spurious ORF for a real gene.

Simple ORF scans are less effective at locating genes in eukaryotic genomes

Although ORF scans work well for bacterial genomes, they are less effective for locating genes in eukaryotic genomes. This is partly because there is substantially more intergenic DNA in a eukaryotic genome, increasing the chances of finding spurious ORFs, but the main problem is the presence of introns (see Figure 11.1). If a gene contains one or more introns, then it does not appear as a continuous ORF in the genome sequence. Many exons are shorter than 100 codons, some fewer than 50 codons, and continuing the reading frame into an intron usually leads to a termination sequence that appears to close the ORF (Figure 12.3). In other words, many of the genes in a eukaryotic genome are not long ORFs, and simple ORF scanning cannot locate them.

Finding ways of locating genes by inspection of a eukaryotic sequence is a major challenge in bioinformatics. Three approaches are being followed:

- **Codon bias** can be taken into account. Not all codons are used equally frequently in the genes of a particular organism. For example, leucine is specified by six codons (TTA, TTG, CTT, CTC, CTA, and CTG), but in human genes leucine is most frequently coded by CTG and is only rarely specified by TTA or CTA. Similarly, of the four valine codons, human genes use GTG four times more frequently than GTA. The biological reason for codon bias is not understood, but all organisms have a bias, which is different in different species. Real exons display this bias, whereas chance series of triplets usually do not.

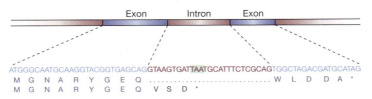

Figure 12.3

ORF scans are complicated by introns. The nucleotide sequence of a short gene containing a single intron is shown. The correct amino acid sequence of the protein translated from the gene is given immediately below the nucleotide sequence, using the single-letter amino acid abbreviations. In this sequence the intron has been left out, because it is removed from the transcript before the mRNA is translated into protein. In the lower line, the sequence has been translated without recognizing that an intron is present. As a result of this error, the amino acid sequence appears to terminate within the intron.

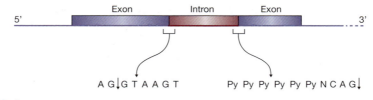

Figure 12.4

The consensus sequences for the upstream and downstream exon–intron boundaries of vertebrate introns. Py = pyrimidine nucleotide (C or T), N = any nucleotide. The arrows indicate the boundary positions.

- **Exon–intron boundaries** can be searched for as these have distinctive sequence features, although unfortunately the distinctiveness of these sequences is not so great as to make their location a trivial task. In vertebrates, the sequence of the upstream exon–intron boundary is usually described as 5′–AG↓GTAAGT–3′ and the downstream one as 5′–PyPyPyPyPyPyNCAG↓–3′, where "Py" means one of the pyrimidine nucleotides (T or C), "N" is any nucleotide, and the arrow shows the precise location of the boundary (Figure 12.4). These are **consensus sequences**, the averages of a large number of related but non-identical sequences, so the search has to include not just the sequences shown but also at least the most common variants. Despite these problems, this type of search can sometimes help delineate the locations of the exons within a region thought to contain a gene.

- **Upstream regulatory sequences** can be used to locate the regions where genes begin. This is because these regulatory sequences, like exon–intron boundaries, have distinctive sequence features that they possess in order to carry out their role as recognition signals for the DNA binding proteins involved in gene expression. As with exon–intron boundaries, the regulatory sequences are variable, and not all genes have the same collection of regulatory sequences. Using these to locate genes is therefore problematic.

These three extensions of simple ORF scanning, despite their limitations, are generally applicable to all higher eukaryotic genomes. Additional strategies are also possible with individual organisms based on the special features of their genomes. For example, vertebrate genomes contain **CpG islands** upstream of many genes, these being sequences of approximately 1 kb in which the GC content is greater than the average for the genome as a whole. Some 40–50% of human genes have an upstream CpG island. These sequences are distinctive and when one is located in vertebrate DNA, a strong assumption can be made that a gene begins in the region immediately downstream.

Gene location is aided by homology searching

Tentative identification of a gene is usually followed by a **homology search**. This is an analysis, carried out by computer, in which the sequence of the gene is compared with all the gene sequences present in the international DNA databases, not just known genes of the organism under study but also genes from all other species. The rationale is that two genes from different organisms that have similar functions have similar sequences, reflecting their common evolutionary histories (Figure 12.5).

To carry out a homology search the nucleotide sequence of the tentative gene is usually translated into an amino acid sequence, as this allows a more sensitive search. This is because there are 20 different amino acids but only 4 nucleotides, so there is less chance of 2 amino acid sequences appearing to be similar purely by chance

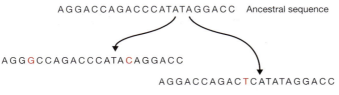

AGGACCAGACCCATATAGGACC   Ancestral sequence

AGGGCCAGACCCATACAGGACC

AGGACCAGACTCATATAGGACC

Two modern sequences

Figure 12.5

Homology between two sequences that share a common ancestor. The two sequences have acquired mutations during their evolutionary histories but their sequence similarities indicate that they are homologs.

```
            G  A  P  G  M  W  L  R  L  A  A  G  S  F  E  H  A  G
Sequence 1 GGTGCACCCGGTATGTGACTGCGATTAGCAGCGGGATCATTTCAGCATGCAGGG
           *  *  * ***** ****  ****  **  ***  ****  *****  ***  **  ****  **  *
Sequence 2 GATACACCCCGTATTTGACAGCAATTTGCAGGGGGATGATTGCACCATGGAGCG
            D  T  P  R  I  W  E  E  P  A  G  G  W  L  H  H  G  A
```

Figure 12.6

Lack of homology between two sequences is often more apparent when comparisons are made at the amino acid level. Two nucleotide sequences are shown, with nucleotides that are identical in the two sequences given in red and non-identities given in blue. The two nucleotide sequences are 76% identical, as indicated by the asterisks. This might be taken as evidence that the sequences are homologous. However, when the sequences are translated into amino acids, the identity decreases to 28%, suggesting that the genes are not homologous, and that the similarity at the nucleotide level was fortuitous. Identical amino acids are shown in brown, and non-identities in green. The amino acid sequences have been written using the one-letter abbreviations.

(Figure 12.6). The analysis is carried out through the internet, by logging on to the web site of one of the DNA databases and using a search program such as **BLAST** (Basic Local Alignment Search Tool). If the test sequence is over 200 amino acids in length and has 30% or greater identity with a sequence in the database (i.e., at 30 out of 100 positions the same amino acid occurs in both sequences), then the two are almost certainly homologous and the ORF under study can be confirmed as a real gene. Further confirmation, if needed, can be obtained by using transcript analysis (p. 186) to show that the gene is transcribed into RNA.

Comparing the sequences of related genomes

A more precise version of homology searching is possible when genome sequences are available for two or more related species. Related species have genomes that share similarities inherited from their common ancestor, overladen with species-specific differences that have arisen since the two species began to evolve independently. Because of natural selection, the sequence similarities between related genomes is greatest within the genes and least in the intergenic regions. Therefore, when related genomes are compared, homologous genes are easily identified because they have high sequence similarity, and any ORF that does not have a clear homolog in the related genome can discounted as almost certainly being a chance sequence and not a genuine gene (Figure 12.7).

This type of homology analysis—called comparative genomics—has proved very valuable for locating genes in the *S. cerevisiae* genome, as complete or partial sequences are now available not only for this yeast but also for several related species, such as *Saccharomyces paradoxus*, *Saccharomyces mikatae*, and *Saccharomyces bayanus*. Comparisons between these genomes has confirmed the authenticity of a number of *S. cerevisiae* ORFs, and also enabled almost 500 putative ORFs to be discounted on the grounds that they have no equivalents in the related genomes. The gene maps of

Figure 12.7

Using comparisons between the genomes of related species to test the authenticity of a short ORF. In this example, the questionable ORF is not present in the related genome and so is probably not a real gene.

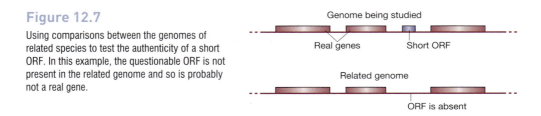

these yeasts are very similar, and although each genome has undergone its own species-specific rearrangements, there are still substantial regions where the gene order in the *S. cerevisiae* genome is the same as in one or more of the related genomes. Conservation of gene order is called synteny, and it makes it very easy to identify homologous genes. More importantly, a spurious ORF, especially a short one, can be discarded with confidence, because its expected location in a related genome can be searched in detail to ensure that no equivalent is present.

12.1.2 Determining the function of an unknown gene

Homology searching serves two purposes. As well as testing the veracity of a tentative gene, identification it can also give an indication of the function of the gene, presuming that the function of the homologous gene is known. Almost 2000 of the genes in the yeast genome were assigned functions in this way. Frequently, however, the matches found by homology searching are to other genes whose functions have yet to be determined. These unassigned genes are called orphans and working out their function is one of the key objectives of post-genomics research.

In future years it will probably be possible to use bioinformatics to gain at least an insight into the function of an orphan gene. It is already possible to use the nucleotide sequence of a gene to predict the positions of α-helices and β-sheets in the encoded protein, albeit with limited accuracy, and the resulting structural information can sometimes be used to make inferences about the function of the protein. Proteins that attach to membranes can often be identified because they possess α-helical arrangements that span the membrane, and DNA binding motifs such as zinc fingers can be recognized. A greater scope and accuracy to this aspect of bioinformatics will be possible when more information is obtained about the relationship between the structure of a protein and its function. In the meantime, functional analysis of orphans depends largely on conventional experiments.

Assigning gene function by experimental analysis requires a reverse approach to genetics

Experimental identification of the function of an unknown gene is proving to be one of the biggest challenges in genomics research. The problem is that the objective—to plot a course from gene to function—is the reverse of the route normally taken by genetic analysis, in which the starting point is a phenotype and the objective is to identify the underlying gene or genes (Figure 12.8). In conventional genetic analysis, the genetic basis of a phenotype is usually studied by searching for mutant organisms in which the phenotype has become altered. The mutants might be obtained experimentally, for example by treating a population of organisms, such as a culture of bacteria, with ultraviolet radiation or a mutagenic chemical, or the mutants might be present in a natural population. The gene or genes that have been altered in the mutant organism are then studied by genetic crosses, which can locate the position of a gene in a genome and also

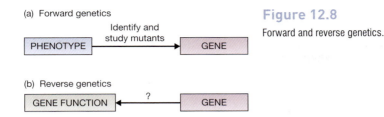

Figure 12.8

Forward and reverse genetics.

determine if the gene is the same as one that has already been characterized. The gene can then be studied further by molecular biology techniques such as cloning and sequencing.

The general principle of this conventional analysis—forward genetics—is that the genes responsible for a phenotype can be identified by determining which genes are inactivated in organisms that display a mutant version of the phenotype. If the starting point is the gene, rather than the phenotype, then the equivalent strategy—reverse genetics —would be to mutate the gene and identify the phenotypic change that results. This is the basis of most of the techniques used to assign functions to unknown genes.

Specific genes can be inactivated by homologous recombination

The easiest way to inactivate a specific gene is to disrupt it with an unrelated segment of DNA (Figure 12.9). This can be achieved by homologous recombination between the chromosomal copy of the gene and a second piece of DNA that shares some sequence identity with the target gene.

How is gene inactivation carried out in practice? We will consider two examples, the first with *S. cerevisiae*. Since completing the genome sequence in 1996, yeast molecular biologists have embarked on a coordinated, international effort to determine the functions of as many of the unknown genes as possible. One technique makes use of a deletion cassette, which carries a gene for antibiotic resistance (Figure 12.10). This gene is not a normal component of the yeast genome but it will work if transferred into a yeast chromosome, giving rise to a transformed yeast cell that is resistant to the antibiotic geneticin. Before using the deletion cassette, new segments of DNA are attached as tails to either end. These segments have sequences identical to parts of the yeast gene that is going to be inactivated. After the modified cassette is introduced into a yeast cell, homologous recombination occurs between the DNA tails and the chromosomal copy of the yeast gene, replacing the latter with the antibiotic resistance gene. The target gene therefore becomes inactivated. Cells which have undergone the replacement are selected by plating the culture onto agar medium containing geneticin, and their phenotypes examined to gain some insight into the function of the gene.

The second example of gene inactivation uses an analogous process, but with mice rather than yeast. The mouse is frequently used as a model organism for humans because

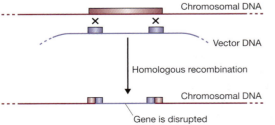

Figure 12.9

Gene disruption by homologous recombination.

Figure 12.10

The use of a yeast deletion cassette.

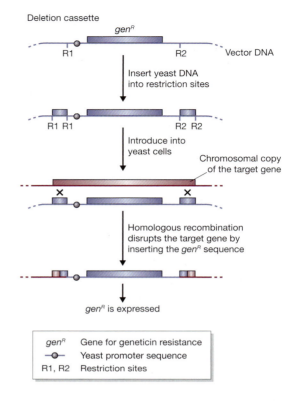

the mouse genome contains many of the same genes. Identifying the functions of unknown human genes is therefore being carried out largely by inactivating the equivalent genes in mice. The homologous recombination part of the procedure is identical to that described for yeast and once again results in a cell in which the target gene has been inactivated. The problem is that we do not want just one mutant cell, we want a whole mutant mouse, as only with the complete organism can we make a full assessment of the effect of the gene inactivation on the phenotype. To achieve this, the vector carrying the deletion cassette is initially microinjected into an embryonic stem cell (p. 124). The eventual result is a knockout mouse, whose phenotype will, with luck, provide the desired information on the function of the gene being studied.

12.2 Studies of the transcriptome and proteome

So far we have considered those aspects of post-genomics research that are concerned with studies of individual genes. The change in emphasis from genes to the genome has resulted in new types of analysis that are aimed at understanding the activity of the genome as a whole. This work has led to the invention of two new terms:

- The transcriptome, which is the messenger RNA (mRNA) content of a cell, and which reflects the overall pattern of gene expression in that cell;
- The proteome, which is the protein content of a cell and which reflects its biochemical capability.

12.2.1 Studying the transcriptome

Transcriptomes can have highly complex compositions, containing hundreds or thousands of different mRNAs, each making up a different fraction of the overall population. To characterize a transcriptome it is therefore necessary to identify the mRNAs that it contains and, ideally, to determine their relative abundances.

Studying a transcriptome by sequence analysis

The most direct way to characterize a transcriptome is to convert its mRNA into cDNA, and then to sequence every clone in the resulting cDNA library. Comparisons between the cDNA sequences and the genome sequence will reveal the identities of the genes whose mRNAs are present in the transcriptome. This approach is feasible but it is laborious, with many different cDNA sequences being needed before a near-complete picture of the composition of the transcriptome begins to emerge. Can any shortcuts be used to obtain the vital sequence information more quickly?

Serial analysis of gene expression (SAGE) provides one possible solution. Rather than studying complete cDNAs, SAGE yields short sequences, as little as 12 bp in length, each of which represents an mRNA present in the transcriptome. The basis of the technique is that these 12 bp sequences, despite their shortness, are sufficient to enable the gene that codes for the mRNA to be identified.

The first step in generating the 12 bp sequences is to immobilize the mRNA in a chromatography column by annealing the poly(A) tails present at the 3′ ends of these molecules to oligo(dT) strands that have been attached to cellulose beads (Figure 12.11). The mRNA is converted into double-stranded cDNA and then treated with a restriction enzyme that recognizes a 4 bp target site, such as *Alu*I, and so cuts frequently in each cDNA. The terminal restriction fragment of each cDNA remains attached to the cellulose beads, enabling all the other fragments to be eluted and discarded. A short linker is now attached to the free end of each cDNA, this linker containing a recognition sequence for *Bsm*FI. This is an unusual restriction enzyme in that rather than cutting within its recognition sequence, it cuts 10–14 nucleotides downstream. Treatment with *Bsm*FI therefore removes a fragment with an average length of 12 bp from the end of each cDNA. The fragments are collected, ligated head-to-tail to produce a catenane, and sequenced. The individual sequences can be identified within the catenane, because they are separated by *Bsm*FI sites.

Studying transcriptomes by microarray or chip analysis

SAGE enables individual mRNAs to be identified in a transcriptome, but provides only approximate information on the relative abundances of those mRNAs. The dominance of one particular type of mRNA in a transcriptome, such as the gliadin mRNAs in wheat seeds (p. 131), will be evident from the frequency of those mRNA sequences among the SAGE fragments, but more subtle variations in mRNA levels will not be apparent.

Techniques that enable more accurate comparisons of the amounts of individual mRNAs were first developed as part of the yeast post-genomics project. In essence, these techniques involve a sophisticated type of hybridization analysis. Every yeast gene—all 6000 of them—was obtained as an individual clone and samples spotted onto glass slides in arrays of 80×80 spots. This is called a microarray. To determine which genes are active in yeast cells grown under particular conditions, mRNA was extracted from the cells, converted to cDNA and the cDNA labeled and hybridized to the microarrays (Figure 12.12). Fluorescent labels were used and hybridization was detected by

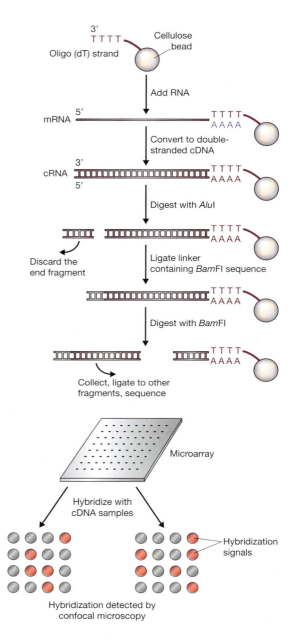

Figure 12.11

Serial analysis of gene expression (SAGE).

Figure 12.12

Microarray analysis. The microarray shown here has been hybridized to two different cDNA preparations, each labelled with a fluorescent marker. The clones which hybridize with the cDNAs are identified by confocal microscopy.

examining the microarrays by confocal microscopy. Those spots that gave a signal indicated genes that were active under the conditions being studied, and the intensities of the hybridization signals revealed the relative amounts of the mRNAs in the transcriptome. Changes in gene expression, when the yeast were transferred to different growth conditions (e.g., oxygen starvation), could be monitored by repeating the experiment with a second cDNA preparation.

Microarrays are now being used to monitor changes in the transcriptomes of many organisms. In some cases the strategy is the same as used with yeast, the microarray representing all the genes in the genome, but this is possible only for those organisms that

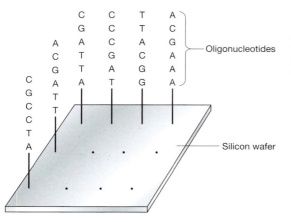

Figure 12.13
A DNA chip. A real chip would carry many more oligonucleotides than those shown here, and each oligonucleotide would be 20–30 nucleotides in length.

Oligonucleotides

Silicon wafer

have relatively few genes. A microarray for all the human genes could be carried by just 10 glass slides of 18 × 18 mm, but preparing clones of every one of the 20,000–30,000 human genes would be a massive task. Fortunately this is not necessary. For example, to study changes in the transcriptome occurring as a result of cancer, a microarray could be prepared with a cDNA library from normal tissue. Hybridization with labeled cDNA from the cancerous tissue would then reveal which genes are up- or down-regulated in response to the cancerous state.

An alternative to microarrays is provided by DNA chips, which are thin wafers of silicon that carry many different oligonucleotides (Figure 12.13). These oligonucleotides are synthesized directly on the surface of the chip and can be prepared at a density of 1 million per cm^2, substantially higher than is possible with a conventional microarray. Hybridization between an oligonucleotide and the probe is detected electronically. Because the oligonucleotides are synthesized *de novo*, using special automated procedures, a chip carrying 20,000–30,000 oligonucleotides, each one specific for a different human gene, is relatively easy to prepare.

12.2.2 Studying the proteome

The proteome is the entire collection of proteins in a cell. Proteome studies (also called proteomics) provide additional information that is not obtainable simply by examining the transcriptome. Examination of the transcriptome gives an accurate indication of which genes are active in a particular cell, but gives a less accurate indication of the proteins that are present. This is because the factors that influence protein content include not only the amount of each mRNA that is available, but also the rate at which an mRNA is translated into protein and the rate at which the protein is degraded.

The method used to identify the constituents of a proteome is called protein profiling. This is based on two techniques—protein electrophoresis and mass spectrometry—both of which have long pedigrees but which were rarely applied together in the pre-genomics era.

Separating the proteins in a proteome

In order to characterize a proteome, it is first necessary to prepare pure samples of its constituent proteins. A mammalian cell may contain 10,000–20,000 different proteins, so a highly discriminating separation system is needed.

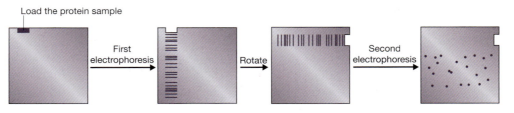

Figure 12.14

Two-dimensional gel electrophoresis.

Polyacrylamide gel electrophoresis is the standard method for separating the proteins in a mixture. Depending on the composition of the gel and the conditions under which the electrophoresis is carried out, different chemical and physical properties of proteins can be used as the basis for their separation. The most popular technique makes use of the detergent called sodium dodecyl sulphate, which denatures proteins and confers a negative charge that is roughly equivalent to the length of the unfolded polypeptide. Under these conditions, the proteins separate according to their molecular masses, the smallest proteins migrating more quickly toward the positive electrode. Alternatively, proteins can be separated by isoelectric focusing in a gel that contains chemicals which establish a pH gradient when the electrical charge is applied. In this type of gel, a protein migrates to its isoelectric point, the position in the gradient where its net charge is zero.

In protein profiling, these methods are combined in two-dimensional gel electrophoresis. In the first dimension, the proteins are separated by isoelectric focusing. The gel is then soaked in sodium dodecyl sulphate, rotated by 90° and a second electrophoresis, separating the proteins according to their sizes, carried out at right angles to the first (Figure 12.14). This approach can separate several thousand proteins in a single gel.

After electrophoresis, staining the gel reveals a complex pattern of spots, each one containing a different protein. When two gels are compared, differences in the pattern and intensities of the spots indicate differences in the identities and relative amounts of individual proteins in the two proteomes that are being studied. Interesting spots can therefore be targeted for the second stage of profiling in which actual protein identities are determined.

Identifying the individual proteins after separation

The second stage of protein profiling is to identify the individual proteins that have been separated from the starting mixture. This used to be a difficult proposition but peptide mass fingerprinting has provided a rapid and accurate identification procedure.

Peptide mass fingerprinting was made possible by advances in mass spectrometry. Mass spectrometry was originally designed as a means of identifying a compound from the mass-to-charge ratios of the ionized forms that are produced when molecules of the compound are exposed to a high-energy field. The standard technique could not be used with proteins because they are too large to be ionized effectively, but matrix-assisted laser desorption ionization time-of-flight (MALDI-TOF), gets around this problem, at least with peptides of up to 50 amino acids in length. Of course, most proteins are much longer than 50 amino acids, and it is therefore necessary to break them into fragments before examining them by MALDI-TOF. The usual approach is to digest the protein with a sequence-specific protease, such as trypsin, which cleaves proteins immediately after arginine or lysine residues. With most proteins, this results in a series of peptides 5–75 amino acids in length.

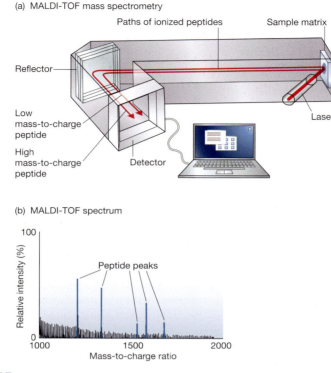

(a) MALDI-TOF mass spectrometry

Paths of ionized peptides

Sample matrix

Reflector

Low mass-to-charge peptide

High mass-to-charge peptide

Detector

Laser

(b) MALDI-TOF spectrum

Relative intensity (%)

Peptide peaks

Mass-to-charge ratio

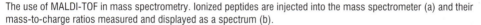

Figure 12.15

The use of MALDI-TOF in mass spectrometry. Ionized peptides are injected into the mass spectrometer (a) and their mass-to-charge ratios measured and displayed as a spectrum (b).

Once ionized, the mass-to-charge ratio of a peptide is determined from its "time of flight" within the mass spectrometer as it passes from the ionization source to the detector (Figure 12.15). The mass-to-charge ratio enables the molecular mass of the peptide to be worked out, which in turn allows its amino acid composition to be deduced. If a number of peptides from a single protein spot in the two-dimensional gel are analyzed, then the resulting compositional information can be related to the genome sequence to identify the gene that specifies that protein.

If two proteomes are being compared, then a key requirement is that proteins that are present in different amounts can be identified. If the differences are relatively large, then they will be apparent simply by looking at the stained gels after two-dimensional electrophoresis. However, important changes in the biochemical properties of a proteome can result from relatively minor changes in the amounts of individual proteins, and methods for detecting small-scale changes are therefore essential. One possibility is to label the constituents of two proteomes with different fluorescent markers, and then run them together in a single two-dimensional gel. Visualization of the two-dimensional gel at different wavelengths enables the intensities of equivalent spots to be judged more easily than is possible when two separate gels are obtained. A more accurate alternative is to label each proteome with an isotope coded affinity tag (ICAT). These markers can be obtained in two forms, one containing normal hydrogen atoms and the other containing deuterium, the heavy isotope of hydrogen. The normal and heavy versions can be distinguished by mass spectrometry, enabling the relative amounts of a protein in two

Figure 12.16

Analyzing two proteomes by ICAT. In the MALDI-TOF spectrum, peaks resulting from peptides containing normal hydrogen atoms are shown in blue, and those from peptides containing deuterium are shown in red. The protein under study is approximately 1.5 times more abundant in the proteome that has been labeled with deuterium.

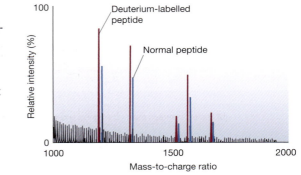

proteomes that have been mixed together to be determined during the MALDI-TOF stage of the profiling procedure (Figure 12.16).

12.2.3 Studying protein–protein interactions

Important information on the activity of a genome can also be obtained by identifying pairs or groups of proteins that work together. Within living cells, few if any proteins act in total isolation. Instead, proteins interact with one another in biochemical pathways and in multiprotein complexes. Two important techniques, **phage display** and the **yeast two hybrid system**, enable these protein–protein interactions to be examined.

Phage display

This technique is called phage display because it involves the "display" of proteins on the surface of a bacteriophage, usually M13 (Figure 12.17a). This is achieved by cloning the gene for the protein in a special type of M13 vector, one that results in the cloned gene becoming fused with a gene for a phage coat protein (Figure 12.17b). After transfection of *E. coli*, this gene fusion directs synthesis of a hybrid protein, made up partly of the coat protein and partly of the product of the cloned gene. With luck this hybrid protein will be inserted into the phage coat so that the product of the cloned gene is now located on the surface of the phage particles.

Normally this technique is carried out with a **phage display library** made up of many recombinant phages, each displaying a different protein. Large libraries can be prepared by cloning a mixture of cDNAs from a particular tissue or, less easily, by cloning genomic DNA fragments. The library consists of phages displaying a range of different proteins and is used to identify those that interact with a test protein. This test protein could be a pure protein or one that is itself displayed on a phage surface. The protein is immobilized in the wells of a microtiter tray or on particles that can be used in an affinity chromatography column, and then mixed with the phage display library (Figure 12.17c). Phages that are retained in the microtiter tray or within the column after a series of washes are ones that display proteins that interact with the immobilized test protein.

The yeast two hybrid system

The yeast two hybrid system is very different to phage display. This procedure is based on the discovery that gene expression in *S. cerevisiae* depends on interactions between pairs of transcription factors (Figure 12.18a). In the two hybrid system, a pair of transcription factors responsible for expression of a yeast gene is replaced by hybrid proteins, each one made partly of transcription factor and partly of test protein. The ability of this pair of hybrids to direct expression of the yeast target gene is then tested.

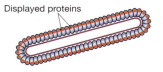

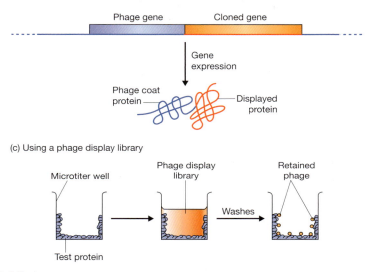

Figure 12.17

Phage display. (a) Display of proteins on the surface of a recombinant filamentous phage. (b) The gene fusion used to display a protein. (c) One way of detecting interactions between a test protein and a phage from within a display library.

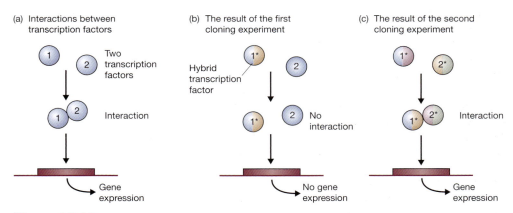

Figure 12.18

The yeast two hybrid system. (a) A pair of transcription factors that must interact in order for a yeast gene to be expressed. (b) Replacement of transcription factor 1 with the hybrid protein 1* abolishes gene expression as 1* cannot interact with transcription factor 2. (c) Replacement of transcription factor 2 with the hybrid protein 2* restores gene expression if the hybrid parts of 1* and 2* are able to interact.

To use the system, two yeast cloning experiments must be carried out. The first cloning experiment involves the gene whose protein product is being studied. This gene is ligated to the gene for one of the pair of transcription factors and the construct inserted into a yeast vector. The recombinant yeasts that are produced are not able to express the target gene, because this modified transcription factor cannot interact with its partner (Figure 12.18b).

In the second cloning experiment, a hybrid version of the partner is made and cloned into the yeast cells. Restoration of expression of the target gene indicates that the two hybrid transcription factors can interact. The fusions are designed in such a way that this can only happen if the interactions occur between the test protein components of the hybrids, not between the transcription factor segments (Figure 12.18c). Pairs of interacting test proteins are therefore identified. The second cloning experiment can involve a library of recombinants representing different proteins, so that one protein can be tested against many others.

Further reading

Altschul, S.F., Gish, W., Miller, W., Myers, E.W. & Lipman, D.J. (1990) Basic local alignment search tool. *Journal of Molecular Biology*, 215, 403–410. [The BLAST program.]

Clackson, T. & Wells, J.A. (1994) *In vitro* selection from protein and peptide libraries. *Trends in Biotechnology*, 12, 173–184. [Phage display.]

Fields, S. & Sternglanz, R. (1994) The two hybrid system: an assay for protein–protein interactions. *Trends in Genetics*, 10, 286–292.

Görg, A., Weiss, W. & Dunn, M.J. (2004) Current two-dimensional electrophoresis technology for proteomics. *Proteomics*, 4, 3665–3685.

Guigó, R., Flicek, P., Abril, J.F. et al. (2006) EGASP: the human ENCODE Genome Annotation Assessment Project. *Genome Biology*, 7, S2, doi:10.1186/gb-2006-7-s1-s2. [Comparison of the accuracy of different computer programs for gene location.]

Kellis, M., Patterson, N., Birren, B. & Lander, E.S. (2003) Sequencing and comparison of yeast species to identify genes and regulatory elements. *Nature*, 423, 241–254. [Using comparative genomics to annotate the yeast genome sequence.]

Ohler, U. & Niemann, H. (2001) Identification and analysis of eukaryotic promoters: recent computational approaches. *Trends in Genetics*, 17, 56–60.

Phizicky, E., Bastiaens, P.I.H., Zhu, H., Snyder, M. & Fields, S. (2003) Protein analysis on a proteomics scale. *Nature* 422: 208–215. [Reviews all aspects of proteomics].

Ramsay, G. (1998) DNA chips: state of the art. *Nature Biotechnology*, 16, 40–44.

Stoughton, R.B. (2005) Applications of DNA microarrays in biology. *Annual Review of Biochemistry*, 74, 53–82.

Velculescu, V.E., Vogelstein, B. & Kinzler, K.W. (2000) Analysing uncharted transcriptomes with SAGE. *Trends in Genetics*, 16, 423–425.

Wach, A., Brachat, A., Pohlmann, R. & Philippsen, P. (1994) New heterologous modules for classical or PCR-based gene disruptions in *Saccharomyces cerevisiae. Yeast*, 10, 1793–1808. [Gene inactivation by homologous recombination.]

PART III

The Applications of Gene Cloning and DNA Analysis in Biotechnology

Chapter 13

Production of Protein from Cloned Genes

Chapter contents

Now that we have covered the basic techniques involved in gene cloning and DNA analysis and examined how these techniques are used in research, we can move on to consider how recombinant DNA technology is being applied in biotechnology. This is not a new subject, although biotechnology has received far more attention during recent years than it ever has in the past. Biotechnology can be defined as the use of biological processes in industry and technology. According to archaeologists, the British biotechnology industry dates back 4000 years, to the late Neolithic period, when fermentation processes that make use of living yeast cells to produce ale and mead were first introduced into this country. Certainly brewing was well established by the time of the Roman invasion.

During the 20th century, biotechnology expanded with the development of a variety of industrial uses for microorganisms. The discovery by Alexander Fleming in 1929 that the mould *Penicillium* synthesizes a potent antibacterial agent led to the use of fungi and bacteria in the large-scale production of antibiotics. At first the microorganisms were grown in large culture vessels from which the antibiotic was purified after the cells had been removed (Figure 13.1a), but this batch culture method has been largely supplanted by continuous culture techniques, making use of a fermenter, from which samples of medium can be continuously drawn off, providing a non-stop supply of the product (Figure 13.1b). This type of process is not limited to antibiotic production and has also been used to obtain large amounts of other compounds produced by microorganisms (Table 13.1).

One of the reasons why biotechnology has received so much attention during the past three decades is because of gene cloning. Although many useful products can be

Gene Cloning and DNA Analysis: An Introduction. 6th edition. By T.A. Brown. Published 2010 by Blackwell Publishing.

Figure 13.1

Two different systems for the growth of microorganisms: (a) batch culture; (b) continuous culture.

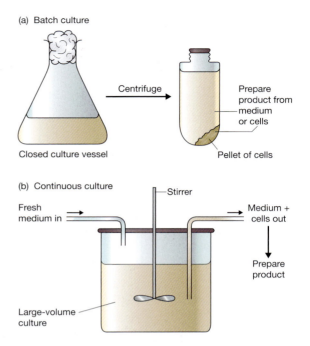

(a) Batch culture

Centrifuge

Prepare product from medium or cells

Closed culture vessel

Pellet of cells

(b) Continuous culture

Stirrer

Fresh medium in

Medium + cells out

Prepare product

Large-volume culture

Table 13.1

Some of the compounds produced by industrial scale culture of microorganisms.

COMPOUND	MICROORGANISM
Antibiotics	
Cephalosporins	*Cephalosporium* spp.
Chloramphenicol, streptomycin	*Streptomyces* spp.
Gramicidins, polymixins	*Bacillus* spp.
Penicillins	*Penicillium* spp.
Enzymes	
Invertase	*Saccharomyces cerevisiae*
Proteases, amylases	*Bacillus* spp., *Aspergillus* spp.
Others	
Acetone, butanol	*Clostridium* spp.
Alcohol	*S. cerevisiae, Saccharomyces carlsbergensis*
Butyric acid	Butyric acid bacteria
Citric acid	*Aspergillus niger*
Dextran	*Leuconostoc* spp.
Glycerol	*S. cerevisiae*
Vinegar	*S. cerevisiae*, acetic acid bacteria

obtained from microbial culture, the list in the past has been limited to those compounds naturally synthesized by microorganisms. Many important pharmaceuticals, which are produced not by microbes but by higher organisms, could not be obtained in this way. This has been changed by the application of gene cloning to biotechnology. The ability to clone genes means that a gene for an important animal or plant protein can now be taken from its normal host, inserted into a cloning vector, and introduced into a bacterium (Figure 13.2). If the manipulations are performed correctly the gene will be

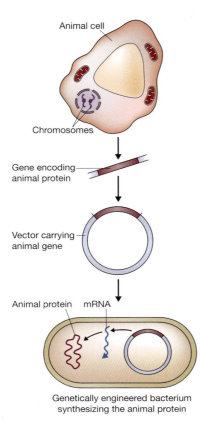

Figure 13.2

A possible scheme for the production of an animal protein by a bacterium. mRNA = messenger RNA.

Animal cell

Chromosomes

Gene encoding animal protein

Vector carrying animal gene

Animal protein   mRNA

Genetically engineered bacterium synthesizing the animal protein

expressed and the recombinant protein synthesized by the bacterial cell. It may then be possible to obtain large amounts of the protein.

Of course, in practice the production of recombinant protein is not as easy as it sounds. Special types of cloning vector are needed, and satisfactory yields of recombinant protein are often difficult to obtain. In this chapter we will look at cloning vectors for recombinant protein synthesis and examine some of the problems associated with their use.

13.1 Special vectors for expression of foreign genes in *E. coli*

If a foreign (i.e., non-bacterial) gene is simply ligated into a standard vector and cloned in *E. coli*, it is very unlikely that a significant amount of recombinant protein will be synthesized. This is because expression is dependent on the gene being surrounded by a collection of signals that can be recognized by the bacterium. These signals, which are short sequences of nucleotides, advertise the presence of the gene and provide instructions for the transcriptional and translational apparatus of the cell. The three most important signals for *E. coli* genes are as follows (Figure 13.3):

- The promoter, which marks the point at which transcription of the gene should start. In *E. coli*, the promoter is recognized by the σ subunit of the transcribing enzyme RNA polymerase.

Figure 13.3

The three most important signals for gene expression in *E. coli*.

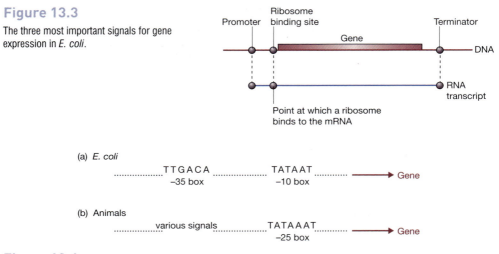

(a) *E. coli*

.... TTGACA .... TATAAT .... ———▶ Gene
                    −35 box   −10 box

(b) Animals

.... various signals .... TATAAAT .... ———▶ Gene
                                                     −25 box

Figure 13.4

Typical promoter sequences for *E. coli* and animal genes.

- The **terminator**, which marks the point at the end of the gene where transcription should stop. A terminator is usually a nucleotide sequence that can base pair with itself to form a **stem–loop** structure.
- The **ribosome binding site**, a short nucleotide sequence recognized by the ribosome as the point at which it should attach to the mRNA molecule. The initiation codon of the gene is always a few nucleotides downstream of this site.

The genes of higher organisms are also surrounded by expression signals, but their nucleotide sequences are not the same as the *E. coli* versions. This is illustrated by comparing the promoters of *E. coli* and human genes (Figure 13.4). There are similarities, but it is unlikely that an *E. coli* RNA polymerase would be able to attach to a human promoter. A foreign gene is inactive in *E. coli*, simply because the bacterium does not recognize its expression signals.

A solution to this problem would be to insert the foreign gene into the vector in such a way that it is placed under control of a set of *E. coli* expression signals. If this can be achieved, then the gene should be transcribed and translated (Figure 13.5). Cloning vectors that provide these signals, and can therefore be used in the production of recombinant protein, are called **expression vectors**.

13.1.1 The promoter is the critical component of an expression vector

The promoter is the most important component of an expression vector. This is because the promoter controls the very first stage of gene expression (attachment of an RNA polymerase enzyme to the DNA) and determines the rate at which mRNA is synthesized. The amount of recombinant protein obtained therefore depends to a great extent on the nature of the promoter carried by the expression vector.

The promoter must be chosen with care

The two sequences shown in Figure 13.4a are consensus sequences, averages of all the *E. coli* promoter sequences that are known. Although most *E. coli* promoters do not

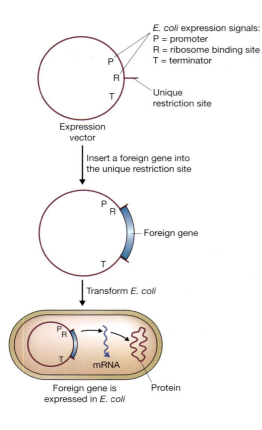

E. coli expression signals:
P = promoter
R = ribosome binding site
T = terminator

Unique
restriction site

Expression
vector

Insert a foreign gene into
the unique restriction site

Foreign gene

Transform E. coli

mRNA

Foreign gene is
expressed in E. coli

Protein

Figure 13.5

The use of an expression vector to achieve expression
of a foreign gene in E. coli.

differ much from these consensus sequences (e.g., TTTACA instead of TTGACA), a small variation may have a major effect on the efficiency with which the promoter can direct transcription. **Strong promoters** are those that can sustain a high rate of transcription; strong promoters usually control genes whose translation products are required in large amounts by the cell (Figure 13.6a). In contrast, **weak promoters**, which are relatively inefficient, direct transcription of genes whose products are needed in only small amounts (Figure 13.6b). Clearly an expression vector should carry a strong promoter, so that the cloned gene is transcribed at the highest possible rate.

A second factor to be considered when constructing an expression vector is whether it will be possible to regulate the promoter in any way. Two major types of gene regulation are recognized in E. coli—**induction** and **repression**. An inducible gene is one whose transcription is switched on by addition of a chemical to the growth medium; often this chemical is one of the substrates for the enzyme coded by the inducible gene (Figure 13.7a). In contrast, a repressible gene is switched off by addition of the regulatory chemical (Figure 13.7b)).

Gene regulation is a complex process that only indirectly involves the promoter itself. However, many of the sequences important for induction and repression lie in the region surrounding the promoter and are therefore also present in an expression vector. It may therefore be possible to extend the regulation to the expression vector, so that the chemical that induces or represses the gene normally controlled by the promoter is also able to regulate expression of the cloned gene. This can be a distinct advantage in the production of recombinant protein. For example, if the recombinant protein has a

Figure 13.6

Strong and weak promoters.

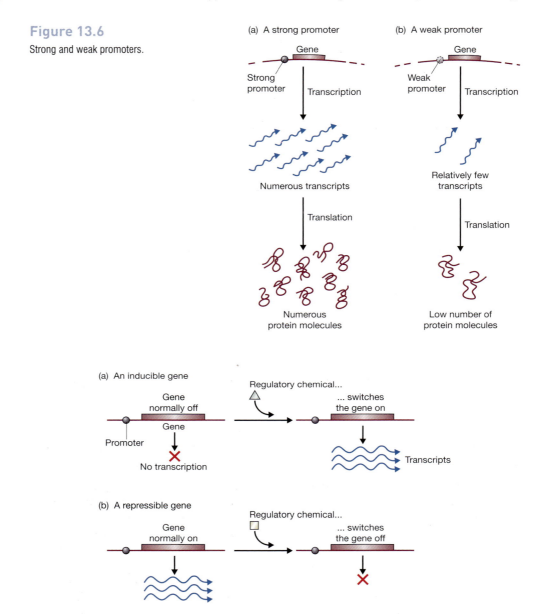

(a) A strong promoter

(b) A weak promoter

Gene

Strong promoter

Transcription

Numerous transcripts

Translation

Numerous protein molecules

Gene

Weak promoter

Transcription

Relatively few transcripts

Translation

Low number of protein molecules

(a) An inducible gene

Gene normally off

Gene

Promoter

✗

No transcription

Regulatory chemical...

... switches the gene on

Transcripts

(b) A repressible gene

Gene normally on

Regulatory chemical...

... switches the gene off

✗

Figure 13.7

Examples of the two major types of gene regulation that occur in bacteria: (a) an inducible gene; (b) a repressible gene.

harmful effect on the bacterium, then its synthesis must be carefully monitored to prevent accumulation of toxic levels: this can be achieved by judicious use of the regulatory chemical to control expression of the cloned gene. Even if the recombinant protein has no harmful effects on the host cell, regulation of the cloned gene is still desirable, as a continuously high level of transcription may affect the ability of the recombinant plasmid to replicate, leading to its eventual loss from the culture.

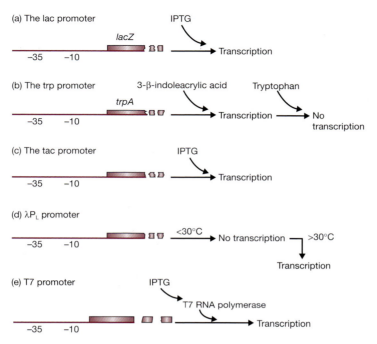

Figure 13.8

Five promoters frequently used in expression vectors. The *lac* and *trp* promoters are shown upstream of the genes that they normally control in *E. coli*.

Examples of promoters used in expression vectors

Several *E. coli* promoters combine the desired features of strength and ease of regulation. Those most frequently used in expression vectors are as follows:

- The *lac* promoter (Figure 13.8a) is the sequence that controls transcription of the *lacZ* gene coding for β-galactosidase (and also the *lacZ'* gene fragment carried by the pUC and M13mp vectors; p. 79). The *lac* promoter is induced by isopropylthiogalactoside (IPTG, p. 80), so addition of this chemical into the growth medium switches on transcription of a gene inserted downstream of the *lac* promoter carried by an expression vector.

- The *trp* promoter (Figure 13.8b) is normally upstream of the cluster of genes coding for several of the enzymes involved in biosynthesis of the amino acid tryptophan. The *trp* promoter is repressed by tryptophan, but is more easily induced by 3-β-indoleacrylic acid.

- The *tac* promoter (Figure 13.8c) is a hybrid between the *trp* and *lac* promoters. It is stronger than either, but still induced by IPTG.

- The λP_L promoter (Figure 13.8d) is one of the promoters responsible for transcription of the λ DNA molecule. λP_L is a very strong promoter that is recognized by the *E. coli* RNA polymerase, which is subverted by λ into transcribing the bacteriophage DNA. The promoter is repressed by the product of the λcI gene. Expression vectors that carry the λP_L promoter are used with a mutant *E. coli* host that synthesizes a temperature-sensitive form of the *c*I protein (p. 40). At a low temperature (less than 30°C) this mutant *c*I protein is able to repress the λP_L promoter, but at higher temperatures the protein is inactivated, resulting in transcription of the cloned gene.

- The **T7 promoter** (Figure 13.8e) is specific for the RNA polymerase coded by T7 bacteriophage. This RNA polymerase is much more active than the *E. coli* RNA polymerase (p. 93), which means that a gene inserted downstream of the T7 promoter will be expressed at a high level. The gene for the T7 RNA polymerase is not normally present in the *E. coli* genome, so a special strain of *E. coli* is needed, one which is lysogenic for T7 phage. Remember that a lysogen contains an inserted copy of the phage DNA in its genome (p. 19). In this particular strain of *E. coli*, the phage DNA has been altered by placing a copy of the *lac* promoter upstream of its gene for the T7 RNA polymerase. Addition of IPTG to the growth medium therefore switches on synthesis of the T7 RNA polymerase, which in turn leads to activation of the gene carried by the T7 expression vector.

13.1.2 Cassettes and gene fusions

An efficient expression vector requires not only a strong, regulatable promoter, but also an *E. coli* ribosome binding sequence and a terminator. In most vectors these expression signals form a **cassette**, so-called because the foreign gene is inserted into a unique restriction site present in the middle of the expression signal cluster (Figure 13.9). Ligation of the foreign gene into the cassette therefore places it in the ideal position relative to the expression signals.

With some cassette vectors the cloning site is not immediately adjacent to the ribosome binding sequence, but instead is preceded by a segment from the beginning of an *E. coli* gene (Figure 13.10). Insertion of the foreign gene into this restriction site must be performed in such a way as to fuse the two reading frames, producing a hybrid gene that starts with the *E. coli* segment and progresses without a break into the codons of the foreign gene. The product of gene expression is therefore a hybrid or **fusion protein**, consisting of the short peptide coded by the *E. coli* reading frame fused to the amino-terminus of the foreign protein. This fusion system has four advantages:

- Efficient translation of the mRNA produced from the cloned gene depends not only on the presence of a ribosome binding site, but is also affected by the nucleotide sequence at the start of the coding region. This is probably because secondary structures resulting from intrastrand base pairs could interfere with attachment of the ribosome to its binding site (Figure 13.11). This possibility is avoided if the pertinent region is made up entirely of natural *E. coli* sequences.

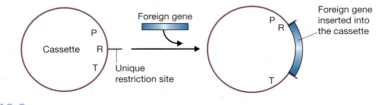

Figure 13.9
A typical cassette vector and the way it is used. P = promoter, R = ribosome binding site, T = terminator.

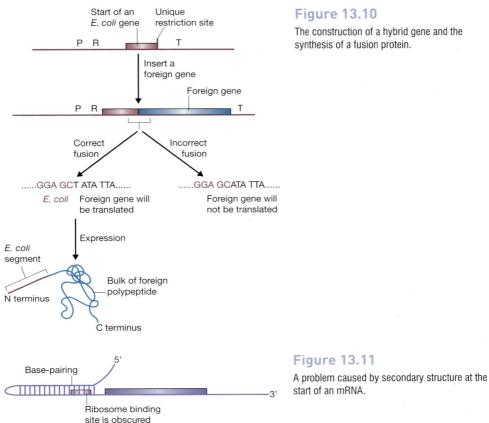

Figure 13.10

The construction of a hybrid gene and the synthesis of a fusion protein.

Figure 13.11

A problem caused by secondary structure at the start of an mRNA.

- The presence of the bacterial peptide at the start of the fusion protein may stabilize the molecule and prevent it from being degraded by the host cell. In contrast, foreign proteins that lack a bacterial segment are often destroyed by the host.
- The bacterial segment may constitute a signal peptide, responsible for directing the *E. coli* protein to its correct position in the cell. If the signal peptide is derived from a protein that is exported by the cell (e.g., the products of the *ompA* or *malE* genes), the recombinant protein may itself be exported, either into the culture medium or into the periplasmic space between the inner and outer cell membranes. Export is desirable as it simplifies the problem of purification of the recombinant protein from the culture.
- The bacterial segment may also aid purification by enabling the fusion protein to be recovered by **affinity chromatography**. For example, fusions involving the *E. coli* glutathione-S-transferase protein can be purified by adsorption onto agarose beads carrying bound glutathione (Figure 13.12).

The disadvantage with a fusion system is that the presence of the *E. coli* segment may alter the properties of the recombinant protein. Methods for removing the bacterial segment are therefore needed. Usually this is achieved by treating the fusion protein with a chemical or enzyme that cleaves the polypeptide chain at or near the junction between the two components. For example, if a methionine is present at the junction, the fusion protein can be cleaved with cyanogen bromide, which cuts polypeptides

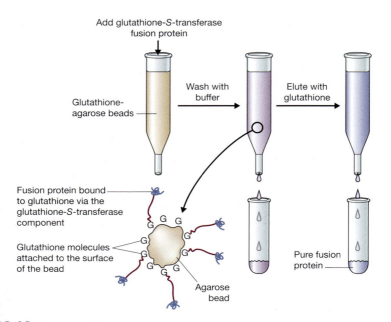

Figure 13.12

The use of affinity chromatography to purify a glutathione-*S*-transferase fusion protein.

Figure 13.13

One method for the recovery of the foreign polypeptide from a fusion protein. The methionine residue at the fusion junction must be the only one present in the entire polypeptide: if others are present cyanogen bromide will cleave the fusion protein into more than two fragments.

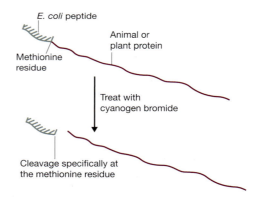

specifically at methionine residues (Figure 13.13). Alternatively, enzymes such as thrombin (which cleaves adjacent to arginine residues) or factor Xa (which cuts after the arginine of Gly–Arg) can be used. The important consideration is that recognition sequences for the cleavage agent must not occur within the recombinant protein.

13.2 General problems with the production of recombinant protein in *E. coli*

Despite the development of sophisticated expression vectors, there are still numerous difficulties associated with the production of protein from foreign genes cloned in *E. coli*. These problems can be grouped into two categories: those that are due to the sequence of the foreign gene, and those that are due to the limitations of *E. coli* as a host for recombinant protein synthesis.

13.2.1 Problems resulting from the sequence of the foreign gene

There are three ways in which the nucleotide sequence might prevent efficient expression of a foreign gene cloned in *E. coli*:

- The foreign gene might contain introns. This would be a major problem, as *E. coli* genes do not contain introns and therefore the bacterium does not possess the necessary machinery for removing introns from transcripts (Figure 13.14a).
- The foreign gene might contain sequences that act as termination signals in *E. coli* (Figure 13.14b). These sequences are perfectly innocuous in the normal host cell, but in the bacterium result in premature termination and a loss of gene expression.
- The codon bias of the gene may not be ideal for translation in *E. coli*. As described on p. 209, although virtually all organisms use the same genetic code, each

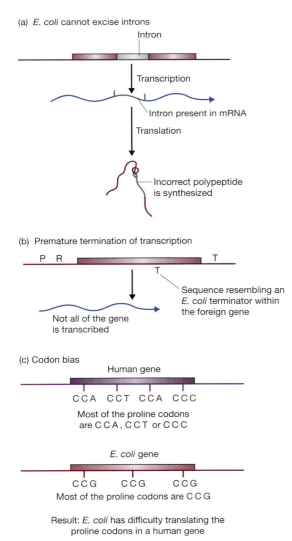

(a) *E. coli* cannot excise introns

Intron

Transcription

Intron present in mRNA

Translation

Incorrect polypeptide is synthesized

(b) Premature termination of transcription

P R   T

T

Sequence resembling an *E. coli* terminator within the foreign gene

Not all of the gene is transcribed

(c) Codon bias

Human gene

CCA   CCT   CCA   CCC

Most of the proline codons are CCA, CCT or CCC

E. coli gene

CCG   CCG   CCG

Most of the proline codons are CCG

Result: *E. coli* has difficulty translating the proline codons in a human gene

Figure 13.14

Three of the problems that could be encountered when foreign genes are expressed in *E. coli*: (a) introns are not removed in *E. coli*; (b) premature termination of transcription; (c) a problem with codon bias.

organism has a bias toward preferred codons. This bias reflects the efficiency with which the tRNA molecules in the organism are able to recognize the different codons. If a cloned gene contains a high proportion of disfavored codons, the *E. coli* tRNAs may encounter difficulties in translating the gene, reducing the amount of protein that is synthesized (Figure 13.14c).

These problems can usually be solved, although the necessary manipulations may be time-consuming and costly (an important consideration in an industrial project). If the gene contains introns then its complementary DNA (cDNA), prepared from the mRNA (p. 133) and so lacking introns, might be used as an alternative. *In vitro* mutagenesis could then be employed to change the sequences of possible terminators and to replace disfavored codons with those preferred by *E. coli*. An alternative with genes that are less than 1 kb in length is to make an artificial version (p. 204). This involves synthesizing a set of overlapping oligonucleotides that are ligated together, the sequences of the oligonucleotides being designed to ensure that the resulting gene contains preferred *E. coli* codons and that terminators are absent.

13.2.2 Problems caused by E. coli

Some of the difficulties encountered when using *E. coli* as the host for recombinant protein synthesis stem from inherent properties of the bacterium. For example:

- *E. coli* might not process the recombinant protein correctly. The proteins of most organisms are processed after translation, by chemical modification of amino acids within the polypeptide. Often these processing events are essential for the correct biological activity of the protein. Unfortunately, the proteins of bacteria and higher organisms are not processed identically. In particular, some animal proteins are glycosylated, meaning that they have sugar groups attached to them after translation. Glycosylation is extremely uncommon in bacteria and recombinant proteins synthesized in *E. coli* are never glycosylated correctly.

- *E. coli* might not fold the recombinant protein correctly, and generally is unable to synthesize the disulphide bonds present in many animal proteins. If the protein does not take up its correctly folded tertiary structure, then usually it is insoluble and forms an **inclusion body** within the bacterium (Figure 13.15). Recovery of the protein from the inclusion body is not a problem, but converting the protein into its correctly folded form can be difficult or impossible in the test tube. Under these circumstances the protein is, of course, inactive.

Figure 13.15
Inclusion bodies.

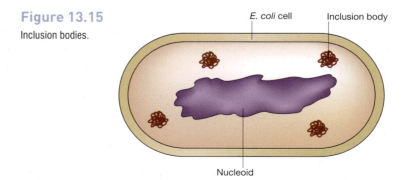

- *E. coli* might degrade the recombinant protein. Exactly how *E. coli* can recognize the foreign protein, and thereby subject it to preferential turnover, is not known.

These problems are less easy to solve than the sequence problems described in the previous section. Degradation of recombinant proteins can be reduced by using as the host a mutant *E. coli* strain that is deficient in one or more of the proteases responsible for protein degradation. Correct folding of recombinant proteins can also be promoted by choosing a special host strain, in this case one that over-synthesizes the chaperone proteins thought to be responsible for protein folding in the cell. But the main problem is the absence of glycosylation. So far this has proved insurmountable, limiting *E. coli* to the synthesis of animal proteins that do not need to be processed in this way.

13.3 Production of recombinant protein by eukaryotic cells

The problems associated with obtaining high yields of active recombinant proteins from genes cloned in *E. coli* have led to the development of expression systems for other organisms. There have been a few attempts to use other bacteria as the hosts for recombinant protein synthesis, and some progress has been made with *Bacillus subtilis*, but the main alternatives to *E. coli* are microbial eukaryotes. The argument is that a microbial eukaryote, such as a yeast or filamentous fungus, is more closely related to an animal, and so may be able to deal with recombinant protein synthesis more efficiently than *E. coli*. Yeasts and fungi can be grown just as easily as bacteria in continuous culture, and might express a cloned gene from a higher organism, and process the resulting protein in a manner more akin to that occurring in the higher organism itself.

13.3.1 Recombinant protein from yeast and filamentous fungi

To a large extent the potential of microbial eukaryotes has been realized and these organisms are now being used for the routine production of several animal proteins. Expression vectors are still required because it turns out that the promoters and other expression signals for animal genes do not, in general, work efficiently in these lower eukaryotes. The vectors themselves are based on those described in Chapter 7.

Saccharomyces cerevisiae *as the host for recombinant protein synthesis*

The yeast *Saccharomyces cerevisiae* is currently the most popular microbial eukaryote for recombinant protein production. Cloned genes are often placed under the control of the *GAL* promoter (Figure 13.16a), which is normally upstream of the gene coding for galactose epimerase, an enzyme involved in the metabolism of galactose. The *GAL* promoter is induced by galactose, providing a straightforward system for regulating expression of a cloned foreign gene. Other useful promoters are *PHO5*, which is regulated by the phosphate level in the growth medium, and *CUP1*, which is induced by copper. Most yeast expression vectors also carry a termination sequence from an *S. cerevisiae* gene, because animal termination signals do not work effectively in yeast.

Yields of recombinant protein are relatively high, but *S. cerevisiae* is unable to glycosylate animal proteins correctly, often adding too many sugar units ("hyperglycosylation"), although this can be prevented or at least reduced by using a mutant host strain. *S. cerevisiae* also lacks an efficient system for secreting proteins into the growth medium.

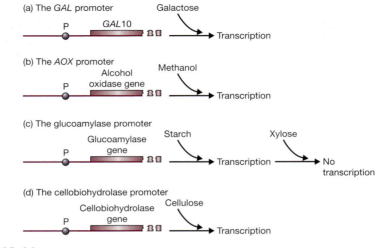

Figure 13.16

Four promoters frequently used in expression vectors for microbial eukaryotes. P = promoter.

In the absence of secretion, recombinant proteins are retained in the cell and consequently are less easy to purify. Codon bias (p. 209) can also be a problem.

Despite these drawbacks, *S. cerevisiae* remains the most frequently used microbial eukaryote for recombinant protein synthesis, partly because it is accepted as a safe organism for production of proteins for use in medicines or in foods, and partly because of the wealth of knowledge built up over the years regarding the biochemistry and genetics of *S. cerevisiae*, which means that it is relatively easy to devise strategies for minimizing the difficulties that arise.

Other yeasts and fungi

Although *S. cerevisiae* retains the loyalty of many molecular biologists, there are other microbial eukaryotes that might be equally if not more effective in recombinant protein synthesis. In particular, *Pichia pastoris*, a second species of yeast, is able to synthesize large amounts of recombinant protein (up to 30% of the total cell protein) and its glycosylation abilities are very similar to those of animal cells. The sugar structures that it synthesizes are not precisely the same as the animal versions (Figure 13.17), but the differences are relatively trivial and would probably not have a significant effect on the activity of a recombinant protein. Importantly, the glycosylated proteins made by *P. pastoris* are unlikely to induce an antigenic reaction if injected into the bloodstream, a problem frequently encountered with the over-glycosylated proteins synthesized by *S. cerevisiae*. Expression vectors for *P. pastoris* make use of the alcohol oxidase (*AOX*) promoter (Figure 13.16b), which is induced by methanol. The only significant problem with *P. pastoris* is that it sometimes degrades recombinant proteins before they can be purified, but this can be controlled by using special growth media. Other yeasts that have been used for recombinant protein synthesis include *Hansenula polymorpha*, *Yarrowia lipolytica*, and *Kluveromyces lactis*. The last of these has the attraction that it can be grown on waste products from the food industry.

The two most popular filamentous fungi are *Aspergillus nidulans* and *Trichoderma reesei*. The advantages of these organisms are their good glycosylation properties and

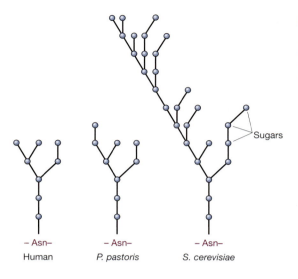

Figure 13.17

Comparison between a typical glycosylation structure found on an animal protein and the structures synthesized by *P. pastoris* and *S. cerevisiae*.

Sugars

– Asn–   – Asn–   – Asn–

Human   *P. pastoris*   *S. cerevisiae*

their ability to secrete proteins into the growth medium. The latter is a particularly strong feature of the wood rot fungus *T. reesei*, which in its natural habitat secretes cellulolytic enzymes that degrade the wood it lives on. The secretion characteristics mean that these fungi are able to produce recombinant proteins in a form that aids purification. Expression vectors for *A. nidulans* usually carry the glucoamylase promoter (Figure 13.16c), induced by starch and repressed by xylose; those for *T. reesei* make use of the cellobiohydrolase promoter (Figure 13.16d), which is induced by cellulose.

13.3.2 Using animal cells for recombinant protein production

The difficulties inherent in synthesis of a fully active animal protein in a microbial host have prompted biotechnologists to explore the possibility of using animal cells for recombinant protein synthesis. For proteins with complex and essential glycosylation structures, an animal cell might be the only type of host within which the active protein can be synthesized.

Protein production in mammalian cells

Culture systems for animal cells have been around since the early 1960s, but only during the past 20 years have methods for large-scale continuous culture become available. A problem with some animal cell lines is that they require a solid surface on which to grow, adding complications to the design of the culture vessels. One solution is to fill the inside of the vessel with plates, providing a large surface area, but this has the disadvantage that complete and continuous mixing of the medium within the vessel becomes very difficult. A second possibility is to use a standard vessel but to provide the cells with small inert particles (e.g., cellulose beads) on which to grow. Rates of growth and maximum cell densities are much less for animal cells compared with microorganisms, limiting the yield of recombinant protein, but this can be tolerated if it is the only way of obtaining the active protein.

Of course, gene cloning may not be necessary in order to obtain an animal protein from an animal cell culture. Nevertheless, expression vectors and cloned genes are still used to maximize yields, by placing the gene under control of a promoter that is stronger

Figure 13.18

Crystalline inclusion bodies in the nuclei of insect cells infected with a baculovirus.

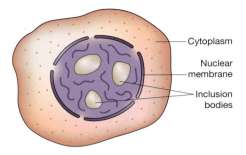

than the one to which it is normally attached. This promoter is often obtained from viruses such as SV40 (p. 123), cytomegalovirus (CMV), or Rous sarcoma virus (RSV). Mammalian cell lines derived from humans or hamsters have been used in synthesis of several recombinant proteins, and in most cases these proteins have been processed correctly and are indistinguishable from the non-recombinant versions. However, this is the most expensive approach to recombinant protein production, especially as the possible co-purification of viruses with the protein means that rigorous quality control procedures must be employed to ensure that the product is safe.

Protein production in insect cells

Insect cells provide an alternative to mammalian cells for animal protein production. Insect cells do not behave in culture any differently to mammalian cells but they have the great advantage that, thanks to a natural expression system, they can provide high yields of recombinant protein.

The expression system is based on the **baculoviruses**, a group of viruses that are common in insects but do not normally infect vertebrates. The baculovirus genome includes the polyhedrin gene, whose product accumulates in the insect cell as large nuclear inclusion bodies toward the end of the infection cycle (Figure 13.18). The product of this single gene frequently makes up over 50% of the total cell protein. Similar levels of protein production also occur if the normal gene is replaced by a foreign one. Baculovirus vectors have been successfully used in production of a number of mammalian proteins, but unfortunately the resulting proteins are not glycosylated correctly. In this regard the baculovirus system does not offer any advantages compared with *S. cerevisiae* or *P. pastoris*. However, the deficiencies in the insect glycosylation process can be circumvented by using a modified baculovirus that carries a mammalian promoter to express genes directly in mammalian cells. The infection is not **productive**, meaning that the virus genome is unable to replicate, but genes cloned into one of the **BacMam** vectors, as they are called, are maintained stably in mammalian cells for enough time for expression to occur. This expression is accompanied by the mammalian cell's own post-translational processing activities, so the recombinant protein is correctly glycosylated and therefore should be fully active.

Of course, in nature baculoviruses infect living insects, not cell cultures. For example, one of the most popular baculoviruses used in cloning is the *Bombyx mori* nucleopolyhedrovirus (BmNPV), which is a natural pathogen of the silkworm. There is a huge conventional industry based on the culturing of silkworms for silk production, and this expertise is now being harnessed for production of recombinant proteins, using expression vectors based on the BmNPV genome. As well as being an easy and cheap means of obtaining proteins, silkworms have the additional advantage of not being infected by viruses that are pathogenic to humans. The possibility that dangerous viruses are co-purified with the recombinant protein is therefore avoided.

13.3.3 Pharming—recombinant protein from live animals and plants

The use of silkworms for recombinant protein production is an example of the process often referred to as **pharming**, where a **transgenic** organism acts as the host for protein synthesis. Pharming is a recent and controversial innovation in gene cloning.

Pharming in animals

A transgenic animal is one that contains a cloned gene in all of its cells. Knockout mice (p. 214), used to study the function of human and other mammalian genes, are examples of transgenic animals. With mice, a transgenic animal can be produced by microinjection of the gene to be cloned into a fertilized egg cell (p. 85). Although this technique works well with mice, injection of fertilized cells is inefficient or impossible with many other mammals, and generation of transgenic animals for recombinant protein production usually involves a more sophisticated procedure called **nuclear transfer** (Figure 13.19). This involves microinjection of the recombinant protein gene into a somatic cell, which is a more efficient process than injection into a fertilized egg. Because the somatic cell will not itself differentiate into an animal, its nucleus, after microinjection, must be transferred to an oocyte whose own nucleus has been removed. After implantation into a foster mother, the engineered cell retains the ability of the original oocyte to divide and differentiate into an animal, one that will contain the transgene in every cell. This is a lengthy procedure and transgenic animals are therefore expensive to produce, but the technology is cost-effective because once a transgenic animal has been made it can reproduce and pass its cloned gene to its offspring according to standard Mendelian principles.

Although proteins have been produced in the blood of transgenic animals, and in the eggs of transgenic chickens, the most successful approach has been to use farm animals such as sheep or pigs, with the cloned gene attached to the promoter for the animal's β-lactoglobulin gene. This promoter is active in the mammary tissue which means that the recombinant protein is secreted in the milk (Figure 13.20). Milk production can be

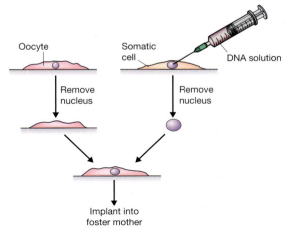

Figure 13.19

Transfer of the nucleus from a transgenic somatic cell to an oocyte.

Figure 13.20

Recombinant protein production in the milk of a transgenic sheep.

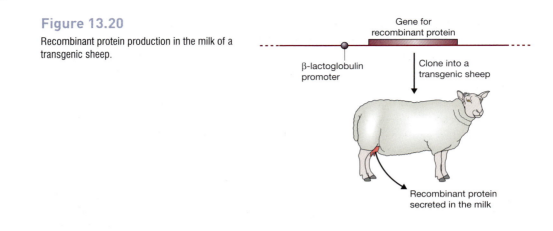

continuous during the animal's adult life, resulting in a high yield of the protein. For example, the average cow produces some 8000 liters of milk per year, yielding 40–80 kg of protein. Because the protein is secreted, purification is relatively easy. Most importantly, sheep and pigs are mammals and so human proteins produced in this way are correctly modified. Production of pharmaceutical proteins in farm animals therefore offers considerable promise for synthesis of correctly modified human proteins for use in medicine.

Recombinant proteins from plants

Plants provide the final possibility for production of recombinant protein. Plants and animals have similar protein processing activities, although there are slight differences in the glycosylation pathways. Plant cell culture is a well established technology that is already used in the commercial synthesis of natural plant products. Alternatively, intact plants can be grown to a high density in fields. The latter approach to recombinant protein production has been used with a variety of crops, such as maize, tobacco, rice, and sugarcane. One possibility is to place the transgene next to the promoter of a seed specific gene such as β-phaseolin, which codes for the main seed protein of the bean *Phaseolus vulgaris*. The recombinant protein is therefore synthesized specifically in the seeds, which naturally accumulate large quantities of proteins and are easy to harvest and to process. Recombinant proteins have also been synthesized in leaves of tobacco and alfalfa and the tubers of potatoes. In all of these cases, the protein has to be purified from the complex biochemical mixture that is produced when the seeds, leaves, or tubers are crushed. One way of avoiding this problem is to express the recombinant protein as a fusion with a signal peptide that directs secretion of the protein by the roots. Although this requires the plants to be grown in hydroponic systems rather than in fields, the decrease in yield is at least partly offset by the low cost of purification.

Whichever production system is used, plants offer a cheap and low-technology means of mass production of recombinant proteins. A range of proteins have been produced in experimental systems, including important pharmaceuticals such as interleukins and antibodies. This is an area of intensive research at the present time, with a number of plant biotechnology companies developing systems that have reached or are nearing commercial production. One very promising possibility is that plants could be used to synthesize vaccines, providing the basis to a cheap and efficient vaccination program (Chapter 14).

Ethical concerns raised by pharming

With our discussion of pharming we have entered one of the areas of gene cloning that causes concern among the public. No student of gene cloning and DNA analysis should ignore the controversies raised by the genetic manipulation of animals and plants but, equally, no textbook on the subject should attempt to teach the "correct" response to these ethical concerns. You must make up your own mind on such matters.

With transgenic animals, one of the fears is that the procedures used might cause suffering. These concerns do not center on the recombinant protein, but on the manipulations that result in production of the transgenic animal. Animals produced by nuclear transfer suffer a relatively high frequency of birth defects, and some of those that survive do not synthesize the required protein adequately, meaning that this type of pharming is accompanied by a high "wastage". Even the healthy animals appear to suffer from premature aging, as was illustrated most famously by "Dolly the sheep" who, although not transgenic, was the first animal to be produced by nuclear transfer. Most sheep of her breed live for up to 12 years, but Dolly developed arthritis at the age of 5 and was put down one year later because she was found to be suffering from terminal lung disease, which is normally found only in old sheep. It has been speculated that this premature aging was related to the age of the somatic cell whose nucleus gave rise to Dolly, as this cell came from a six-year-old sheep and so Dolly was effectively six when she was born. Although the technology has moved on dramatically since Dolly was born in 1997, the welfare issues regarding transgenic animals have not been resolved, and the broader issues concerning the use of nuclear transfer to "clone" animals (i.e., to produce identical offspring, rather than to clone individual genes) remain at the forefront of public awareness.

Pharming in plants raises a completely different set of ethical concerns, relating in part to the impact that genetically manipulated crops might have on the environment. These concerns apply to all GM crops, not just those used for pharming, and we will return to them in Chapter 15 after we have examined the more general uses of gene cloning in agriculture.

Further reading

de Boer, H.A., Comstock, L.J. & Vasser, M. (1983) The *tac* promoter: a functional hybrid derived from the *trp* and *lac* promoters. *Proceedings of the National Academy of Sciences of the USA*, 80, 21–25.

Borisjuk, N.V., Borisjuk, L.G., Logendra, S. et al. (1999) Production of recombinant proteins in plant root exudates. *Nature Biotechnology*, 17, 466–469.

Gellissen, G. & Hollenberg, C.P. (1997) Applications of yeasts in gene expression studies: a comparison of *Saccharomyces cerevisiae*, *Hansenula polymorpha* and *Kluveromyces lactis*—a review. *Gene*, 190, 87–97.

Hannig, G. & Makrides, S.C. (1998) Strategies for optimizing heterologous protein expression in *Escherichia coli*. *Trends in Biotechnology*, 16, 54–60.

Hellwig, S., Drossard, J., Twyman, R.M. & Fischer, R. (2004) Plant cell cultures for the production of recombinant proteins. *Nature Biotechnology*, 22, 1415–1422.

Houdebine, L.-M. (2009) Production of pharmaceutical proteins by transgenic animals. *Comparative Immunology, Microbiology and Infectious Diseases*, 32, 107–121.

Ikonomou, L., Schneider, Y.J. & Agathos, S.N. (2003) Insect cell culture for industrial production of recombinant proteins. *Applied Microbiology and Biotechnology*, 62, 1–20.

Kaiser, J. (2008) Is the drought over for pharming? *Science*, 320, 473–475.

Kind, A. & Schnieke, A. (2008) Animal pharming, two decades on. *Transgenic Research*, 17, 1025–1033. [Reviews the progress and controversies with animal pharming.]

Kost, T.A., Condreay, J.P. & Jarvis, D.L. (2005) Baculovirus as versatile vectors for protein expression in insect and mammalian cells. *Nature Biotechnology*, 23, 567–575.

Remaut, E., Stanssens, P. & Fiers, W. (1981) Plasmid vectors for high-efficiency expression controlled by the P_L promoter of coliphage. *Gene*, 15, 81–93. [Construction of an expression vector.]

Robinson, M., Lilley, R., Little, S. et al. (1984) Codon usage can affect efficiency of translation of genes in *Escherichia coli*. *Nucleic Acids Research*, 12, 6663–6671.

Sørensen, H.P. & Mortensen, K.K. (2004) Advanced genetic strategies for recombinant protein expression in *Escherichia coli*. *Journal of Biotechnology*, 115, 113–128. [A general review of the use of *E. coli* for recombinant protein production.]

Sreekrishna, K., Brankamp, R.G., Kropp, K.E. et al. (1997) Strategies for optimal synthesis and secretion of heterologous proteins in the methylotrophic yeast *Pichia pastoris*. *Gene*, 190, 55–62.

Stoger, E., Ma, J.K.-C., Fischer, R. & Christou, P. (2005) Sowing the seeds of success: pharmaceutical proteins from plants. *Current Opinion in Biotechnology*, 16, 167–173.

Thomson, A.J. & McWhir, J. (2004) Biomedical and agricultural applications of animal transgenesis. *Molecular Biotechnology*, 27, 231–244.

Wiebe, M.G. (2003) Stable production of recombinant proteins in filamentous fungi—problems and improvements. *Mycologist*, 17, 140–144.

Wilmut, I., Schnieke, A.E., McWhir, J. et al. (1997) Viable offspring derived from fetal and adult mammalian cells. *Nature*, 385, 810–813. [The method used to produce Dolly the sheep.]

Wurm, F. (2004) Production of recombinant protein therapeutics in cultivated mammalian cells. *Nature Biotechnology*, 22, 1393–1398.

Chapter **14**

Gene Cloning and DNA Analysis in Medicine

Chapter contents

Medicine has been and will continue to be a major beneficiary of the recombinant DNA revolution, and an entire book could be written on this topic. Later in this chapter we will see how recombinant DNA techniques are being used to identify genes responsible for inherited diseases and to devise new therapies for these disorders. First, we will continue the theme developed in the last chapter and examine the ways in which cloned genes are being used in the production of recombinant pharmaceuticals.

14.1 Production of recombinant pharmaceuticals

A number of human disorders can be traced to the absence or malfunction of a protein normally synthesized in the body. Most of these disorders can be treated by supplying the patient with the correct version of the protein, but for this to be possible the relevant protein must be available in relatively large amounts. If the defect can be corrected only by administering the human protein, then obtaining sufficient quantities will be a major problem unless donated blood can be used as the source. Animal proteins are therefore used whenever these are effective, but there are not many disorders that can be treated with animal proteins, and there is always the possibility of side effects such as an allergenic response.

We learned in Chapter 13 that gene cloning can be used to obtain large amounts of recombinant human proteins. How are these techniques being applied to the production of proteins for use as pharmaceuticals?

Gene Cloning and DNA Analysis: An Introduction. 6th edition. By T.A. Brown. Published 2010 by Blackwell Publishing.

14.1.1 Recombinant insulin

Insulin, synthesized by the β-cells of the islets of Langerhans in the pancreas, controls the level of glucose in the blood. An insulin deficiency manifests itself as diabetes mellitus, a complex of symptoms which may lead to death if untreated. Fortunately, many forms of diabetes can be alleviated by a continuing program of insulin injections, thereby supplementing the limited amount of hormone synthesized by the patient's pancreas. The insulin used in this treatment was originally obtained from the pancreas of pigs and cows slaughtered for meat production. Although animal insulin is generally satisfactory, problems may arise in its use to treat human diabetes. One problem is that the slight differences between the animal and the human proteins can lead to side effects in some patients. Another is that the purification procedures are difficult, and potentially dangerous contaminants cannot always be completely removed.

Insulin displays two features that facilitate its production by recombinant DNA techniques. The first is that the human protein is not modified after translation by the addition of sugar molecules (p. 236): recombinant insulin synthesized by a bacterium should therefore be active. The second advantage concerns the size of the molecule. Insulin is a relatively small protein, comprising two polypeptides, one of 21 amino acids (the A chain) and one of 30 amino acids (the B chain; Figure 14.1). In humans these chains are

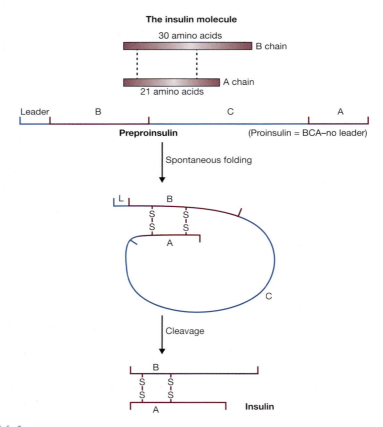

Figure 14.1

The structure of the insulin molecule and a summary of its synthesis by processing from preproinsulin.

synthesized as a precursor called preproinsulin, which contains the A and B segments linked by a third chain (C) and preceded by a leader sequence. The leader sequence is removed after translation and the C chain excised, leaving the A and B polypeptides linked to each other by two disulphide bonds.

Several strategies have been used to obtain recombinant insulin. One of the first projects, involving synthesis of artificial genes for the A and B chains followed by production of fusion proteins in *E. coli*, illustrates a number of the general techniques used in recombinant protein production.

Synthesis and expression of artificial insulin genes

In the late 1970s, the idea of making an artificial gene was extremely innovative. Oligonucleotide synthesis was in its infancy at that time, and the available methods for making artificial DNA molecules were much more cumbersome than the present-day automated techniques. Nevertheless, genes coding for the A and B chains of insulin were synthesized as early as 1978.

The procedure used was to synthesize trinucleotides representing all the possible codons and then join these together in the order dictated by the amino acid sequences of the A and B chains. The artificial genes would not necessarily have the same nucleotide sequences as the real gene segments coding for the A and B chains, but they would still specify the correct polypeptides. Two recombinant plasmids were constructed, one carrying the artificial gene for the A chain, and one the gene for the B chain.

In each case the artificial gene was ligated to a *lacZ′* reading frame present in a pBR322-type vector (Figure 14.2a). The insulin genes were therefore under the control of the strong *lac* promoter (p. 231), and were expressed as fusion proteins, consisting of the first few amino acids of β-galactosidase followed by the A or B polypeptides (Figure 14.2b). Each gene was designed so that its β-galactosidase and insulin segments were separated by a methionine residue, so that the insulin polypeptides could be cleaved from the β-galactosidase segments by treatment with cyanogen bromide (p. 233). The purified A and B chains were then attached to each other by disulphide bond formation in the test tube.

The final step, involving disulphide bond formation, is actually rather inefficient. A subsequent improvement was to synthesize not the individual A and B genes, but the entire proinsulin reading frame, specifying B chain–C chain–A chain (see Figure 14.1). Although this is a more daunting proposition in terms of DNA synthesis, the prohormone has the big advantage of folding spontaneously into the correct disulphide-bonded structure. The C chain segment can then be excised relatively easily by proteolytic cleavage.

14.1.2 Synthesis of human growth hormones in *E. coli*

At about the same time that recombinant insulin was first being made in *E. coli*, other researchers were working on similar projects with the human growth hormones somatostatin and somatotrophin. These two proteins act in conjunction to control growth processes in the human body, their malfunction leading to painful and disabling disorders such as acromegaly (uncontrolled bone growth) and dwarfism.

Somatostatin was the first human protein to be synthesized in *E. coli*. Being a very short protein, only 14 amino acids in length, it was ideally suited for artificial gene synthesis. The strategy used was the same as described for recombinant insulin, involving insertion of the artificial gene into a *lacZ′* vector (Figure 14.3), synthesis of a fusion protein, and cleavage with cyanogen bromide.

Figure 14.2

The synthesis of recombinant insulin from artificial A and B chain genes.

Figure 14.3

Production of recombinant somatostatin.

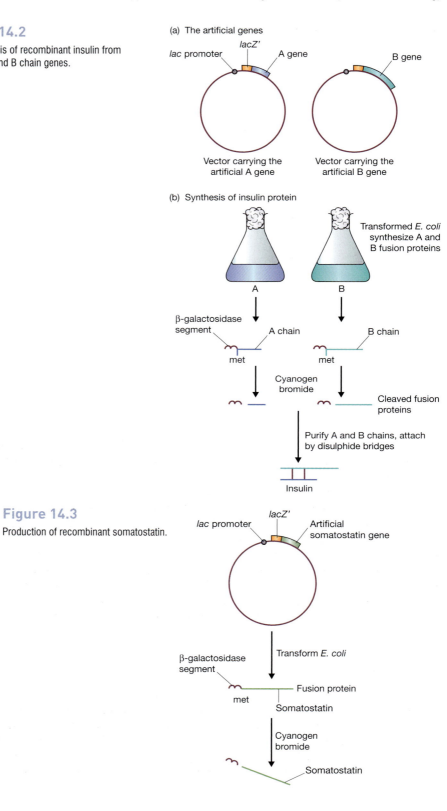

(a) Preparation of the somatotrophin cDNA fragment

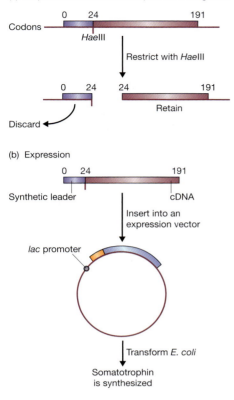

Figure 14.4

Production of recombinant somatotrophin.

Somatotrophin presented a more difficult problem. This protein is 191 amino acids in length, equivalent to almost 600 bp, an impossible prospect for the DNA synthesis capabilities of the late 1970s. In fact, a combination of artificial gene synthesis and complementary DNA (cDNA) cloning was used to obtain a somatotrophin-producing *E. coli* strain. Messenger RNA was obtained from the pituitary, the gland that produces somatotrophin in the human body, and a cDNA library prepared. The somatotrophin cDNA contained a single site for the restriction endonuclease *Hae*III, which therefore cuts the cDNA into two segments (Figure 14.4a). The longer segment, consisting of codons 24–191, was retained for use in construction of the recombinant plasmid. The smaller segment was replaced by an artificial DNA molecule that reproduced the start of the somatotrophin gene and provided the correct signals for translation in *E. coli* (Figure 14.4b). The modified gene was then ligated into an expression vector carrying the *lac* promoter.

14.1.3 Recombinant factor VIII

Although a number of important pharmaceutical compounds have been obtained from genes cloned in *E. coli*, the general problems associated with using bacteria to synthesize foreign proteins (p. 234) have led in many cases to these organisms being replaced by eukaryotes. An example of a recombinant pharmaceutical produced in eukaryotic cells is human factor VIII, a protein that plays a central role in blood clotting. The commonest form of hemophilia in humans results from an inability to synthesize factor VIII,

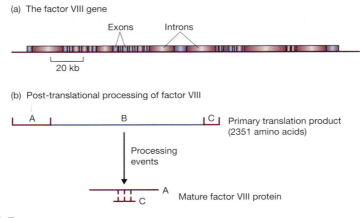

Figure 14.5

The factor VIII gene and its translation product.

leading to a breakdown in the blood clotting pathway and the well-known symptoms associated with the disease.

Until recently the only way to treat hemophilia was by injection of purified factor VIII protein, obtained from human blood provided by donors. Purification of factor VIII is a complex procedure and the treatment is expensive. More critically, the purification is beset with difficulties, in particular in removing virus particles that may be present in the blood. Hepatitis and acquired immune deficiency syndrome (AIDS) can be and have been passed on to hemophiliacs via factor VIII injections. Recombinant factor VIII, free from contamination problems, would be a significant achievement for biotechnology.

The factor VIII gene is very large, over 186 kb in length, and is split into 26 exons and 25 introns (Figure 14.5a). The mRNA codes for a large polypeptide (2351 amino acids), which undergoes a complex series of post-translational processing events, eventually resulting in a dimeric protein consisting of a large subunit, derived from the upstream region of the initial polypeptide, and a small subunit from the downstream segment (Figure 14.5b). The two subunits contain a total of 17 disulphide bonds and a number of glycosylated sites. As might be anticipated for such a large and complex protein, it has not been possible to synthesize an active version in *E. coli*.

Initial attempts to obtain recombinant factor VIII therefore involved mammalian cells. In the first experiments to be carried out the entire cDNA was cloned in hamster cells, but yields of protein were disappointingly low. This was probably because the post-translational events, although carried out correctly in hamster cells, did not convert all the initial product into an active form, limiting the overall yield. As an alternative, two separate fragments from the cDNA were used, one fragment coding for the large subunit polypeptide, the second for the small subunit. Each cDNA fragment was ligated into an expression vector, downstream of the Ag promoter (a hybrid between the chicken β-actin and rabbit β-globin sequences) and upstream of a polyadenylation signal from SV40 virus (Figure 14.6). The plasmid was introduced into a hamster cell line and recombinant protein obtained. The yields were over ten times greater than those from cells containing the complete cDNA, and the resulting factor VIII protein was indistinguishable in terms of function from the native form.

Pharming (p. 241) has also been used for production of recombinant factor VIII. The complete human cDNA has been attached to the promoter for the whey acidic protein

Factor VIII cDNA

Ag promoter   SV40 polyadenylation
sequence

Figure 14.6

The expression signals used in production of recombinant factor VIII. The promoter is an artificial hybrid of the chicken β-actin and rabbit β-globin sequences, and the polyadenylation signal (needed for correct processing of the mRNA before translation into protein) is obtained from SV40 virus.

Table 14.1

Some of the human proteins that have been synthesized from genes cloned in bacteria and/or eukaryotic cells or by pharming.

PROTEIN	USED IN THE TREATMENT OF
α_1-Antitrypsin	Emphysema
Deoxyribonuclease	Cystic fibrosis
Epidermal growth factor	Ulcers
Erythropoietin	Anemia
Factor VIII	Hemophilia
Factor IX	Christmas disease
Fibroblast growth factor	Ulcers
Follicle stimulating hormone	Infertility treatment
Granulocyte colony stimulating factor	Cancers
Insulin	Diabetes
Insulin-like growth factor 1	Growth disorders
Interferon-α	Leukemia and other cancers
Interferon-β	Cancers, AIDS
Interferon-γ	Cancers, rheumatoid arthritis
Interleukins	Cancers, immune disorders
Lung surfactant protein	Respiratory distress
Relaxin	Used to aid childbirth
Serum albumin	Used as a plasma supplement
Somatostatin	Growth disorders
Somatotrophin	Growth disorders
Superoxide dismutase	Free radical damage in kidney transplants
Tissue plasminogen activator	Heart attack
Tumor necrosis factor	Cancers

gene of pig, leading to synthesis of human factor VIII in pig mammary tissue and subsequent secretion of the protein in the milk. The factor VIII produced in this way appears to be exactly the same as the native protein and is fully functional in blood clotting assays.

14.1.4 Synthesis of other recombinant human proteins

The list of human proteins synthesized by recombinant technology continues to grow (Table 14.1). As well as proteins used to treat disorders by replacement or supplementation of the dysfunctional versions, the list includes a number of growth factors (e.g., interferons and interleukins) with potential uses in cancer therapy. These proteins are synthesized in very limited amounts in the body, so recombinant technology is the only viable means of obtaining them in the quantities needed for clinical purposes. Other proteins, such as serum albumin, are more easily obtained, but are needed in such large quantities that production in microorganisms is still a more attractive option.

Figure 14.7

The principle behind the use of a preparation of isolated virus coat proteins as a vaccine.

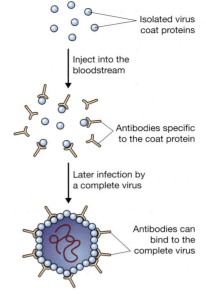

Isolated virus coat proteins

Inject into the bloodstream

Antibodies specific to the coat protein

Later infection by a complete virus

Antibodies can bind to the complete virus

14.1.5 Recombinant vaccines

The final category of recombinant protein is slightly different from the examples given in Table 14.1. A vaccine is an antigenic preparation that, after injection into the bloodstream, stimulates the immune system to synthesize antibodies that protect the body against infection. The antigenic material present in a vaccine is normally an inactivated form of the infectious agent. For example, antiviral vaccines often consist of virus particles that have been attenuated by heating or a similar treatment. In the past, two problems have hindered the preparation of attenuated viral vaccines:

- The inactivation process must be 100% efficient, as the presence in a vaccine of just one live virus particle could result in infection. This has been a problem with vaccines for the cattle disease foot-and-mouth.
- The large amounts of virus particles needed for vaccine production are usually obtained from tissue cultures. Unfortunately some viruses, notably hepatitis B virus, do not grow in tissue culture.

Producing vaccines as recombinant proteins

The use of gene cloning in this field centers on the discovery that virus-specific antibodies are sometimes synthesized in response not only to the whole virus particle, but also to isolated components of the virus. This is particularly true of purified preparations of the proteins present in the virus coat (Figure 14.7). If the genes coding for the antigenic proteins of a particular virus could be identified and inserted into an expression vector, the methods described above for the synthesis of animal proteins could be employed in the production of recombinant proteins that might be used as vaccines. These vaccines would have the advantages that they would be free of intact virus particles and they could be obtained in large quantities.

The greatest success with this approach has been with hepatitis B virus. Hepatitis B is endemic in many tropical parts of the world and leads to liver disease and possibly, after chronic infection, to cancer of the liver. A person who recovers from hepatitis B is immune to future infection because their blood contains antibodies to the hepatitis B surface antigen (HBsAg), which is one of the virus coat proteins. This protein has been synthesized in both *Saccharomyces cerevisiae*, using a vector based on the 2 μm plasmid (p. 105), and in Chinese hamster ovary (CHO) cells. In both cases, the protein was obtained in reasonably high quantities, and when injected into test animals provided protection against hepatitis B.

The key to the success of recombinant HbsAg as a vaccine is provided by an unusual feature of the natural infection process for the virus. The bloodstream of infected individuals contains not only intact hepatitis B virus particles, which are 42 nm in diameter, but also smaller, 22 nm spheres made up entirely of HBsAg protein molecules. Assembly of these 22 nm spheres occurs during HbsAg synthesis in both yeast and hamster cells and it is almost certainly these spheres, rather than single HbsAg molecules, that are the effective component of the recombinant vaccine. The recombinant vaccine therefore mimics part of the natural infection process and stimulates antibody production, but as the spheres are not viable viruses the vaccine does not itself cause the disease. Both the yeast and hamster cell vaccines have been approved for use in humans, and the World Health Organization is promoting their use in national vaccination programmes.

Recombinant vaccines in transgenic plants

The advent of pharming (p. 241) has led to the possibility of using transgenic plants as the hosts for synthesis of recombinant vaccines. The ease with which plants can be grown and harvested means that this technology might be applicable for developing parts of the world where the more expensive approaches to recombinant protein production might be difficult to sustain. If the recombinant vaccine is effective after oral administration, then immunity could be acquired simply by eating part or all of the transgenic plant. A simpler and cheaper means of carrying out a mass vaccination program would be hard to imagine.

The feasibility of this approach has been demonstrated by trials with vaccines such as HbsAg and the coat proteins of measles virus and respiratory syncytial virus. In each case, immunity was conferred by feeding the transgenic plant to test animals. Attempts are also being made to engineer plants that express a variety of vaccines, so that immunity against a range of diseases can be acquired from a single source. The main problem currently faced by the companies developing this technology is that the amount of recombinant protein synthesized by the plant is often insufficient to stimulate complete immunity against the target disease. To be completely effective the yield of the vaccine needs to make up 8–10% of the soluble protein content of the part of the plant which is eaten, but in practice yields are much less than this, usually not more than 0.5%. Variability in the yields between different plants in a single crop is also a concern. A partial solution is provided by placing the cloned gene in the chloroplast genome rather than the plant nucleus (p. 119), as this generally results in much higher yields of recombinant protein. However, proteins made in the chloroplast are not glycosylated and so those vaccines that require post-translational modification will be inactive if produced in this way. These include most of the relevant viral coat proteins, but not bacterial surface proteins such as the *Vibrio cholerae* B subunit, which can be used to confer immunity against diseases such as cholera. This protein has been synthesized in transgenic tobacco, tomato, and rice plants and shown to elicit an anti-cholera immune response when leaves, fruits, or seeds are fed to mice.

Live recombinant virus vaccines

The use of live vaccinia virus as a vaccine for smallpox dates back to 1796, when Edward Jenner first realized that this virus, harmless to humans, could stimulate immunity

Figure 14.8

The rationale behind the potential use of a recombinant vaccinia virus.

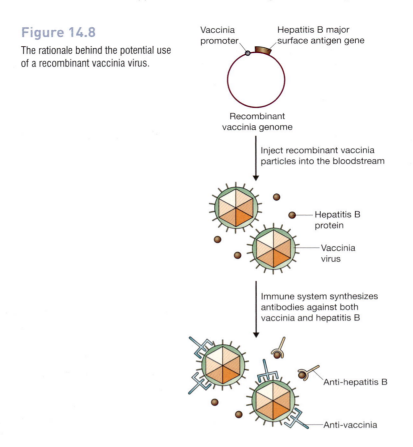

against the much more dangerous smallpox virus. The term "vaccine" comes from vaccinia; its use resulted in the worldwide eradication of smallpox in 1980.

A more recent idea is that recombinant vaccinia viruses could be used as live vaccines against other diseases. If a gene coding for a virus coat protein, for example HBsAg, is ligated into the vaccinia genome under control of a vaccinia promoter, then the gene will be expressed (Figure 14.8). After injection into the bloodstream, replication of the recombinant virus results not only in new vaccinia particles, but also in significant quantities of the major surface antigen. Immunity against both smallpox and hepatitis B would result.

This remarkable technique has considerable potential. Recombinant vaccinia viruses expressing a number of foreign genes have been constructed and shown to confer immunity against the relevant diseases in experimental animals (Table 14.2). The possibility of broad-spectrum vaccines is raised by the demonstration that a single recombinant vaccinia, expressing the genes for influenza virus hemagglutinin, HBsAg, and herpes simplex virus glycoprotein, confers immunity against each disease in monkeys. Studies of vaccinia viruses expressing the rabies glycoprotein have shown that deletion of the vaccinia gene for the enzyme thymidine kinase prevents the virus from replicating. This avoids the possibility that animals treated with the live vaccine will develop any form of cowpox, the disease caused by normal vaccinia. This particular live virus vaccine is now being used in rabies control in Europe and North America.

Table 14.2

Some of the foreign genes that have been expressed in recombinant vaccinia viruses.

GENE
Plasmodium falciparum (malaria parasite) surface antigen
Influenza virus coat proteins
Rabies virus G protein
Hepatitis B surface antigen
Herpes simplex glycoproteins
Human immunodeficiency virus (HIV) envelope proteins
Vesicular somatitis coat proteins
Sindbis virus proteins

14.2 Identification of genes responsible for human diseases

A second major area of medical research in which gene cloning is having an impact is in the identification and isolation of genes responsible for human diseases. A genetic or inherited disease is one that is caused by a defect in a specific gene (Table 14.3), individuals carrying the defective gene being predisposed toward developing the disease at some stage of their lives. With some inherited diseases, such as hemophilia, the gene is present on the X chromosome, so all males carrying the gene express the disease state; females with one defective gene and one correct gene are healthy but can pass the disease on to their male offspring. Genes for other diseases are present on autosomes and in most cases are recessive, so both chromosomes of the pair must carry a defective version for the disease to occur. A few diseases, including Huntington's chorea, are autosomal dominant, so a single copy of the defective gene is enough to lead to the disease state.

With some genetic diseases, the symptoms manifest themselves early in life, but with others the disease may not be expressed until the individual is middle-aged or elderly.

Table 14.3

Some of the commonest genetic diseases in the UK.

DISEASE	SYMPTOMS	FREQUENCY (BIRTHS PER YEAR)
Inherited breast cancer	Cancer	1 in 300 females
Cystic fibrosis	Lung disease	1 in 2000
Huntington's chorea	Neurodegeneration	1 in 2000
Duchenne muscular dystrophy	Progressive muscle weakness	1 in 3000 males
Hemophilia A	Blood disorder	1 in 4000 males
Sickle cell anemia	Blood disorder	1 in 10,000
Phenylketonuria	Mental retardation	1 in 12,000
β-Thalassaemia	Blood disorder	1 in 20,000
Retinoblastoma	Cancer of the eye	1 in 20,000
Hemophilia B	Blood disorder	1 in 25,000 males
Tay–Sachs disease	Blindness, loss of motor control	1 in 200,000

Cystic fibrosis is an example of the former, and neurodegenerative diseases such as Alzheimer's and Huntington's are examples of the latter. With a number of diseases that appear to have a genetic component, cancers in particular, the overall syndrome is complex with the disease remaining dormant until triggered by some metabolic or environmental stimulus. If predisposition to these diseases can be diagnosed, the risk factor can be reduced by careful management of the patient's lifestyle to minimize the chances of the disease being triggered.

Genetic diseases have always been present in the human population but their importance has increased in recent decades. This is because vaccination programs, antibiotics, and improved sanitation have reduced the prevalence of infectious diseases such as smallpox, tuberculosis, and cholera, which were major killers up to the mid-20th century. The result is that a greater proportion of the population now dies from a disease that has a genetic component, especially the late-onset diseases that are now more common because of increased life expectancies. Medical research has been successful in controlling many infectious diseases: can it be equally successful with genetic disease?

There are a number of reasons why identifying the gene responsible for a genetic disease is important:

- Gene identification may provide an indication of the biochemical basis to the disease, enabling therapies to be designed.
- Identification of the mutation present in a defective gene can be used to devise a screening programme so that the mutant gene can be identified in individuals who are carriers or who have not yet developed the disease. Carriers can receive counseling regarding the chances of their children inheriting the disease. Early identification in individuals who have not yet developed the disease allows appropriate precautions to be taken to reduce the risk of the disease becoming expressed.
- Identification of the gene is a prerequisite for gene therapy (p. 259).

14.2.1 How to identify a gene for a genetic disease

There is no single strategy for identification of genes that cause diseases, the best approach depending on the information that is available about the disease. To gain an understanding of the principles of this type of work, we will consider the most common and most difficult scenario. This is when all that is known about the disease is that certain people have it. Even with such an unpromising starting point, DNA techniques can locate the relevant gene.

Locating the approximate position of the gene in the human genome

If there is no information about the desired gene, how can it be located in the human genome? The answer is to use basic genetics to determine the approximate position of the gene on the human genetic map. Genetic mapping is usually carried out by **linkage analysis**, in which the inheritance pattern for the target gene is compared with the inheritance patterns for genetic loci whose map positions are already known. If two loci are inherited together, they must be very close on the same chromosome. If they are not close together, then recombination events and the random segregation of chromosomes during meiosis will result in the loci displaying different inheritance patterns (Figure 14.9). Demonstration of linkage with one or more mapped genetic loci is therefore the key to understanding the chromosomal position of an unmapped gene.

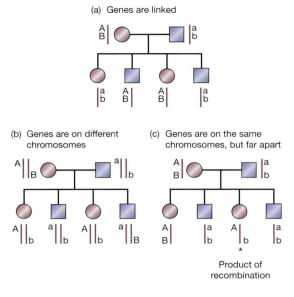

(a) Genes are linked

(b) Genes are on different chromosomes

(c) Genes are on the same chromosomes, but far apart

Product of recombination

Figure 14.9

Inheritance patterns for linked and unlinked genes. Three families are shown, circles representing females and squares representing males. (a) Two closely linked genes are almost always inherited together. (b) Two genes on different chromosomes display random segregation. (c) Two genes that are far apart on a single chromosome are often inherited together, but recombination may unlink them.

With humans it is not possible to carry out directed breeding programs aimed at determining the map position of a desired gene. Instead, mapping of disease genes must make use of data available from **pedigree analysis**, in which inheritance of the gene is examined in families with a high incidence of the disease being studied. It is important to be able to obtain DNA samples from at least three generations of each family, and the more family members there are the better, but unless the disease is very uncommon it is usually possible to find suitable pedigrees. Linkage between the presence/absence of the disease and the inheritance of other genes could be studied, but as DNA samples are being analyzed, linkage to DNA markers is more usually tested (p. 180).

To illustrate how linkage analysis is used we will look briefly at the way in which one of the genes conferring susceptibility to human breast cancer was mapped. The first breakthrough in this project occurred in 1990 as a result of **restriction fragment length polymorphism (RFLP) linkage analyses** carried out by a group at the University of California at Berkeley. This study showed that in families with a high incidence of breast cancer, a significant number of the women who suffered from the disease all possessed the same version of an RFLP called *D17S74*. This RFLP had previously been mapped to the long arm of chromosome 17 (Figure 14.10): the gene being sought—*BRCA1*—must therefore also be located on the long arm of chromosome 17.

This initial linkage result was extremely important, as it indicated in which region of the human genome this breast cancer susceptibility gene was to be found, but it was far from the end of the story. In fact, over 1000 genes are thought to lie in this particular 20 Mb stretch of chromosome 17. The next objective was therefore to carry out more linkage studies to try to pinpoint *BRCA1* more accurately. This was achieved by first examining the region containing *BRCA1* for short tandem repeats (STRs) (p. 181), which are useful for fine scale mapping because many of them exist in three or more allelic forms, rather than just the two alleles that are possible for an RFLP. Several alleles of an STR might therefore be present within a single pedigree, enabling more detailed mapping to be carried out. STR linkage mapping reduced the size of the *BRCA1* region from 20 Mb down to just 600 kb (Figure 14.10). This approach to locating a gene is called **positional cloning**.

Figure 14.10

Mapping the breast cancer gene. Initially the gene was mapped to a 20 Mb segment of chromosome 17 (highlighted region in the left drawing). Additional mapping experiments narrowed this down to a 600 kb region flanked by two previously mapped loci, *D17S1321* and *D17S1325* (middle drawing). After examination of expressed sequences, a strong candidate for *BRCA1* was eventually identified (right drawing).

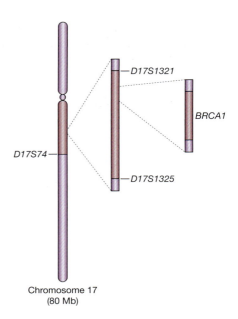

D17S74

D17S1321

BRCA1

D17S1325

Chromosome 17
(80 Mb)

Identification of candidates for the disease gene

One might imagine that once the map location of the disease gene has been determined the next step is simply to refer to the genome sequence in order to identify the gene. Unfortunately a great deal of work still has to be done. Genetic mapping, even in its most precise form, only gives an approximate indication of the location of a gene. In the breast cancer project the researchers were fortunate in being able to narrow the search area down to just 600 kb—often 10 Mb or more of DNA sequence has to be examined. Such large stretches of DNA could contain many genes: the 600 kb breast cancer region contained over 60 genes, any one of which could have been *BRCA1*.

A variety of approaches can be used to identify which of the genes in a mapped region is the disease gene:

- The expression profiles of the **candidate genes** can be examined by hybridization analysis or reverse transcription–polymerase chain reaction (RT–PCR) (p. 161) of RNA from different tissues. For example, *BRCA1* would be expected to hybridize to RNA prepared from breast tissue, and also to ovary tissue RNA, ovarian cancer frequently being associated with inherited breast cancer.

- Southern hybridization analysis (p. 142) can be carried out with DNA from different species (these are called **zoo blots**). The rationale is that an important human gene will almost certainly have homologs in other mammals, and that this homolog, although having a slightly different sequence from the human version, will be detectable by hybridization with a suitable probe.

- The gene sequences could be examined in individuals with and without the disease to see if the genes from affected individuals contain mutations that might explain why they have the disease.

- To confirm the identity of a candidate gene, it might be possible to prepare a knockout mouse (p. 214) that has an inactive version of the equivalent mouse gene. If the knockout mouse displays symptoms compatible with the human disease, then the candidate gene is almost certainly the correct one.

When applied to the breast cancer region, these analyses resulted in identification of an approximately 100 kb gene, made up of 22 exons and coding for a 1863 amino acid protein, that was a strong candidate for *BRCA1*. Transcripts of the gene were detectable in breast and ovary tissues, and homologs were present in mice, rats, rabbits, sheep, and pigs, but not chickens. Most importantly, the genes from five susceptible families contained mutations (such as frameshift and nonsense mutations) likely to lead to a non-functioning protein. Although circumstantial, the evidence in support of the candidate was sufficiently overwhelming for this gene to be identified as *BRCA1*. Subsequent research has shown that both this gene and *BRCA2*, a second gene associated with susceptibility to breast cancer, are involved in transcription regulation and DNA repair, and that both act as tumor suppressor genes, inhibiting abnormal cell division.

14.3 Gene therapy

The final application of recombinant DNA technology in medicine that we will consider is gene therapy. This is the name originally given to methods that aim to cure an inherited disease by providing the patient with a correct copy of the defective gene. Gene therapy has now been extended to include attempts to cure any disease by introduction of a cloned gene into the patient. First we will examine the techniques used in gene therapy, and then we will attempt to address the ethical issues.

14.3.1 Gene therapy for inherited diseases

There are two basic approaches to gene therapy: germline therapy and somatic cell therapy. In germline therapy, a fertilized egg is provided with a copy of the correct version of the relevant gene and re-implanted into the mother. If successful, the gene is present and expressed in all cells of the resulting individual. Germline therapy is usually carried out by microinjection of a somatic cell followed by nuclear transfer into an oocyte (p. 241), and theoretically could be used to treat any inherited disease.

Somatic cell therapy involves manipulation of cells, which either can be removed from the organism, transfected, and then placed back in the body, or transfected *in situ* without removal. The technique has most promise for inherited blood diseases (e.g., hemophilia and thalassaemia), with genes being introduced into stem cells from the bone marrow, which give rise to all the specialized cell types in the blood. The strategy is to prepare a bone extract containing several billion cells, transfect these with a retrovirus-based vector, and then re-implant the cells. Subsequent replication and differentiation of transfectants leads to the added gene being present in all the mature blood cells (Figure 14.11). The advantage of a retrovirus is that this type of vector has an extremely high transfection frequency, enabling a large proportion of the stem cells in a bone marrow extract to receive the new gene.

Somatic cell therapy also has potential in the treatment of lung diseases such as cystic fibrosis, as DNA cloned in adenovirus vectors (p. 123) or contained in liposomes (p. 85) is taken up by the epithelial cells in the lungs after introduction into the respiratory tract via an inhaler. However, gene expression occurs for only a few weeks, and as yet this has not been developed into an effective means of treating cystic fibrosis.

With those genetic diseases where the defect arises because the mutated gene does not code for a functional protein, all that is necessary is to provide the cell with the correct version of the gene: removal of the defective genes is unnecessary. The situation

Figure 14.11

Differentiation of a transfected stem cell leads to the new gene being present in all the mature blood cells.

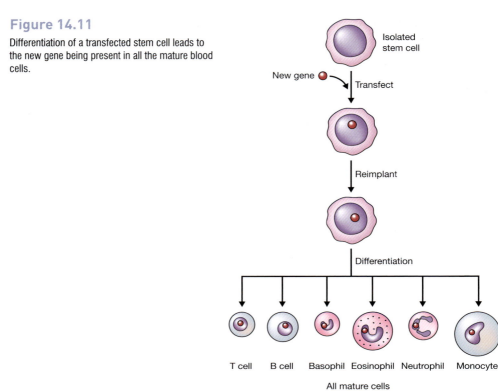

All mature cells contain the new gene

is less easy with dominant genetic diseases (p. 255), as with these it is the defective gene product itself that is responsible for the disease state, and so the therapy must include not only addition of the correct gene but also removal of the defective version. This requires a gene delivery system that promotes recombination between the chromosomal and vector-borne versions of the gene, so that the defective chromosomal copy is replaced by the gene from the vector. The technique is complex and unreliable, and broadly applicable procedures have not yet been developed.

14.3.2 Gene therapy and cancer

The clinical uses of gene therapy are not limited to treatment of inherited diseases. There have also been attempts to use gene cloning to disrupt the infection cycles of human pathogens such as the AIDS virus. However, the most intensive area of current research concerns the potential use of gene therapy as a treatment for cancer.

Most cancers result from activation of an oncogene that leads to tumor formation, or inactivation of a gene that normally suppresses formation of a tumor. In both cases a gene therapy could be envisaged to treat the cancer. Inactivation of a tumor suppressor gene could be reversed by introduction of the correct version of the gene by one of the methods described above for inherited disease. Inactivation of an oncogene would, however, require a more subtle approach, as the objective would be to prevent expression of the oncogene, not to replace it with a non-defective copy. One possible way of doing this would be to introduce into a tumor a gene specifying an **antisense** version of the

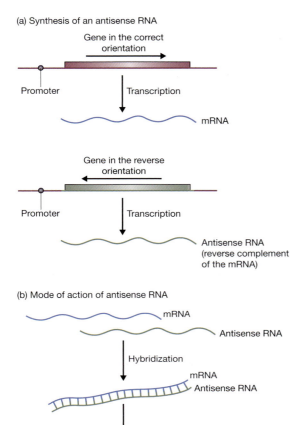

(a) Synthesis of an antisense RNA

Gene in the correct
orientation

Promoter

Transcription

mRNA

Gene in the reverse
orientation

Promoter

Transcription

Antisense RNA
(reverse complement
of the mRNA)

(b) Mode of action of antisense RNA

mRNA

Antisense RNA

Hybridization

mRNA
Antisense RNA

Degraded by cellular
ribonucleases

Figure 14.12

Antisense RNA can be used to silence a cellular
mRNA.

mRNA transcribed from the oncogene (Figure 14.12a). An antisense RNA is the reverse complement of a normal RNA, and can prevent synthesis of the protein coded by the gene it is directed against, probably by hybridizing to the mRNA producing a double-stranded RNA molecule that is rapidly degraded by cellular ribonucleases (Figure 14.12b). The target is therefore inactivated.

An alternative would be to introduce a gene that selectively kills cancer cells or promotes their destruction by drugs administered in a conventional fashion. This is called **suicide gene therapy** and is looked on as an effective general approach to cancer treatment, because it does not require a detailed understanding of the genetic basis of the particular disease being treated. Many genes that code for toxic proteins are known, and there are also examples of enzymes that convert non-toxic precursors of drugs into the toxic form. Introduction of the gene for one of these toxic proteins or enzymes into a tumor should result in the death of the cancer cells, either immediately or after drug administration. It is obviously important that the introduced gene is targeted accurately at the cancer cells, so that healthy cells are not killed. This requires a very precise delivery system, such as direct inoculation into the tumor, or some other means of ensuring that the gene is expressed only in the cancer cells. One possibility is to place the gene under control of the human telomerase promoter, which is active only in cancerous tissues.

Another approach is to use gene therapy to improve the natural killing of cancer cells by the patient's immune system, perhaps with a gene that causes the tumor cells to synthesize strong antigens that are efficiently recognized by the immune system. All of these approaches, and many not based on gene therapy, are currently being tested in the fight against cancer.

14.3.3 The ethical issues raised by gene therapy

Should gene therapy be used to cure human disease? As with many ethical questions, there is no simple answer. On the one hand, there could surely be no justifiable objection to the routine application via a respiratory inhaler of correct versions of the cystic fibrosis gene as a means of managing this disease. Similarly, if bone marrow transplants are acceptable, then it is difficult to argue against gene therapies aimed at correction of blood disorders via stem cell transfection. And cancer is such a terrible disease that the withholding of effective treatment regimens on moral grounds could itself be criticized as immoral.

Germline therapy is a more difficult issue. The problem is that the techniques used for germline correction of inherited diseases are exactly the same techniques that could be used for germline manipulation of other inherited characteristics. Indeed, the development of this technique with animals has not been prompted by any desire to cure genetic diseases, the aims being to "improve" farm animals, for example by making genetic changes that result in lower fat content. This type of manipulation, where the genetic constitution of an organism is changed in a directed, heritable fashion, is clearly unacceptable with humans. At present, technical problems mean that human germline manipulation would be difficult. Before these problems are solved we should ensure that the desire to do good should not raise the possibility of doing tremendous harm.

Further reading

Brocher, B., Kieny, M.P., Costy, F. et al. (1991) Large-scale eradication of rabies using recombinant vaccinia-rabies vaccine. *Nature*, 354, 520–522.

Broder, C.C. & Earl, P.L. (1999) Recombinant vaccinia viruses—design, generation, and isolation. *Molecular Biotechnology*, 13, 223–245.

Goeddel, D.V., Heyneker, H.L., Hozumi, T. et al. (1979) Direct expression in *Escherichia coli* of a DNA sequence coding for human growth hormone. *Nature*, 281, 544–548. [Production of recombinant somatotrophin.]

Goeddel, D.V., Kleid, D.G., Bolivar, F. et al. (1979) Expression in *Escherichia coli* of chemically synthesized genes for human insulin. *Proceedings of the National Academy of Sciences of the USA*, 76, 106–110.

Itakura, K., Hirose, T., Crea, R. et al. (1977) Expression in *Escherichia coli* of a chemically synthesized gene for the hormone somatostatin. *Science*, 198, 1056–1063.

Kaufman, R.J., Wasley, L.C. & Dorner, A.J. (1988) Synthesis, processing, and secretion of recombinant human factor VIII expressed in mammalian cells. *Journal of Biological Chemistry*, 263, 6352–6362.

Liu, M.A. (1998) Vaccine developments. *Nature Medicine*, 4, 515–519. [Describes the development of recombinant vaccines.]

Miki, Y., Swensen, J., Shattuck Eidens, D. et al. (1994) A strong candidate for the breast and ovarian cancer susceptibility gene *BRCA1*. *Science*, 266, 66–71.

Paleyanda, R.K., Velander, W.H., Lee, T.K. et al. (1997) Transgenic pigs produce functional human factor VIII in milk. *Nature Biotechnology*, 15, 971–975.

Schepelmann, S., Ogilvie, L.M., Hedley, D. et al. (2007) Suicide gene therapy of human colon carcinoma xenografts using an armed oncolytic adenovirus expressing carboxypeptidase G2. *Cancer Research*, 67, 4949–4955. [An example of suicide gene therapy using the telomerase promoter and an enzyme that converts a non-toxic pro-drug into the toxic form.]

Smith, K.R. (2003) Gene therapy: theoretical and bioethical concepts. *Archives of Medical Research*, 34, 247–268.

Tiwari, S., Verma, P.C., Sing, P.K. & Tuli, R. (2009) Plants as bioreactors for the production of vaccine antigens. *Biotechnology Advances*, 27, 449–467.

van Deutekom, J.C.T. & van Ommen, G.J.B. (2003) Advances in Duchenne muscular dystrophy gene therapy. *Nature Reviews Genetics*, 4, 774–783. [A example of the use of gene therapy.]

Chapter 15

Gene Cloning and DNA Analysis in Agriculture

Chapter contents

Agriculture, or more specifically the cultivation of plants, is the world's oldest biotechnology, with an unbroken history that stretches back at least 10,000 years. Throughout this period humans have constantly searched for improved varieties of their crop plants: varieties with better nutritional qualities, higher yields, or features that aid cultivation and harvesting. During the first few millennia, crop improvements occurred in a sporadic fashion, but in recent centuries new varieties have been obtained by breeding programs of ever increasing sophistication. However, the most sophisticated breeding program still retains an element of chance, dependent as it is on the random merging of parental characteristics in the hybrid offspring that are produced. The development of a new variety of crop plant, displaying a precise combination of desired characteristics, is a lengthy and difficult process.

Gene cloning provides a new dimension to crop breeding by enabling directed changes to be made to the genotype of a plant, circumventing the random processes inherent in conventional breeding. Two general strategies have been used:

- **Gene addition**, in which cloning is used to alter the characteristics of a plant by providing it with one or more new genes;
- **Gene subtraction**, in which genetic engineering techniques are used to inactivate one or more of the plant's existing genes.

A number of projects are being carried out around the world, many by biotechnology companies, aimed at exploiting the potential of gene addition and gene subtraction in crop improvement. In this chapter we will investigate a representative selection of

Gene Cloning and DNA Analysis: An Introduction. 6th edition. By T.A. Brown. Published 2010 by Blackwell Publishing.

these projects, and look at some of the problems that must be solved if plant genetic engineering is to gain widespread acceptance in agriculture.

15.1 The gene addition approach to plant genetic engineering

Gene addition involves the use of cloning techniques to introduce into a plant one or more new genes coding for a useful characteristic that the plant lacks. A good example of the technique is provided by the development of plants that resist insect attack by synthesizing insecticides coded by cloned genes.

15.1.1 Plants that make their own insecticides

Plants are subject to predation by virtually all other types of organism—viruses, bacteria, fungi, and animals—but in agricultural settings the greatest problems are caused by insects. To reduce losses, crops are regularly sprayed with insecticides. Most conventional insecticides (e.g., pyrethroids and organophosphates) are relatively non-specific poisons that kill a broad spectrum of insects, not just the ones eating the crop. Because of their high toxicity, several of these insecticides also have potentially harmful side effects for other members of the local biosphere, including in some cases humans. These problems are exacerbated by the need to apply conventional insecticides to the surfaces of plants by spraying, which means that subsequent movement of the chemicals in the ecosystem cannot be controlled. Furthermore, insects that live within the plant, or on the undersurfaces of leaves, can sometimes avoid the toxic effects altogether.

What features would be displayed by the ideal insecticide? Clearly it must be toxic to the insects against which it is targeted, but if possible this toxicity should be highly selective, so that the insecticide is harmless to other insects and is not poisonous to animals and to humans. The insecticide should be biodegradable, so that any residues that remain after the crop is harvested, or which are carried out of the field by rainwater, do not persist long enough to damage the environment. And it should be possible to apply the insecticide in such a way that all parts of the crop, not just the upper surfaces of the plants, are protected against insect attack.

The ideal insecticide has not yet been discovered. The closest we have are the δ-endotoxins produced by the soil bacterium *Bacillus thuringiensis*.

The δ-endotoxins of Bacillus thuringiensis

Insects not only eat plants: bacteria also form an occasional part of their diet. In response, several types of bacteria have evolved defense mechanisms against insect predation, an example being *B. thuringiensis* which, during sporulation, forms intracellular crystalline bodies that contain an insecticidal protein called the δ-endotoxin. The activated protein is highly poisonous to insects, some 80,000 times more toxic than organophosphate insecticides, and is relatively selective, different strains of the bacterium synthesizing proteins effective against the larvae of different groups of insects (Table 15.1).

The δ-endotoxin protein that accumulates in the bacterium is an inactive precursor. After ingestion by the insect, this protoxin is cleaved by proteases, resulting in shorter versions of the protein that display the toxic activity, by binding to the inside of the

Table 15.1

The range of insects poisoned by the various types of *B. thuringiensis* δ-endotoxins.

δ-ENDOTOXIN TYPE	EFFECTIVE AGAINST
CryI	Lepidoptera (moth and butterfly) larvae
CryII	Lepidoptera and Diptera (two-winged fly) larvae
CryIII	Lepidoptera larvae
CryIV	Diptera larvae
CryV	Nematode worms
CryVI	Nematode worms

Figure 15.1

Mode of action of a δ-endotoxin.

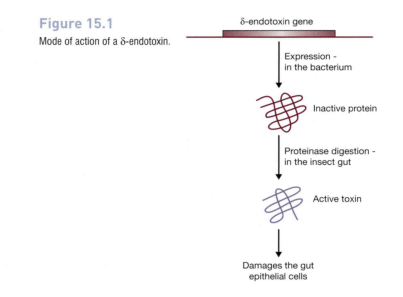

insect's gut and damaging the surface epithelium, so that the insect is unable to feed and consequently starves to death (Figure 15.1). Variation in the structure of these binding sites in different groups of insects is probably the underlying cause of the high specificities displayed by the different types of δ-endotoxin.

    B. thuringiensis toxins are not recent discoveries, the first patent for their use in crop protection having been granted in 1904. Over the years there have been several attempts to market them as environmentally friendly insecticides, but their biodegradability acts as a disadvantage because it means that they must be reapplied at regular intervals during the growing season, increasing the farmer's costs. Research has therefore been aimed at developing δ-endotoxins that do not require regular application. One approach is via protein engineering (p. 206), modifying the structure of the toxin so that it is more stable. A second approach is to engineer the crop to synthesize its own toxin.

Cloning a δ-endotoxin gene in maize

Maize is an example of a crop plant that is not served well by conventional insecticides. A major pest is the European corn borer (*Ostrinia nubilalis*), which tunnels into the plant from eggs laid on the undersurfaces of leaves, thereby evading the effects of insecticides applied by spraying. The first attempt at countering this pest by engineering

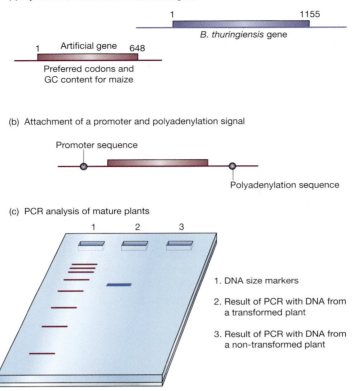

(a) Synthesis of an artificial δ-endotoxin gene

1   1155
B. thuringiensis gene

1   Artificial gene   648
Preferred codons and
GC content for maize

(b) Attachment of a promoter and polyadenylation signal

Promoter sequence

Polyadenylation sequence

(c) PCR analysis of mature plants

1   2   3

1. DNA size markers

2. Result of PCR with DNA from
   a transformed plant

3. Result of PCR with DNA from
   a non-transformed plant

Figure 15.2
Important steps in the procedure used to obtain genetically engineered maize plants expressing an artificial δ-endotoxin gene.

maize plants to synthesize δ-endotoxin was made by plant biotechnologists in 1993, working with the CryIA(b) version of the toxin. The CryIA(b) protein is 1155 amino acids in length, with the toxic activity residing in the segment from amino acids 29–607. Rather than isolating the natural gene, a shortened version containing the first 648 codons was made by artificial gene synthesis. This strategy enabled modifications to be introduced into the gene to improve its expression in maize plants. For example, the codons that were used in the artificial gene were those known to be preferred by maize, and the overall GC content of the gene was set at 65%, compared with the 38% GC content of the native bacterial version of the gene (Figure 15.2a). The artificial gene was ligated into a cassette vector (p. 232) between a promoter and polyadenylation signal from cauliflower mosaic virus (Figure 15.2b), and introduced into maize embryos by bombardment with DNA-coated microprojectiles (p. 85). The embryos were grown into mature plants, and transformants identified by PCR analysis of DNA extracts, using primers specific for a segment of the artificial gene (Figure 15.2c).

The next step was to use an immunological test to determine if δ-endotoxin was being synthesized by the transformed plants. The results showed that the artificial gene was indeed active, but that the amounts of δ-endotoxin being produced varied from plant to plant, from about 250–1750 ng of toxin per mg of total protein. These

Figure 15.3

Positional effects.

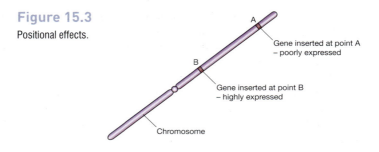

Gene inserted at point A
– poorly expressed

Gene inserted at point B
– highly expressed

Chromosome

differences were probably due to **positional effects**, the level of expression of a gene cloned in a plant (or animal) often being influenced by the exact location of the gene in the host chromosomes (Figure 15.3).

Were the transformed plants able to resist the attentions of the corn borers? This was assessed by field trials in which transformed and normal maize plants were artificially infested with larvae and the effects of predation measured over a period of 6 weeks. The criteria that were used were the amount of damage suffered by the foliage of the infested plants, and the lengths of the tunnels produced by the larvae boring into the plants. In both respects the transformed plants gave better results than the normal ones. In particular, the average length of the larval tunnels was reduced from 40.7 cm for the controls to just 6.3 cm for the engineered plants. In real terms, this is a very significant level of resistance.

Cloning δ-endotoxin genes in chloroplasts

One objection that has been raised to the use of GM crops is the possibility that the cloned gene might escape from the engineered plant and become established in a weed species. From a biological viewpoint, this is an unlikely scenario as the pollen produced by a plant is usually only able to fertilize the ovary of a plant of the same species, so transfer of a gene from a crop to a weed is highly unlikely. However, one way of making such transfer totally impossible would be to place the cloned gene not in the nucleus but in the plant's chloroplasts. A transgene located in the chloroplast genome cannot escape via pollen for the simple reason that pollen does not contain chloroplasts.

Synthesis of δ-endotoxin protein in transgenic chloroplasts has been achieved in an experimental system with tobacco. The CryIIA(a2) gene was used, which codes for a protein that has a broader toxicity spectrum than the CryIA toxins, killing the larvae of two-winged flies as well as lepidopterans (see Table 15.1). In the *B. thuringiensis* genome, the CryIIA(a2) gene is the third gene in a short operon, the first two genes of which code for proteins that help to fold and process the δ-endotoxin (Figure 15.4). One advantage of using chloroplasts as the sites of recombinant protein synthesis is that the gene expression machinery of chloroplasts, being related to that of bacteria (because chloroplasts were once free-living prokaryotes), is able to express all the genes in an operon. In contrast, each gene that is placed in a plant (or animal) nuclear genome must be cloned individually, with its own promoter and other expression signals, which makes it very difficult to introduce two or more genes at the same time.

Biolistics (p. 85) was used to introduce the CryIIA(a2) operon into tobacco leaf cells. Insertion into the chloroplast genome was ensured by attaching chloroplast DNA

Figure 15.4

The CryIIA(a2) operon.

Helper proteins   CryIIA (a2)

sequences to the operon (p. 119), and by rigorously selecting for the kanamycin resistance marker by placing leaf segments on agar containing kanamycin for up to 13 weeks. Transgenic shoots growing out of the leaf segments were then placed on a medium that induced root formation, and plants grown.

The amounts of CryIIA(a2) protein produced in the tissues of these GM plants was quite remarkable, the toxin making up over 45% of the total soluble protein, more than previously achieved in any plant cloning experiment. This high level of expression almost certainly results from the combined effects of the high copy number for the transgene (there being many chloroplast genomes per cell, compared with just two copies of the nuclear genome) and the presence in the chloroplasts of the two helper proteins coded by the other genes in the CryIIA(a2) operon. As might be anticipated, the plants proved to be extremely toxic to susceptible insect larvae. Five days after being placed on the GM plants, all cotton bollworm and beet armyworm larvae were dead, with appreciable damage being visible only on the leaves of the plants exposed to armyworms, which have a relatively high natural resistance to δ-endotoxins. The presence of large amounts of toxin in the leaf tissues appeared not to affect the plants themselves, the GM tobacco being undistinguishable from non-GM plants when factors such as growth rates, chlorophyll content, and level of photosynthesis were considered. Attempts to repeat this experiment with maize, cotton, and other more useful crops have been hampered by the difficulties in achieving chloroplast transformation with plants other than tobacco (see p. 119).

Countering insect resistance to δ-endotoxin crops

It has long been recognized that crops synthesizing δ-endotoxins might become ineffective after a few seasons due to the build-up of resistance among the insect populations feeding on the crops. This would be a natural consequence of exposing these populations to high amounts of toxins and, of course, could render the GM plants no better than the non-GM versions after a just a few years. Various strategies have been proposed to prevent the development of δ-endotoxin resistant insects. One of the first to be suggested was to develop crops expressing both the CryI and CryII genes, the rationale being that as these toxins are quite different it would be difficult for an insect population to develop resistance to both types (Figure 15.5a). Whether or not this is a sound argument is not yet clear. Most examples of δ-endotoxin resistance that have been documented have not been broad spectrum: for example, the CryIIA(a2) tobacco plants described above were equally poisonous to cotton budworms that were or were not resistant to CryIA(b). However, some strains of meal moth larvae exposed to plants containing the CryIA(c) toxin have acquired a resistance that also provides protection against the CryII toxins. In any case, it would be risky to base a counter-resistance strategy on assumed limitations to the genetic potential of the insect pests.

An alternative might be to engineer toxin production in such a way that synthesis occurs only in those parts of the plant that need protection. For example, in a crop such as maize, some damage to the non-fruiting parts of the plant could be tolerated if this did not affect the production of cobs (Figure 15.5b). If expression of the toxin only occurred late in the plant life cycle, when the cobs are developing, then overall exposure of the insects to the toxin might be reduced without any decrease in the value of the crop. However, this strategy might delay the onset of resistance, but it is unlikely to avoid it altogether.

A third strategy is to mix GM plants with non-GM ones, so that each field contains plants that the insects can feed on without being exposed to the toxin produced by the

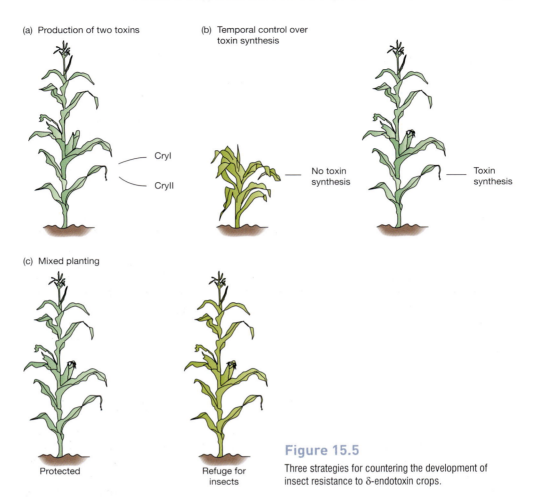

(a) Production of two toxins

CryI

CryII

(b) Temporal control over toxin synthesis

No toxin synthesis

Toxin synthesis

(c) Mixed planting

Protected

Refuge for insects

Figure 15.5

Three strategies for countering the development of insect resistance to δ-endotoxin crops.

engineered versions (Figure 15.5c). These non-GM plants would act as a refuge for the insects, ensuring that the insect population continually includes a high proportion of non-resistant individuals. As all the δ-endotoxin resistance phenotypes so far encountered are recessive, heterozygotes arising from a mating between a susceptible insect and a resistant partner would themselves be susceptible, continually diluting the proportion of resistant insects in the population. Trials have been carried out, and theoretical models have been examined, to identify the most effective mixed planting strategies. In practice, success or failure would depend to a very large extent on the farmers who grow the crops, these farmers having to adhere to the precise planting strategy dictated by the scientists, despite the resulting loss in productivity due to the damage suffered by the non-GM plants. Again, this introduces an element of risk. The success of GM projects with plants clearly depends on much more that the cleverness of the genetic engineers.

15.1.2 Herbicide resistant crops

Although δ-endotoxin production has been engineered in crops as diverse as maize, cotton, rice, potato, and tomato, these plants are not the most widespread GM crops

Figure 15.6

Glyphosate competes with phosphoenol pyruvate in the EPSPS catalyzed synthesis of enolpyruvylshikimate-3-phosphate, and hence inhibits synthesis of tryptophan, tyrosine, and phenylalanine.

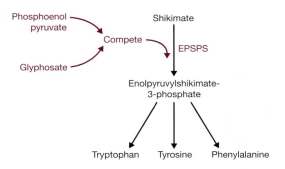

grown today. In commercial terms the most important transgenic plants are those that have been engineered to withstand the herbicide glyphosate. This herbicide, which is widely used by farmers and horticulturists, is environmentally friendly, as it is non-toxic to insects and to animals and has a short residence time in soils, breaking down over a period of a few days into harmless products. However, glyphosate kills all plants, both weeds and crop species, and so has to be applied to fields very carefully in order to prevent the growth of weeds without harming the crop itself. GM crops that are able to withstand the effects of glyphosate are therefore desirable as they would enable a less rigorous and hence less expensive herbicide application regime to be followed.

"Roundup Ready" crops

The first crops to be engineered for glyphosate resistance were produced by Monsanto Co. and called "Roundup Ready", reflecting the trade name of the herbicide. These plants contain modified genes for the enzyme enolpyruvylshikimate-3-phosphate synthase (EPSPS), which converts shikimate and phosphoenol pyruvate (PEP) into enolpyruvylshikimate-3-phosphate, an essential precursor for synthesis of the aromatic amino acids tryptophan, tyrosine, and phenylalanine (Figure 15.6). Glyphosate competes with PEP for binding to the enzyme surface, thereby inhibiting synthesis of enolpyruvylshikimate-3-phosphate and preventing the plant from making the three amino acids. Without these amino acids, the plant quickly dies.

Initially, genetic engineering was used to generate plants that made greater than normal amounts of EPSPS, in the expectation that these would be able to withstand higher doses of glyphosate than non-engineered plants. However, this approach was unsuccessful because, although engineered plants that made up to 80 times the normal amount of EPSPS were obtained, the resulting increase in glyphosate tolerance was not sufficient to protect these plants from herbicide application in the field.

A search was therefore carried out for an organism whose EPSPS enzyme is resistant to glyphosate inhibition and whose EPSPS gene might therefore be used to confer resistance on a crop plant. After testing the genes from various bacteria, as well as mutant forms of *Petunia* that displayed glyphosate resistance, the EPSPS gene from *Agrobacterium* strain CP4 was chosen, because of its combination of high catalytic activity and high resistance to the herbicide. EPSPS is located in the plant chloroplasts, so the *Agrobacterium* EPSPS gene was cloned in a Ti vector as a fusion protein with a leader sequence that would direct the enzyme across the chloroplast membrane and into the organelle. Biolistics was used to introduce the recombinant vector into soybean callus culture. After regeneration, the GM plants were found to have a threefold increase in herbicide resistance.

(a) Detoxification of glyphosate by GAT

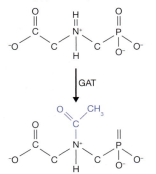

(b) Directed evolution to produce a highly active GAT enzyme

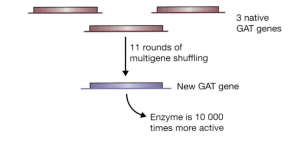

Figure 15.7

Use of glyphosate *N*-acetyltransferase to generate plants that detoxify glyphosate. (a) GAT detoxifies glyphosate by adding an acetyl group (shown in blue). (b) Creation of a highly active GAT enzyme by multigene shuffling.

A new generation of glyphosate resistant crops

Roundup Ready versions of a variety of crops have been produced in recent years, and several of these, in particular soybean and maize, are grown routinely in the USA and other parts of the world. However, these plants do not actually destroy glyphosate, which means that the herbicide can accumulate in the plant tissues. Glyphosate is not poisonous to humans or other animals, so the use of such plants as food or forage should not be a concern, but accumulation of the herbicide can interfere with reproduction of the plant. The degree of resistance displayed by Roundup Ready crops has also been found to be too low to provide a major economic benefit with some crops, notably wheat.

Until recently, there had been only a few scattered reports of organisms capable of actively degrading glyphosate. However, searches of microbial collections have revealed that this property is relatively common among bacteria of the genus *Bacillus*, which possess an enzyme, now called glyphosate *N*-acetyltransferase (GAT), which detoxifies glyphosate by attaching an acetyl group to the herbicide molecule (Figure 15.7a). The most active detoxifier known is a strain of *B. licheniformis*, but even this bacterium detoxifies glyphosate at rates that are too low to be of value if transferred to a GM crop.

Is it possible to increase the activity of the GAT synthesized by *B. licheniformis*? The discovery that the bacterium possesses three related genes for this enzyme pointed a way forward. A type of **directed evolution** called **multigene shuffling** was used. Multigene shuffling involves taking parts of each member of a multigene family and reassembling these parts to create new gene variants. At each stage of the process, the most active genes are identified by cloning all variants in *E. coli* and assaying the recombinant colonies for GAT activity. The most active genes are then used as the substrates for the next round of shuffling. After 11 rounds, a gene specifying a GAT with 10,000 times the activity of the enzymes present in the original *B. licheniformis* strain was obtained (Figure 15.7b). This gene was introduced into maize, and the resulting GM plants were found to tolerate levels of glyphosate six times higher than the amount normally used by farmers to control weeds, without any reduction in the productivity of the plant. This new way of engineering glyphosate resistance is currently being examined in greater detail to determine if it presents a real alternative to Roundup Ready crops.

Table 15.2

Examples of gene addition projects with plants.

GENE FOR	SOURCE ORGANISM	CHARACTERISTIC CONFERRED ON MODIFIED PLANTS
δ-Endotoxin	*B. thuringiensis*	Insect resistance
Proteinase inhibitors	Various legumes	Insect resistance
Chitinase	Rice	Fungal resistance
Glucanase	Alfalfa	Fungal resistance
Ribosome-inactivating protein	Barley	Fungal resistance
Ornithine carbamyltransferase	*Pseudomonas syringae*	Bacterial resistance
RNA polymerase, helicase	Potato leafroll luteovirus	Virus resistance
Satellite RNAs	Various viruses	Virus resistance
Virus coat proteins	Various viruses	Virus resistance
2′–5′-Oligoadenylate synthetase	Rat	Virus resistance
Acetolactate synthase	*Nicotiana tabacum*	Herbicide resistance
Enolpyruvylshikimate-3-phosphate synthase	*Agrobacterium* spp.	Herbicide resistance
Glyphosate oxidoreductase	*Ochrobactrum anthropi*	Herbicide resistance
Glyphosate *N*-acetyltransferase	*B. licheniformis*	Herbicide resistance
Nitrilase	*Klebsiella ozaenae*	Herbicide resistance
Phosphinothricin acetyltransferase	*Streptomyces* spp.	Herbicide resistance
Phosphatidylinositol-specific phospholipase C	Maize	Drought tolerance
Barnase ribonuclease inhibitor	*Bacillus amyloliquefaciens*	Male sterility
DNA adenine methylase	*E. coli*	Male sterility
Methionine-rich protein	Brazil nuts	Improved sulphur content
1-Aminocyclopropane-1-carboxylic acid deaminase	Various	Modified fruit ripening
S-Adenosylmethionine hydrolase	Bacteriophage T3	Modified fruit ripening
Monellin	*Thaumatococcus danielli*	Sweetness
Thaumatin	*T. danielli*	Sweetness
Acyl carrier protein thioesterase	*Umbellularia californica*	Modified fat/oil content
Delta-12 desaturase	*Glycine max*	Modified fat/oil content
Dihydroflavanol reductase	Various flowering plants	Modified flower color
Flavonoid hydroxylase	Various flowering plants	Modified flower color

15.1.3 Other gene addition projects

GM crops that synthesize δ-endotoxins or glyphosate resistance enzymes are by no means the only examples of plants engineered by gene addition. Examples of other gene addition projects are listed in Table 15.2. These projects include an alternative means of conferring insect resistance, using genes coding for proteinase inhibitors, small polypeptides that disrupt the activities of enzymes in the insect gut, preventing or slowing growth. Proteinase inhibitors are produced naturally by several types of plant, notably legumes such as cowpeas and common beans, and their genes have been successfully transferred to other crops which do not normally make significant amounts of these proteins. The inhibitors are particularly effective against beetle larvae that feed on seeds, and so may be a better alternative than δ-endotoxin for plants whose seeds are stored for long periods. Other projects are exploring the use of genetic modification to improve the nutritional quality of crop plants, for example by increasing the content of essential amino acids or by changing the plant biochemistry so that more of the available nutrients can be utilized during digestion by humans or animals. Finally, in a different sphere of commercial activity, ornamental plants with unusual flower colors are being produced by transferring genes for enzymes involved in pigment production from one species to another.

15.2 Gene subtraction

The second way of changing the genotype of a plant is by gene subtraction. This term is a misnomer, as the modification does not involve the actual removal of a gene, merely its inactivation. There are several possible strategies for inactivating a single, chosen gene in a living plant, the most successful so far in practical terms being the use of antisense RNA (p. 260).

15.2.1 Antisense RNA and the engineering of fruit ripening in tomato

To illustrate how antisense RNA has been used in plant genetic engineering, we will examine how tomatoes with delayed ripening have been produced. This is an important example of plant genetic modification as it resulted in one of the first GM foodstuffs to be approved for sale to the general public.

Commercially grown tomatoes and other soft fruits are usually picked before they are completely ripe, to allow time for the fruits to be transported to the marketplace before they begin to spoil. This is essential if the process is to be economically viable, but there is a problem in that most immature fruits do not develop their full flavor if they are removed from the plant before they are fully ripe. The result is that mass-produced tomatoes often have a bland taste, which makes them less attractive to the consumer. Antisense technology has been used in two ways to genetically engineer tomato plants so that the fruit ripening process is slowed down. This enables the grower to leave the fruits on the plant until they ripen to the stage where the flavor has fully developed, there still being time to transport and market the crop before spoilage sets in.

Using antisense RNA to inactivate the polygalacturonase gene
The timescale for development of a fruit is measured as the number of days or weeks after flowering. In tomato, this process takes approximately eight weeks from start to finish, with the color and flavor changes associated with ripening beginning after about six weeks. At this time a number of genes involved in the later stages of ripening are switched on, including one coding for the polygalacturonase enzyme (Figure 15.8). This enzyme slowly breaks down the polygalacturonic acid component of the cell walls in the fruit pericarp, resulting in a gradual softening. The softening makes the fruit palatable, but if taken too far results in a squashy, spoilt tomato, attractive only to students with limited financial resources.

Partial inactivation of the polygalacturonase gene should increase the time between flavor development and spoilage of the fruit. To test this hypothesis, a 730 bp restriction fragment was obtained from the 5′ region of the normal polygalacturonase gene, representing just under half of the coding sequence (Figure 15.9). The orientation of the fragment was reversed, a cauliflower mosaic virus promoter was ligated to the start of the sequence, and a plant polyadenylation signal attached to the end. The construction was then inserted into the Ti plasmid vector pBIN19 (p. 116). Once inside the plant, transcription from the cauliflower mosaic virus promoter should result in synthesis of an antisense RNA complementary to the first half of the polygalacturonase mRNA.

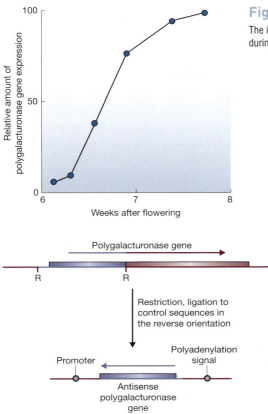

Figure 15.8

The increase in polygalacturonase gene expression seen during the later stages of tomato fruit ripening.

Figure 15.9

Construction of an antisense polygalacturonase "gene". R = restriction site.

Previous experiments with antisense RNA had suggested that this would be sufficient to reduce or even prevent translation of the target mRNA.

Transformation was carried out by introducing the recombinant pBIN19 molecules into *Agrobacterium tumefaciens* bacteria and then allowing the bacteria to infect tomato stem segments. Small amounts of callus material collected from the surfaces of these segments were tested for their ability to grow on an agar medium containing kanamycin (remember that pBIN19 carries a gene for kanamycin resistance; see Figure 7.14). Resistant transformants were identified and allowed to develop into mature plants.

The effect of antisense RNA synthesis on the amount of polygalacturonase mRNA in the cells of ripening fruit was determined by northern hybridization with a single-stranded DNA probe specific for the sense mRNA. These experiments showed that ripening fruit from transformed plants contained less polygalacturonase mRNA than the fruits from normal plants. The amounts of polygalacturonase enzyme produced in the ripening fruits of transformed plants were then estimated from the intensities of the relevant bands after separation of fruit proteins by polyacrylamide gel electrophoresis, and by directly measuring the enzyme activities in the fruits. The results showed that less enzyme was synthesized in transformed fruits (Figure 15.10). Most importantly, the transformed fruits, although undergoing a gradual softening, could be stored for a prolonged period before beginning to spoil. This indicated that the antisense RNA had not completely inactivated the polygalacturonase gene, but had nonetheless produced a sufficient reduction in gene expression to delay the ripening process as desired. The

Figure 15.10

The differences in polygalacturonase activity in normal tomato fruits and in fruits expressing the antisense polygalacturonase gene.

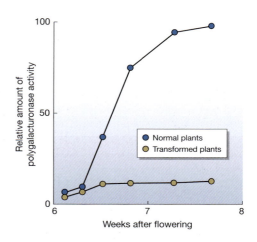

GM tomatoes—marketed under the trade name "FlavrSavr"—were one of the first genetically engineered plants to be approved for sale to the public, first appearing in supermarkets in 1994.

Using antisense RNA to inactivate ethylene synthesis

The main trigger that switches on the genes involved in the later stages of tomato ripening is ethylene which, despite being a gas, acts as a hormone in many plants. A second way of delaying fruit ripening would therefore be to engineer plants so that they do not synthesize ethylene. Fruits on these plants would develop as normal for the first six weeks, but would be unable to complete the ripening process. The unripe fruit could therefore be transported to the marketplace without any danger of the crop spoiling. Before selling to the consumer, or conversion into paste or some other product, artificial ripening would be induced by spraying the tomatoes with ethylene.

The penultimate step in the ethylene synthesis pathway is conversion of S-adenosyl-methionine to 1-aminocyclopropane-1-carboxylic acid (ACC), which is the immediate precursor for ethylene. This step is catalyzed by an enzyme called ACC synthase. As with polygalacturonase, ACC synthase inactivation was achieved by cloning into tomato a truncated version of the normal ACC synthase gene, inserted into the cloning vector in the reverse orientation, so that the construct would direct synthesis of an antisense version of the ACC synthase mRNA. After regeneration, the engineered plants were grown to the fruiting stage and found to make only 2% of the amount of ethylene produced by non-engineered plants. This reduction was more than sufficient to prevent the fruit from completing the ripening process. These tomatoes have been marketed as the "Endless Summer" variety.

15.2.2 Other examples of the use of antisense RNA in plant genetic engineering

In general terms, the applications of gene subtraction in plant genetic engineering are probably less broad than those of gene addition. It is easier to think of useful characteristics that a plant lacks and which might be introduced by gene addition, than it is to identify disadvantageous traits that the plant already possesses and which could be removed by gene subtraction. There are, however, a growing number of plant

Table 15.3

Examples of gene subtraction projects with plants.

TARGET GENE	MODIFIED CHARACTERISTIC
Polygalacturonase	Delay of fruit spoilage in tomato
1-Aminocyclopropane-1-carboxylic acid synthase	Modified fruit ripening in tomato
Polyphenol oxidase	Prevention of discoloration in fruits and vegetables
Starch synthase	Reduction of starch content in vegetables
Delta-12 desaturase	High oleic acid content in soybean
Chalcone synthase	Modification of flower color in various decorative plants
1D-*myo*-inositol 3-phosphate synthase	Reduction of indigestible phosphorus content of rice grains

biotechnology projects based on gene subtraction (Table 15.3), and the approach is likely to increase in importance as the uncertainties that surround the underlying principles of antisense technology are gradually resolved.

15.3 Problems with genetically modified plants

Ripening-delayed tomatoes produced by gene subtraction were among the first genetically modified whole foods to be approved for marketing. Partly because of this, plant genetic engineering has provided the battleground on which biotechnologists and other interested parties have fought over the safety and ethical issues that arise from our ability to alter the genetic make-up of living organisms. A number of the most important questions do not directly concern genes and the expertise needed to answer them will not be found in this book. For example, we cannot discuss in an authoritative fashion the possible impact, good or otherwise, that GM crops might have on local farming practices in the developing world. However, we can, and should, look at the biological issues.

15.3.1 Safety concerns with selectable markers

One of the main areas of concern to emerge from the debate over genetically modified tomatoes is the possible harmful effects of the marker genes used with plant cloning vectors. Most plant vectors carry a copy of a gene for kanamycin resistance, enabling transformed plants to be identified during the cloning process. The *kan*^{R} gene, also called *nptII*, is bacterial in origin and codes for the enzyme neomycin phosphotransferase II. This gene and its enzyme product are present in all cells of an engineered plant. The fear that neomycin phosphotransferase might be toxic to humans has been allayed by tests with animal models, but two other safety issues remain:

- Could the *kan*^{R} gene contained in a genetically modified foodstuff be passed to bacteria in the human gut, making these resistant to kanamycin and related antibiotics?
- Could the *kan*^{R} gene be passed to other organisms in the environment, and would this result in damage to the ecosystem?

Neither question can be fully answered with our current knowledge. It can be argued that digestive processes would destroy all the *kan*^{R} genes in a genetically modified food before they could reach the bacterial flora of the gut, and that, even if a gene did avoid

Figure 15.11

DNA excision by the Cre recombinase enzyme.

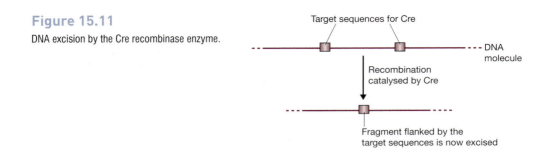

destruction, the chances of it being transferred to a bacterium would be very small. Nevertheless, the risk factor is not zero. Similarly, although experiments suggest that growth of genetically modified plants would have a negligible effect on the environment, as kan^R genes are already common in natural ecosystems, the future occurrence of some unforeseen and damaging event cannot be considered an absolute impossibility.

The fears surrounding the use of kan^R and other marker genes have prompted biotechnologists to devise ways of removing these genes from plant DNA after the transformation event has been verified. One of the strategies makes use of an enzyme from bacteriophage P1, called Cre, which catalyzes a recombination event that excises DNA fragments flanked by specific 34 bp recognition sequences (Figure 15.11). To use this system the plant is transformed with two cloning vectors, the first carrying the gene being added to the plant along with its kan^R selectable marker gene surrounded by the Cre target sequences, and the second carrying the Cre gene. After transformation, expression of the Cre gene results in excision of the kan^R gene from the plant DNA.

What if the Cre gene is itself hazardous in some way? This is immaterial as the two vectors used in the transformation would probably integrate their DNA fragments into different chromosomes, so random segregation during sexual reproduction would result in first generation plants that contained one integrated fragment but not the other. A plant that contains neither the Cre gene nor the kan^R selectable marker, but does contain the important gene that we wished to add to the plant's genome, can therefore be obtained.

15.3.2 The terminator technology

The Cre recombination system also underlies one of the most controversial aspects of plant genetic engineering, the so-called terminator technology. This is one of the processes by which the companies who market GM crops attempt to protect their financial investment by ensuring that farmers must buy new seed every year, rather than simply collecting seed from the crop and sowing this second generation seed the following year. In reality, even with conventional crops, mechanisms have been devised to ensure that second generation seed cannot be grown by farmers, but the general controversies surrounding GM crops have placed the terminator technology in the public eye.

The terminator technology centers on the gene for ribosome inactivating protein (RIP). The ribosome inactivating protein blocks protein synthesis by cutting one of the ribosomal RNA molecules into two segments (Figure 15.12a). Any cell in which the ribosome inactivating protein is active will quickly die. In GM plants that utilize the terminator system, the RIP gene is placed under control of a promoter that is active only during embryo development. The plants therefore grow normally but the seeds that they produce are sterile.

(a) The RIP gene

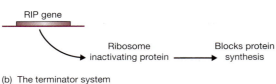

(b) The terminator system

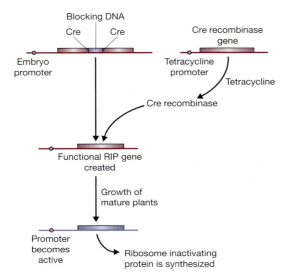

Figure 15.12

The terminator technology. (a) The RIP gene codes for a protein that blocks protein synthesis. (b) The system that is used to allow first generation seeds to be produced.

How are the first generation seeds, those sold to farmers, obtained? To begin with, the RIP gene is non-functional because it is disrupted by a segment of non-RIP DNA (Figure 15.12b). However, this DNA is flanked by the 34 bp recognition sequences for the Cre recombinase. In these plants the gene for the Cre recombinase is placed under control of a promoter that is switched on by tetracycline. Once seeds have been obtained, the supplier activates the Cre recombinase by placing the seeds in a tetracycline solution. This removes the blocking DNA from the RIP gene, which becomes functional but remains silent until its own promoter becomes active during embryogenesis.

15.3.3 The possibility of harmful effects on the environment

A second area of concern regarding genetically modified plants is that their new gene combinations might harm the environment in some way. These concerns have to be addressed individually for each type of GM crop, as different engineered genes might have different impacts. We will examine the work that has been carried out to assess whether it is possible that herbicide resistant plants, one of the two examples of gene addition that we studied earlier in this chapter, can have a harmful effect. As these are the most widely grown GM crops, they have been subject to some of the most comprehensive environmental studies. In particular, in 1999, the UK Government commissioned an independent investigation into how herbicide resistant crops, whose growth in the UK was not at that time permitted, might affect the abundance and diversity of farmland wildlife.

After delays due to activists attempting to prevent the work from being carried out, the UK research team reported their findings in 2003. The study involved 273 field trials throughout England, Wales, and Scotland, and included glyphosate resistant sugar beet as well as maize and spring rape engineered for resistance to a second herbicide, glufosinate-ammonium. The results, as summarized in the official report (see Burke (2003) in *Further Reading*), were as follows:

> The team found that there were differences in the abundance of wildlife between GM crop fields and conventional crop fields. Growing conventional beet and spring rape was better for many groups of wildlife than growing GM beet and spring rape. There were more insects, such as butterflies and bees, in and around the conventional crops because there were more weeds to provide food and cover. There were also more weed seeds in conventional beet and spring rape crops than in their GM counterparts. Such seeds are important in the diets of some animals, particularly some birds. In contrast, growing GM maize was better for many groups of wildlife than conventional maize. There were more weeds in and around the GM crops, more butterflies and bees around at certain times of the year, and more weed seeds. The researchers stress that the differences they found do not arise just because the crops have been genetically modified. They arise because these GM crops give farmers new options for weed control. That is, they use different herbicides and apply them differently. The results of this study suggest that growing such GM crops could have implications for wider farmland biodiversity. However, other issues will affect the medium- and long-term impacts, such as the areas and distribution of land involved, how the land is cultivated and how crop rotations are managed. These make it hard for researchers to predict the medium- and large-scale effects of GM cropping with any certainty. In addition, other management decisions taken by farmers growing conventional crops will continue to impact on wildlife.

Further reading

Burke, M. (2003) GM crops: effects on farmland wildlife. Produced by the Farmscale Evaluations Research Team and the Scientific Steering Committee. ISBN: 0-85521-035-4. [For a more detailed description of this work see *Philosophical Transactions of the Royal Society, Biological Sciences*, 358, 1775–1889 (2003).]

Castle, L.A., Siehl, D.L. & Gorton, R. (2004) Discovery and directed evolution of a glyphosate tolerance gene. *Science*, 304, 1151–1154. [Cloning the GAT gene in maize.]

De Cosa, B., Moar, W., Lee, S.B. et al. (2001) Overexpression of the Bt *cry2Aa2* operon in chloroplasts leads to formation of insecticidal crystals. *Nature Biotechnology*, 19, 71–74.

Feitelson, J.S., Payne, J. & Kim, L. (1992) *Bacillus thuringiensis*: insects and beyond. *Biotechnology*, 10, 271–275. [Details of δ-endotoxins and their potential as conventional insecticides and in genetic engineering.]

Fischhoff, D.A., Bowdish, K.S., Perlak, F.J. et al. (1987) Insect-tolerant transgenic tomato plants. *Biotechnology*, 5, 807–813. [The first transfer of a δ-endotoxin gene to a plant.]

Groot, A.T. & Dicke, M. (2002) Insect-resistant transgenic plants in a multi-trophic context. *Plant Journal*, 31, 387–406. [Examines the impact of δ-endotoxin plants on the ecological food chain.]

Koziel, M.G., Beland, G.L., Bowman, C. et al. (1993) Field performance of elite transgenic maize plants expressing an insecticidal protein derived from *Bacillus thuringiensis*. *Biotechnology*, 11, 194–200. [Cloning a δ-endotoxin gene in maize.]

Matas, A.J., Gapper, N.E., Chung, M.-Y., Giovannoni, J.J. & Rose, J.K.C. (2009) Biology and genetic engineering of fruit maturation for enhanced quality and shelf-life. *Current Opinion in Biotechnology*, 20, 197–203.

Miki, B. & McHugh, S. (2003) Selectable marker genes in transgenic plants: applications, alternatives and biosafety. *Journal of Biotechnology*, 107, 193–232.

Shade, R.E., Schroeder, H.E., Pueyo, J.J. et al. (1994) Transgenic pea seeds expressing the α-amylase inhibitor of the common bean are resistant to bruchid beetles. *Biotechnology*, 12, 793–796. [A second approach to insect-resistant plants.]

Shelton, A.M., Zhao, J.Z. & Roush, R.T. (2002) Economic, ecological, food safety, and social consequences of the deployment of Bt transgenic plants. *Annual Review of Entomology*, 47, 845–881. [Various issues relating to δ-endotoxin plants.]

Smith, C.J.S., Watson, C.F., Ray, J. et al. (1988) Antisense RNA inhibition of polygalacturonase gene expression in transgenic tomatoes. *Nature*, 334, 724–726.

Tabashnik, B.E., Gassmann, A.J., Crowder, D.W. & Carriére, Y. (2008) Insect resistance to *Bt* crops: evidence versus theory. *Nature Biotechnology*, 26, 199–202. [Discusses the effectiveness of refuges in reducing insect resistance to δ-endotoxin crops.]

Chapter 16

Gene Cloning and DNA Analysis in Forensic Science and Archaeology

Forensic science is the final area of biotechnology that we will consider. Hardly a week goes by without a report in the national press of another high profile crime that has been solved thanks to DNA analysis. The applications of molecular biology in forensics center largely on the ability of DNA analysis to identify an individual from hairs, bloodstains, and other items recovered from the crime scene. In the popular media, these techniques are called genetic fingerprinting, though the more accurate term for the procedures used today is DNA profiling. We begin this chapter by examining the methods used in genetic fingerprinting and DNA profiling, including their use both in identification of individuals and in establishing if individuals are members of a single family. This will lead us into an exploration of the ways in which genetic techniques are being used in archaeology.

Gene Cloning and DNA Analysis: An Introduction. 6th edition. By T.A. Brown. Published 2010 by Blackwell Publishing.

16.1 DNA analysis in the identification of crime suspects

It is probably impossible for a person to commit a crime without leaving behind a trace of his or her DNA. Hairs, spots of blood, and even conventional fingerprints contain traces of DNA, enough to be studied by the polymerase chain reaction (PCR). The analysis does not have to be done immediately, and in recent years a number of past crimes—so-called "cold cases"—have been solved and the criminal brought to justice because of DNA testing that has been carried out on archived material. So how do these powerful methods work?

The basis to genetic fingerprinting and DNA profiling is that identical twins are the only individuals who have identical copies of the human genome. Of course, the human genome is more or less the same in everybody—the same genes will be in the same order with the same stretches of intergenic DNA between them. But the human genome, as well as those of other organisms, contains many polymorphisms, positions where the nucleotide sequence is not the same in every member of the population. We have already encountered the most important of these polymorphic sites earlier, because these variable sequences are the same ones that are used as DNA markers in genome mapping (p. 180). They include restriction fragment length polymorphisms (RFLPs), short tandem repeats (STRs), and single nucleotide polymorphisms (SNPs). All three can occur within genes as well as in intergenic regions, and altogether there are several million of these polymorphic sites in the human genome, with SNPs being the most common.

16.1.1 Genetic fingerprinting by hybridization probing

The first method for using DNA analysis to identify individuals was developed in the mid-1980s by Sir Alec Jeffreys of Leicester University. This technique was not based on any of the types of polymorphic site listed above, but on a different kind of variation in the human genome called a hypervariable dispersed repetitive sequence. As the name indicates, this is a repeated sequence that occurs at various places ("dispersed") in the human genome. The key feature of these sequences is that their genomic positions are variable: they are located at different positions in the genomes of different people (Figure 16.1a).

The particular repeat that was initially used in genetic fingerprinting contains the sequence GGGCAGGANG (where N is any nucleotide). To prepare a fingerprint, a sample of DNA is digested with a restriction endonuclease, the fragments separated by agarose gel electrophoresis, and a Southern blot prepared (p. 142). Hybridization to the blot of a labeled probe containing the repeat sequence reveals a series of bands, each one representing a restriction fragment that contains the repeat (Figure 16.1b). Because the insertion sites of the repeat sequence are variable, the same procedure carried out with a DNA sample from a second person will give a different pattern of bands. These are the genetic fingerprints for those individuals.

16.1.2 DNA profiling by PCR of short tandem repeats

Strictly speaking, genetic fingerprinting refers only to hybridization analysis of dispersed repeat sequences. This technique has been valuable in forensic work but suffers from three limitations:

Figure 16.1

Genetic fingerprinting. (a) The positions of polymorphic repeats, such as hypervariable dispersed repetitive sequences, in the genomes of two individuals. In the chromosome segments shown, the second person has an additional repeat sequence. (b) An autoradiograph showing the genetic fingerprints of two individuals.

(a) Polymorphic repeat sequences in the human genome

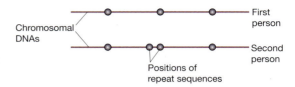

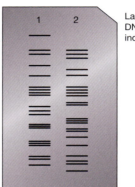

(b) Two genetic fingerprints

Lanes 1 and 2: DNA from two individuals

- A relatively large amount of DNA is needed because the technique depends on hybridization analysis. Fingerprinting cannot be used with the minute amounts of DNA in hair and bloodstains.
- Interpretation of the fingerprint can be difficult because of variations in the intensities of the hybridization signals. In a court of law, minor differences in band intensity between a test fingerprint and that of a suspect can be sufficient for the suspect to be acquitted.
- Although the insertion sites of the repeat sequences are hypervariable, there is a limit to this variability and therefore a small chance that two unrelated individuals could have the same, or at least very similar, fingerprints. Again, this consideration can lead to acquittal when a case is brought to court.

The more powerful technique of DNA profiling avoids these problems. Profiling makes use of the polymorphic sequences called STRs. As described on p. 181, an STR is a short sequence, 1–13 nucleotides in length, which is repeated several times in a tandem array. In the human genome, the most common type of STR is the dinucleotide repeat $[CA]_n$, where "n", the number of repeats, is usually between 5 and 20 (Figure 16.2a).

The number of repeats in a particular STR is variable because repeats can be added or, less frequently, removed by errors that occur during DNA replication. In the population as a whole, there might be as many as ten different versions of a particular STR, each of the alleles characterized by a different number of repeats. In DNA profiling, the alleles of a selected number of different STRs are identified. This can be achieved quickly and with very small amounts of DNA by PCRs with primers that anneal to the DNA sequences either side of a repeat (see Figure 10.19). After the PCR, the products can be examined by agarose gel electrophoresis. The size of the band or bands that are seen in the gel indicate the allele or alleles present in the DNA sample that has been tested (Figure 16.2b). Two alleles of an STR can be present in a single DNA sample because

(a) Two alleles of an STR

....CACACACACA.... $n = 5$

....CACACACACACA.... $n = 6$

(b) The results of PCR

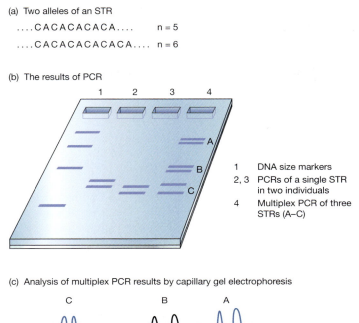

1   DNA size markers

2, 3   PCRs of a single STR
        in two individuals

4   Multiplex PCR of three
        STRs (A–C)

(c) Analysis of multiplex PCR results by capillary gel electrophoresis

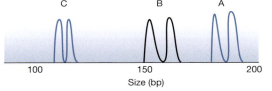

Size (bp)

Figure 16.2

DNA profiling. (a) DNA profiling makes use of STRs which have variable repeat units. (b) A gel obtained after DNA profiling. In lanes 2 and 3 the same STR has been examined in two individuals. These two people have different profiles, but have a band in common. Lane 4 shows the result of a multiplex PCR in which three STRs have been typed in a single PCR. (c) Capillary gel electrophoresis can be used to determine the sizes of multiplex PCR products.

there are two copies of each STR, one on the chromosome inherited from the mother and one on the chromosome from the father.

Because PCR is used, DNA profiling is very sensitive and enables results to be obtained with hairs and other specimens that contain trace amounts of DNA. The results are unambiguous, and a match between DNA profiles is usually accepted as evidence in a trial. The current methodology, called CODIS (Combined DNA Index System), makes use of 12 STRs with sufficient variability to give only a one in 10^{15} chance that two individuals, other than identical twins, have the same profile. As the world population is around 6×10^9, the statistical likelihood of two individuals on the planet sharing the same profile is so low as to be considered implausible when DNA evidence is presented in a court of law. Each STR is typed by PCRs with primers that are fluorescently labeled and which anneal either side of the variable repeat region. The alleles present at the STR are then typed by determining the sizes of the amplicons by capillary gel electrophoresis. Two or more STRs can be typed together in a **multiplex PCR** if their product sizes do not overlap, or if the individual primer pairs are labeled with different fluorescent markers, enabling the products to be distinguished in the capillary gel (Figure 16.2c).

Figure 16.3

Inheritance of STR alleles within a family.

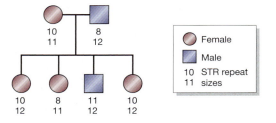

16.2 Studying kinship by DNA profiling

As well as identification of criminals, DNA profiling can also be used to infer if two or more individuals are members of the same family. This type of study is called **kinship analysis** and its main day-to-day application is in paternity testing.

16.2.1 Related individuals have similar DNA profiles

Your DNA profile, like all other aspects of your genome, is inherited partly from your mother and partly from your father. Relationships within a family therefore become apparent when the alleles of a particular STR are marked on the family pedigree (Figure 16.3). In this example, we see that 3 of the 4 children have inherited the 12-repeat allele from the father. This observation in itself is not sufficient to deduce that these three children are siblings, though the statistical chance would be quite high if the 12-repeat allele was uncommon in the population as a whole. To increase the degree of certainty, more STRs would have to be typed but, as with identification of individuals, the analysis need not be endless, because a comparison of 12 STRs gives an acceptable probability that relationships that are observed are real.

16.2.2 DNA profiling and the remains of the Romanovs

An interesting example of the use of DNA profiling in a kinship study is provided by work carried out during the 1990s on the bones of the Romanovs, the last members of the Russian ruling family. The Romanovs and their descendents ruled Russia from the early 17th century until the time of the Russian Revolution, when Tsar Nicholas II was deposed and he and his wife, the Tsarina Alexandra, and their five children imprisoned. On 17 July 1918, all seven, along with their doctor and three servants, were murdered and their bodies disposed of in a shallow roadside grave near Yekaterinburg in the Urals. In 1991, after the fall of communism, the remains were recovered with the intention that they should be given a more fitting burial.

STR analysis of the Romanov bones

Although it was suspected that the bones recovered were indeed those of the Romanovs, the possibility that they belonged to some other unfortunate group of people could not be discounted. Nine skeletons had been found in the grave—six adults and three children—examination of the bones suggesting that four of the adults were male and two female, and the three children were all female. If these were the remains of the Romanovs, then their son Alexei and one their daughters were, for some, reason, absent. The bodies showed signs of violence, consistent with reports of their treatment during

(a) The Romanov family tree

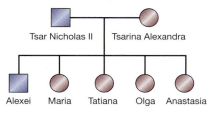

(b) The STR analysis

	STRs				
	VWA/31	THO1	F13A1	FES/FPS	ACTBP2
Child 1	15, 16	8, 10	5, 7	12, 13	11, 32
Child 2	15, 16	7, 8	5, 7	12, 13	11, 36
Child 3	15, 16	8, 10	3, 7	12, 13	32, 36
Female adult 1	15, 16	8, 8	3, 5	12, 13	32, 36
Female adult 2	16, 17	6, 6	6, 7	11, 12	not done
Male adult 1	14, 20	9, 10	6, 16	10, 11	not done
Male adult 2	17, 17	6, 10	5, 7	10, 11	11, 30
Male adult 3	15, 16	7, 10	7, 7	12, 12	11, 32
Male adult 4	15, 17	6, 9	5, 7	8, 10	not done

Figure 16.4

Short tandem repeat analysis of the Romanov bones. (a) The Romanov family tree. (b) The results of the STR analysis. Data taken from Gill *et al.* (1994) (see *Further Reading*).

and after death, and at least some of the remains were clearly aristocratic as their teeth were filled with porcelain, silver, and gold, dentistry well beyond the means of the average Russian of the early 20th century.

DNA was extracted from bones from each individual and five STRs typed by PCR to test the hypothesis that the three children were siblings and that two of the adults were their parents, as would be the case if indeed these were the Romanovs. The results immediately showed that the three children could be siblings, as they have identical genotypes for the STRs called VWA/31 and FES/FPS and share alleles at each of the three other loci (Figure 16.4). The THO1 data show that female adult 2 cannot be the mother of the children because she only possesses allele 6, which none of the children have. Female adult 1, however, has allele 8, which all three children have. Examination of the other STRs confirms that she could be the mother of each of the children, and so she is identified as the Tsarina. The THO1 data exclude male adult 4 as a possible father of the children, and the VWA/31 results exclude male adults 1 and 2. When all the STRs are taken into account, male adult 3 could be the father of the children and therefore is identified as the Tsar. Note that all these conclusions can be drawn simply from the TH01 and VWA/31 results: the other STR data simply provide corroborating evidence.

Mitochondrial DNA was used to link the Romanov skeletons with living relatives
The STR analysis showed that the skeletons included a family group as expected, if these were indeed the bones of the Romanovs. But could they be the remains of some

Figure 16.5

Family tree showing the matrilineal relationship between Prince Philip, Duke of Edinburgh and Princess Victoria of Hesse, the Tsarina's sister. Males are shown as blue boxes and females as red circles.

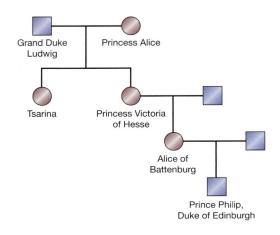

other unfortunate group of people? To address this problem, the DNA from the bones was compared with DNA samples from living relatives of the Romanovs. This work included studies of **mitochondrial DNA**, the small 16 kb circles of DNA contained in the energy-generating mitochondria of cells. Mitochondrial DNA contains polymorphisms that can be used to infer relationships between individuals, but the degree of variability is not as great as displayed by STRs, so mitochondrial DNA is rarely used for kinship studies among closely related individuals such as those of a single family group. But mitochondrial DNA has the important property of being inherited solely through the female line, the father's mitochondrial DNA being lost during fertilization and not contributing to the son or daughter's DNA content. This maternal inheritance pattern makes it easier to distinguish relationships when the individuals being compared are more distantly related, as was the case with the living relatives of the Romanovs.

Comparisons were therefore made between mitochondrial DNA sequences obtained from the skeletons and that of Prince Philip, the Duke of Edinburgh, whose maternal grandmother was Princess Victoria of Hesse, the Tsarina Alexandra's sister (Figure 16.5). The mitochondrial DNA sequences from four of the female skeletons—the three children and the adult female identified as the Tsarina—were exactly the same as that of Prince Philip, strong evidence that the four females were members of the same lineage. Comparisons were also made with of two living matrilineal descendants of Tsar Nicholas's grandmother, Louise of Hesse-Cassel. This analysis was more complicated, as two sequences were present among the clones of the PCR product obtained from the adult male thought to be the Tsar. These sequences differed at a single position which was either a C or a T, the former four times more frequent than the latter. This could indicate that the sample was contaminated with somebody else's DNA but instead was interpreted as showing that the Tsar's mitochondrial DNA was **heteroplasmic**, an infrequent situation where two different mitochondrial DNAs co-exist within the same cells. The two descendents of the Tsar's grandmother both had the mitochondrial DNA version with a T at this position, suggesting that the mutation producing the C variant had occurred very recently in the Tsar's lineage. Support for this hypothesis was subsequently provided by analysis of DNA from the Tsar's brother, Grand Duke George Alexandrovich, who died in 1899, which showed that he also displayed heteroplasmy at the same position in his mitochondrial DNA. On balance, the evidence suggested that the Tsar's remains had been correctly identified.

The missing children

Only three children were found in the Romanovs' grave. Alexei, the only boy, and one of the four girls were missing. During the middle decades of the 20th century, several women claimed to be a Romanov princess, because even before the bones were recovered there had been rumors that one of the girls, Anastasia, had escaped the clutches of the Bolsheviks and fled to the West. One of the most famous of these claimants was Anna Anderson, whose case was first widely publicised in the 1920s. Anna Anderson died in 1984 but she left an archived tissue sample whose mitochondrial DNA does not match the Tsarina's and suggests a woman of Polish origin. There have also been various people claiming to be descended from Tsarevich Alexei. But these stories are almost certainly romances, as the partially cremated bodies of two other children found near Yekaterinburg in 2007 have now been shown to have mitochondrial DNA sequences that suggest that they are missing Romanov children.

16.3 Sex identification by DNA analysis

DNA analysis can also be used to identify the sex of an individual. The genetic difference between the sexes is the possession of a Y chromosome by males, so detection of DNA specific for the Y chromosome would enable males and females to be distinguished. Forensic scientists occasionally have to deal with bodies that are so badly damaged that DNA analysis is the only way of identifying their sex.

DNA tests can also be used to identify the sex of an unborn child. Finding out if a fetus is a boy or a girl is usually delayed until the anatomical differences have developed and the sex can be identified by scanning, but under some circumstances an earlier indication of sex is desirable. An example is when the pedigree of the family indicates that an unborn male might suffer from an inherited disease and the parents wish to make an early decision about whether to continue the pregnancy.

A third application of DNA-based sex identification, and the one that has been responsible for many of the developments in this field, is in the analysis of archaeological specimens. Male and female skeletons can be distinguished if key bones such as the skull or the pelvis are intact, but with fragmentary remains, or those of young children, there are not enough sex-specific anatomical differences for a confident identification to be made. If ancient DNA is preserved in the bones, a DNA-based method can tell the archaeologists if they are dealing with a male or a female.

16.3.1 PCRs directed at Y chromosome-specific sequences

The simplest way to use DNA analysis to identify sex is to design a PCR specific for a region of the Y chromosome. The PCR has to be designed with care, because the X and Y chromosomes are not completely different, some segments being shared between the two. But there are many unique regions within the Y chromosome. In particular, there are several repeated sequences that are only located in the Y chromosome, these repeated sequences acting as multiple targets for the PCR and hence giving greater sensitivity, an important consideration if you are dealing with a badly damaged body or an ancient bone.

A PCR directed at Y-specific DNA sequences would give a product with male DNA but no band if the sample comes from a female (Figure 16.6). This is a clear distinction between the two alternatives and hence a perfectly satisfactory system for most

Figure 16.6

Sex identification by PCR of a Y-specific DNA sequence. Male DNA gives a PCR product (lane 2), but female DNA does not (lane 3). The problem is that a failed PCR (lane 4) gives the same result as female DNA.

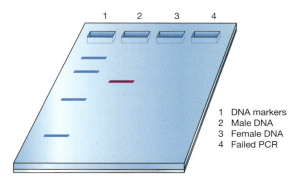

1   DNA markers
2   Male DNA
3   Female DNA
4   Failed PCR

applications. But what if the sample did not contain any DNA, or if the DNA was too degraded to work in the PCR, or if the sample also contained inhibitors of *Taq* polymerase that prevented the enzyme from carrying out the PCR? All of these possibilities could occur with archaeological specimens, especially those that have been buried in the ground and become contaminated with humic acids and other compounds known to inhibit many of the enzymes used in molecular biology research. Now the test becomes ambiguous because a specimen that is unable to give a PCR product for one of these reasons could mistakenly be identified as female. The result would be exactly the same: no band on the gel.

16.3.2 PCR of the amelogenin gene

The lack of discrimination between "female" and "failed PCR" that occurs when Y-specific sequences are studied has led to the development of more sophisticated DNA tests for sex identification, ones that give unambiguous results for both males and females. The most widely used of these involves PCRs that amplify the amelogenin gene.

The amelogenin gene codes for a protein found in tooth enamel. It is one of the few genes that are present on the Y chromosome and, like many of these genes, there is also a copy on the X chromosome. But the two copies are far from identical, and when the nucleotide sequences are aligned a number of indels, positions where a segment of DNA has either been inserted into one sequence or deleted from the other sequence, are seen (Figure 16.7a). If the primers for a PCR anneal either side of an indel, the products obtained from the X and Y chromosomes would have different sizes. Female DNA would give a single band when the products are examined, because females only have the X chromosome, whereas males would give two bands, one from the X chromosome and one from the Y (Figure 16.7b). If the sample contains no DNA or the PCR fails for some other reason, no band will be obtained: there is no confusion between the failure and a male or female result.

The development of the amelogenin system for sex identification is having an important impact in archaeology. No longer is it necessary to assign sex to buried bones on the basis of vague differences in the structures of the bones. The greater confidence that DNA-based sex testing allows is resulting in some unexpected discoveries. In particular, archaeologists are now reviewing their preconceptions about the meaning of the objects buried in a grave along with the body. It was thought that if a body was accompanied by a sword then it must be male, or if the grave contained beads then the body was female. DNA testing has shown that these stereotypes are not always correct and that archaeologists must take a broader view of the link between grave goods and sex.

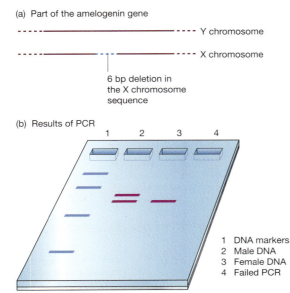

(a) Part of the amelogenin gene

Y chromosome

X chromosome

6 bp deletion in
the X chromosome
sequence

(b) Results of PCR

1   2   3   4

1 DNA markers
2 Male DNA
3 Female DNA
4 Failed PCR

Figure 16.7

Sex identification by PCR of part of the amelogenin gene. (a) An indel in the amelogenin gene. (b) The results of PCRs spanning the indel. Male DNA gives two PCR products, of 106 and 112 bp in the standard system used in forensics and biomolecular archaeology. Female DNA gives just the smaller product. A failed PCR gives no products and so is clearly distinguishable from the two types of positive result.

16.4 Archaeogenetics—using DNA to study human prehistory

Sex identification and kinship studies are not the only ways in which gene cloning and DNA analysis are being applied in archaeology. By examining DNA sequences in living and dead humans, archaeologists have begun to understand the evolutionary origins of modern humans, and the routes followed by prehistoric people as they colonized the planet. This area of research is called archaeogenetics.

16.4.1 The origins of modern humans

Paleontologists believe that humans originated in Africa because it is here that all of the oldest pre-human fossils have been found. The fossil evidence reveals that hominids first migrated out of Africa over one million years ago, but these were not modern humans. Instead it was an earlier species called *Homo erectus*, who were the first hominids to become geographically dispersed, eventually spreading to all parts of the Old World.

The events that followed the dispersal of *H. erectus* are controversial. From studies of fossils, many paleontologists believe that the *H. erectus* populations which became located in different parts of the Old World gave rise to the modern *Homo sapiens* populations found in those areas today (Figure 16.8a). This process is called multi-regional evolution. There may have been a certain amount of interbreeding between humans from different geographical regions but, to a large extent, these various populations remained separate throughout their evolutionary history.

DNA analysis has challenged the multiregional hypothesis
Doubts about the multiregional hypothesis were raised in 1987 when geneticists first started using DNA analysis to ask questions about human evolution. In one of the very

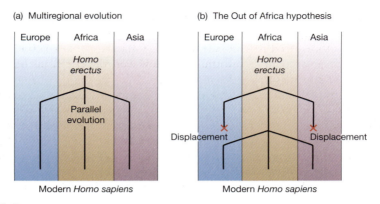

Figure 16.8

Two hypotheses for the origins of modern humans: (a) multiregional evolution; (b) the Out of Africa hypothesis.

first archaeogenetics projects, restriction fragment length polymorphisms (RFLPs) were measured in mitochondrial DNA samples taken from 147 humans, from all parts of the world. The resulting data were then used to construct a phylogenetic tree showing the evolutionary relationships between different human populations. From this tree, various deductions were made:

- The root of the tree represents a woman (remember, mitochondrial DNA is inherited only through the female line) whose mitochondrial genome is ancestral to all the 147 modern mitochondrial DNAs that were tested. This woman has been called **mitochondrial Eve**. Of course, she was not equivalent to the Biblical character and was by no means the only woman alive at the time: she simply was the person who carried the ancestral mitochondrial DNA that gave rise to all the mitochondrial DNAs in existence today.
- Mitochondrial Eve lived in Africa. This was deduced because the ancestral sequence split the tree into two segments, one of which was composed solely of African mitochondrial DNAs. Because of this split, it was inferred that the ancestor was also located in Africa.
- Mitochondrial Eve lived between 140,000 and 290,000 years ago. This conclusion was drawn by applying the **molecular clock** to the phylogenetic tree. The molecular clock is a measure of the speed at which evolutionary change occurs in mitochondrial DNA sequences, and is calibrated from the rate at which mutations are known to accumulate in mitochondrial DNA. By comparing the sequence inferred for Eve's mitochondrial DNA with the sequences of the 147 modern DNAs, the number of years needed for all of the necessary evolutionary changes to take place was calculated.

The key finding that mitochondrial Eve lived in Africa no earlier than 290,000 years ago does not agree with the suggestion that we are all descended from *H. erectus* populations who left Africa over a million years ago. A new hypothesis for human origins was therefore devised, called **Out of Africa**. According to this hypothesis, modern humans—*H. sapiens*—evolved specifically from those *H. erectus* populations that remained in Africa. Modern humans then moved into the rest of the Old World between 100,000 and 50,000 years ago, displacing the descendents of *H. erectus* that they encountered (Figure 16.8b).

At first, the mitochondrial Eve results were heavily criticized. It became apparent that the computer analysis used to construct the phylogenetic tree was flawed, mainly because the algorithms used to compare the RFLP data were not sufficiently robust to deal with this huge amount of information. However, the criticisms have now died away as the results of more extensive mitochondrial DNA studies, using actual DNA sequences rather than RFLPs and analyzed using modern, powerful computers, have all confirmed the findings of the first project. To take one example, when the complete mitochondrial genome sequences of 53 people, again from all over the world, were compared, a date of 220,000–120,000 years for mitochondrial Eve was obtained. An interesting complement has been provided by studies of the Y chromosome which, of course, descends exclusively through the male line. This work has revealing that "Y chromosome Adam" also lived in Africa between 40,000 and 140,000 years ago.

DNA analysis shows that Neanderthals are not the ancestors of modern Europeans

Neanderthals are extinct hominids who lived in Europe between 300,000 and 30,000 years ago. They were descended from the *H. erectus* populations who left Africa about one million years ago and, according to the Out of Africa hypothesis, were displaced when modern humans reached Europe about 50,000 years ago. Therefore, one prediction of the Out of Africa hypothesis is that Neanderthals are not the ancestors of modern Europeans. Analysis of ancient DNA from Neanderthal bones has been used to test this prediction.

The first Neanderthal specimen selected for study was the type specimen, which had been found in Germany in the 19th century. This fossil has not been precisely dated but is between 30,000 and 100,000 years old. This places it at the very limits of ancient DNA survival, as natural degradation processes are thought to break down the DNA surviving in bones so that very little remains after about 50,000 years, even in specimens that are kept very cold, for example by burial in permafrost. The Neanderthal specimen had not been preserved under particularly cold conditions, but nonetheless it was possible to obtain a short part of the mitochondrial DNA sequence of this individual. This was achieved by carrying out nine overlapping PCRs, each one amplifying less than 170 bp of DNA but together giving a total length of 377 bp.

A phylogenetic tree was constructed to compare the sequence obtained from the Neanderthal bone with the sequences of six of the main mitochondrial DNA variants (called **haplogroups**) present in modern Europeans. The Neanderthal sequence was positioned on a branch of its own, connected to the root of the tree but not linked directly to any of the modern human sequences (Figure 16.9). This was the first evidence suggesting that Neanderthals are not ancestral to modern Europeans.

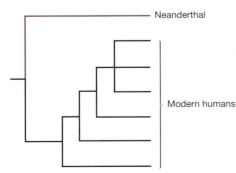

Figure 16.9

Phylogenetic analysis of ancient DNA suggests that Neanderthals are not directly related to modern humans.

Next, the Neanderthal sequence was aligned with the equivalent sequences from 994 modern humans. The differences were striking. The Neanderthal sequence differed from the modern sequences at an average of 27.2 ± 2.2 nucleotide positions, whereas the modern sequences, which came from all over the world, not just Europe, differed from each other at only 8.0 ± 3.1 positions. This degree of difference is incompatible with the notion that modern Europeans are descended from Neanderthals. Similar results were obtained when mitochondrial DNA from a second Neanderthal skeleton was examined. The results therefore provide an independent proof of the Out of Africa hypothesis, and show that, at least for Europe, the multiregional model is incorrect.

16.4.2 DNA can also be used to study prehistoric human migrations

The modern humans that displaced Neanderthals arrived in Europe about 40,000 years ago. This is clear from the fossil and archaeological records. But were these humans themselves displaced by newer populations who migrated into Europe more recently?

The spread of agriculture into Europe

Some archaeologists have suggested that new populations of humans might have moved into Europe during the past 10,000 years, and that these humans brought agriculture into the continent. The transition from hunting and gathering to farming occurred in southwest Asia about 10,000 years ago, when early Neolithic villagers began to cultivate crops such as wheat and barley. After becoming established, farming spread into Asia, Europe, and North Africa. By searching archaeological sites for the remains of cultivated plants or for implements used in farming, two routes by which agriculture spread into Europe have been traced. One of these trajectories followed the Mediterranean coast to Spain and eventually to Britain, and the second passed along the Danube and Rhine valleys to northern Europe (Figure 16.10).

One explanation for the spread of agriculture is that farmers moved from one place to another, taking with them their implements, animals, and crops, and displacing the indigenous, pre-agricultural communities that were present in Europe at that time. This **wave of advance** model was initially supported by archaeogeneticists, as it agrees with the results of a large phylogenetic study, carried out in the 1990s, of the allele frequencies for 95 nuclear genes in populations from across Europe. This dataset was analyzed by a technique often used in population genetics, called **principal component analysis**, which attempts to identify patterns in the geographical distribution of alleles, these patterns possibly being indicative of past population migrations.

The most striking pattern within the European dataset, accounting for about 28% of the total genetic variation, is a gradation of allele frequencies running from southeast to northwest across the continent (Figure 16.11). This pattern suggests that a migration of people occurred either from southeast to northwest Europe, or in the opposite direction. Because the former migration coincides with the spread of farming, as revealed by the archaeological record, this first principal component was looked on as providing strong support for the wave of advance model.

Using mitochondrial DNA to study past human migrations into Europe

Principal component analysis has one weakness when applied to past human migrations. It is difficult to determine when a migration identified in this way took place. This means that the link between the first principal component and the spread of

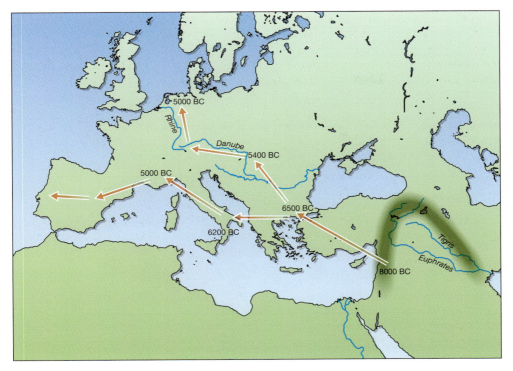

Figure 16.10

The spread of agriculture from southwest Asia into Europe. The shaded area in southwest Asia is the Fertile Crescent, where the wild versions of wheat and barley grow, and where these plants were first cultivated by farmers some 10,000 years ago.

agriculture is based solely on the pattern of the allele gradation, not on any complementary evidence relating to the period when this gradation was set up.

A second study of European human populations, one which does include a time dimension, has been carried out using mitochondrial DNA. To begin with, the distribution of mitochondrial DNA sequence variations in 821 individuals from populations across Europe were compared. The data gave no evidence for a gradation of allele frequencies, and instead suggested that European populations have remained relatively static over the past 20,000 years. This result raised important doubts about the wave of advance model. How then did agriculture spread into Europe?

More sophisticated studies of mitochondrial DNA variations in modern European populations is pointing toward a new model for the spread of agriculture. It has been discovered that the mitochondrial genomes present among modern Europeans can be grouped into 11 major sequence classes, or haplogroups, each one displaying distinctive nucleotide sequence variations. For each of these haplogroups, the molecular clock can be used to deduce a date of origin, which is thought to correspond with the date at which the haplogroup entered Europe (Figure 16.12). The most ancient haplogroup, called U, first appeared in Europe about 50,000 years ago, coinciding with the period when, according to the archaeological record, the first modern humans moved into the continent as the ice sheets withdrew to the north at the end of the last major glaciation.

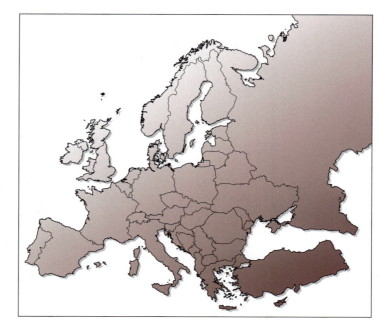

Figure 16.11

Principal component analysis reveals a southeast to northwest gradient of human allele frequencies across Europe.

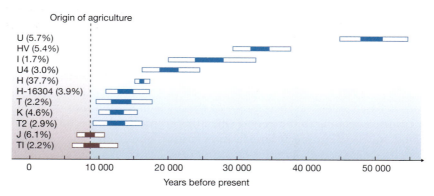

Figure 16.12

The deduced times of arrival in Europe of the 11 major mitochondrial DNA haplogroups found in modern populations. Those haplogroups whose arrivals coincide with the spread of agriculture are shown in red.

The youngest haplogroups, J and T1, which at 9000 years in age could correspond to the origins of agriculture, are possessed by just 8.3% of the modern European population, confirming that the spread of farming into Europe was not the huge wave of advance indicated by the principal component study. Instead, it is now thought that farming was brought into Europe by a smaller group of "pioneers" who interbred with the existing pre-farming communities rather than displacing them.

Archaeogenetics truly illustrates how broad ranging an impact gene cloning and DNA analysis have had on science.

Further reading

Cann, R.L., Stoneking, M. & Wilson, A.C. (1987) Mitochondrial DNA and human evolution. *Nature*, 325, 31–36. [The first paper to propose the mitochondrial Eve hypothesis.]

Cavalli-Sforza, L.L. (1998) The DNA revolution in population genetics. *Trends in Genetics*, 14, 60–65. [The use of principal component analysis to study nuclear allele frequencies in Europe.]

Coble, M.D., Loreille, O.M., Wadhams, M.J. et al. (2009) Mystery solved: the identification of the two missing Romanov children using DNA analysis. *PLoS ONE*, 4, e4838.

Gill, P., Ivanov, P.L., Kimpton, C. et al. (1994) Identification of the remains of the Romanov family by DNA analysis. *Nature Genetics*, 6, 130–135.

Jeffreys, A.J., Wilson, V. & Thein, L.S. (1985) Individual-specific fingerprints of human DNA. *Nature*, 314, 67–73. [Genetic fingerprinting by analysis of dispersed repeats.]

Jobling, M.A. & Gill, P. (2004) Encoded evidence: DNA in forensic analysis. *Nature Reviews Genetics*, 5, 739–751.

Krings, M., Stone, A., Schmitz, R.W. et al. (1997) Neandertal DNA sequences and the origin of modern humans. *Cell*, 90, 19–30.

Nakahori, Y., Hamano, K., Iwaya, M. & Nakagome, Y. (1991) Sex identification by polymerase chain reaction using X–Y homologous primer. *American Journal of Medical Genetics*, 39, 472–473. [The amelogenin method.]

Richards, M., Macauley, V., Hickey, E. et al. (2000) Tracing European founder lineages in the Near Eastern mtDNA pool. *American Journal of Human Genetics*, 67, 1251–1276. [Using mitochondrial DNA to study human migrations into Europe.]

Glossary

GLOSSARY

2 μm circle A plasmid found in the yeast *Saccharomyces cerevisiae* and used as the basis for a series of cloning vectors.

3′ terminus One of the two ends of a polynucleotide: that which carries the hydroxyl group attached to the 3′ position of the sugar.

5′ terminus One of the two ends of a polynucleotide: that which carries the phosphate group attached to the 5′ position of the sugar.

Adaptor A synthetic, double-stranded oligonucleotide used to attach sticky ends to a blunt-ended molecule.

Adeno-associated virus (AAV) A virus that is unrelated to adenovirus but which is often found in the same infected tissues, because AAV makes use of some of the proteins synthesized by adenovirus in order to complete its replication cycle.

Adenovirus An animal virus, derivatives of which have been used to clone genes in mammalian cells.

Affinity chromatography A chromatography method that makes use of a ligand that binds a specific protein and which can therefore be used to aid purification of that protein.

Agrobacterium tumefaciens The soil bacterium which, when containing the Ti plasmid, is able to form crown galls on a number of dicotyledonous plant species.

Ancient DNA Preserved DNA from an archaeological or fossil specimen.

Annealing Attachment of an oligonucleotide to a single-stranded DNA molecule by hybridization.

Antisense RNA An RNA molecule that is the reverse complement of a naturally occurring mRNA, and which can be used to prevent translation of that mRNA in a transformed cell.

Archaeogenetics The use of DNA analysis to study the human past.

Artificial gene synthesis Construction of an artificial gene from a series of overlapping oligonucleotides.

Autoradiography A method of detecting radioactively labeled molecules through exposure of an X-ray-sensitive photographic film.

Auxotroph A mutant microorganism that grows only when supplied with a nutrient not required by the wild type.

Avidin A protein that has a high affinity for biotin and is used in a detection system for biotinylated probes.

BacMam vector A modified baculovirus that carries a mammalian promoter and so is able to express a cloned gene directly in a mammalian cell.

Bacterial artificial chromosome (BAC) A cloning vector based on the F plasmid, used for cloning relatively large fragments of DNA in *E. coli*.

Bacteriophage or **phage** A virus whose host is a bacterium. Bacteriophage DNA molecules are often used as cloning vectors.

Baculovirus A virus that has been used as a cloning vector for the production of recombinant protein in insect cells.

Batch culture Growth of bacteria in a fixed volume of liquid medium in a closed vessel, with no additions or removals made during the period of incubation.

Bioinformatics The use of computer methods in studies of genomes.

Biolistics A means of introducing DNA into cells that involves bombardment with high velocity microprojectiles coated with DNA.

Biological containment One of the precautionary measures taken to prevent the replication of recombinant DNA molecules in microorganisms in the natural environment. Biological containment involves the use of vectors and host organisms that have been modified so that they will not survive outside the laboratory.

Biotechnology The use of biological processes in industry and technology.

Biotin A molecule that can be incorporated into dUTP and used as a non-radioactive label for a DNA probe.

BLAST An algorithm frequently used in homology searching.

Blunt end or **flush end** An end of a DNA molecule at which both strands terminate at the same nucleotide position with no single-stranded extension.

Broad host range plasmid A plasmid that can replicate in a variety of host species.

Broth culture Growth of microorganisms in a liquid medium.

Buoyant density The density possessed by a molecule or particle when suspended in an aqueous salt or sugar solution.

Candidate gene A gene, identified by positional cloning, that might be a disease-causing gene.

Capsid The protein coat that encloses the DNA or RNA molecule of a bacteriophage or virus.

Cassette A DNA sequence consisting of promoter–ribosome binding site–unique restriction site–terminator (or for a eukaryotic host, promoter–unique restriction site–polyadenylation sequence) carried by certain types of expression vector. A foreign gene inserted into the unique restriction site is placed under control of the expression signals.

Cauliflower mosaic virus (CaMV) The best studied of the caulimoviruses, used in the past as a cloning vector for some species of higher plant. Cauliflower mosaic virus is the source of strong promoters used in other types of plant cloning vector.

Caulimoviruses One of the two groups of DNA viruses to infect plants, the members of which have potential as cloning vectors for some species of higher plant.

Cell extract A preparation consisting of a large number of broken cells and their released contents.

Cell-free translation system A cell extract containing all the components required for protein synthesis (i.e., ribosomal subunits, tRNAs, amino acids, enzymes and cofactors) and able to translate added mRNA molecules.

Chimera (1) A recombinant DNA molecule made up of DNA fragments from more than one organism, named after the mythological beast. (2) The initial product of cloning using embryonic stem cells: an animal made up of a mixture of cells with different genotypes.

Chromosome One of the DNA–protein structures that contains part of the nuclear genome of a eukaryote. Less accurately, the DNA molecule(s) that contains a prokaryotic genome.

Chromosome walking A technique that can be used to construct a clone contig by identifying overlapping fragments of cloned DNA.

Cleared lysate A cell extract that has been centrifuged to remove cell debris, subcellular particles and possibly chromosomal DNA.

Clone A population of identical cells, generally those containing identical recombinant DNA molecules.

Clone contig approach A genome sequencing strategy in which the molecules to be sequenced are broken into manageable segments, each a few hundred kb or a few Mb in length, which are sequenced individually.

Clone fingerprinting Any one of a variety of techniques that compares cloned DNA fragments in order to identify ones that overlap.

Codon bias The fact that not all codons are used equally frequently in the genes of a particular organism.

Combinatorial screening A technique that reduces the number of PCRs or other analyses that must be performed by combining samples in an ordered fashion, so that a sample giving a particular result can be identified even though that sample is not individually examined.

Comparative genomics A research strategy that uses information obtained from the study of one genome to make inferences about the map positions and functions of genes in a second genome.

Compatibility The ability of two different types of plasmid to coexist in the same cell.

Competent A culture of bacteria that has been treated to enhance their ability to take up DNA molecules.

Complementary Two polynucleotides that can base pair to form a double-stranded molecule.

Complementary DNA (cDNA) cloning A cloning technique involving conversion of purified mRNA to DNA before insertion into a vector.

Conformation The spatial organization of a molecule. Linear and circular are two possible conformations of a polynucleotide.

Conjugation Physical contact between two bacteria, usually associated with transfer of DNA from one cell to the other.

Consensus sequence A nucleotide sequence used to describe a large number of related though non-identical sequences. Each position of the consensus sequence represents the nucleotide most often found at that position in the real sequences.

Contig A contiguous segment of DNA sequence obtained as part of a genome sequencing project.

Continuous culture The culture of microorganisms in liquid medium under controlled conditions, with additions to and removals from the medium over a lengthy period of time.

Contour clamped homogeneous electric fields (CHEF) An electrophoresis technique for the separation of large DNA molecules.

Copy number The number of molecules of a plasmid contained in a single cell.

Cosmid A cloning vector consisting of the λ *cos* site inserted into a plasmid, used to clone DNA fragments up to 40 kb in size.

***cos* site** One of the cohesive, single-stranded extensions present at the ends of the DNA molecules of certain strains of λ phage.

Covalently closed-circular DNA (cccDNA) A completely double-stranded circular DNA molecule, with no nicks or discontinuities, usually with a supercoiled conformation.

CpG island A GC-rich DNA region located upstream of approximately 56% of the genes in the human genome.

Defined medium A bacterial growth medium in which all the components are known.

Deletion analysis The identification of control sequences for a gene by determining the effects on gene expression of specific deletions in the upstream region.

Deletion cassette A segment of DNA that is transferred to a yeast chromosome by homologous recombination in order to create a deleted version of a target gene, in order to inactivate that gene and identify its function.

Denaturation Of nucleic acid molecules: breakdown by chemical or physical means of the hydrogen bonds involved in base pairing.

Density gradient centrifugation Separation of molecules and particles on the basis of buoyant density, by centrifugation in a concentrated sucrose or caesium chloride solution.

Deoxyribonuclease An enzyme that degrades DNA.

Dideoxynucleotide A modified nucleotide that lacks the 3′ hydroxyl group and so prevents further chain elongation when incorporated into a growing polynucleotide.

Directed evolution A set of experimental techniques that is used to obtain novel genes with improved products.

Directed shotgun approach A genome-sequencing strategy which combines random shotgun sequencing with a genome map, the latter used to aid assembly of the master sequence.

Direct gene transfer A cloning process that involves transfer of a gene into a chromosome without the use of a cloning vector able to replicate in the host organism.

Disarmed plasmid A Ti plasmid that has had some or all of the T-DNA genes removed, so it is no longer able to promote cancerous growth of plant cells.

DNA chip A wafer of silicon carrying a high density array of oligonucleotides used in transcriptome and other studies.

DNA ladder A mixture of DNA fragments, whose sizes are multiples of 100 bp or of 1 kb, used as size markers.

DNA marker A DNA sequence that exists as two or more alleles and which can therefore be used in genetic mapping.

DNA polymerase An enzyme that synthesizes DNA on a DNA or RNA template.

DNA profiling A PCR technique that determines the alleles present at different STR loci within a genome in order to use DNA information to identify individuals.

DNA sequencing Determination of the order of nucleotides in a DNA molecule.

Double digestion Cleavage of a DNA molecule with two different restriction endonucleases, either concurrently or consecutively.

Electrophoresis Separation of molecules on the basis of their charge-to-mass ratio.

Electroporation A method for increasing DNA uptake by protoplasts through prior exposure to a high voltage, which results in the temporary formation of small pores in the cell membrane.

Elution The unbinding of a molecule from a chromatography column.

Embryonic stem (ES) cell A totipotent cell from the embryo of a mouse or other organism, used in construction of a transgenic animal such as a knockout mouse.

End filling Conversion of a sticky end to a blunt end by enzymatic synthesis of the complement to the single-stranded extension.

Endonuclease An enzyme that breaks phosphodiester bonds within a nucleic acid molecule.

Episome A plasmid capable of integration into the host cell's chromosome.

Ethanol precipitation Precipitation of nucleic acid molecules by ethanol plus salt, used primarily as a means of concentrating DNA.

Ethidium bromide A fluorescent chemical that intercalates between base pairs in a double-stranded DNA molecule, used in the detection of DNA.

Exonuclease An enzyme that sequentially removes nucleotides from the ends of a nucleic acid molecule.

Expressed sequence tag (EST) A partial or complete cDNA sequence.

Expression vector A cloning vector designed so that a foreign gene inserted into the vector is expressed in the host organism.

Fermenter A vessel used for the large scale culture of microorganisms.

Field inversion gel electrophoresis (FIGE) An electrophoresis technique for the separation of large DNA molecules.

Fluorescence *in situ* hybridization (FISH) A hybridization technique that uses fluorochromes of different colors to enable two or more genes to be located within a chromosome preparation in a single *in situ* experiment.

Footprinting The identification of a protein binding site on a DNA molecule by determining which phosphodiester bonds are protected from cleavage by DNase I.

Forward genetics The strategy by which the genes responsible for a phenotype are identified by determining which genes are inactivated in organisms that display a mutant version of that phenotype.

Functional genomics Studies aimed at identifying all the genes in a genome and determining their expression patterns and functions.

Fusion protein A recombinant protein that carries a short peptide from the host organism at its amino or, less commonly, carboxyl terminus.

Gel electrophoresis Electrophoresis performed in a gel matrix so that molecules of similar electric charge can be separated on the basis of size.

Gel retardation A technique that identifies a DNA fragment that has a bound protein by virtue of its decreased mobility during gel electrophoresis.

Geminivirus One of the two groups of DNA viruses that infect plants, the members of which have potential as cloning vectors for some species of higher plants.

Gene A segment of DNA that codes for an RNA and/or polypeptide molecule.

Gene addition A genetic engineering strategy that involves the introduction of a new gene or group of genes into an organism.

Gene cloning Insertion of a fragment of DNA, carrying a gene, into a cloning vector, and subsequent propagation of the recombinant DNA molecule in a host organism. Also used to describe those techniques that achieve the same result without the use of a cloning vector (e.g., direct gene transfer).

Gene knockout A technique that results in inactivation of a gene, as a means of determining the function of that gene.

Gene mapping Determination of the relative positions of different genes on a DNA molecule.

Gene subtraction A genetic engineering strategy that involves the inactivation of one or more of an organism's genes.

Gene therapy A clinical procedure in which a gene or other DNA sequence is used to treat a disease.

Genetic engineering The use of experimental techniques to produce DNA molecules containing new genes or new combinations of genes.

Genetic fingerprinting A hybridization technique that determines the genomic distribution of a hypervariable dispersed repetitive sequence and results in a banding pattern that is specific for each individual.

Genetic map A genome map that has been obtained by analysing the results of genetic crosses.

Genetics The branch of biology devoted to the study of genes.

Genome The complete set of genes of an organism.

Genome annotation The process by which the genes, control sequences and other interesting features are identified in a genome sequence.

Genomic DNA Consists of all the DNA present in a single cell or group of cells.

Genomic library A collection of clones sufficient in number to include all the genes of a particular organism.

Genomics The study of a genome, in particular the complete sequencing of a genome.

GM (genetically modified) crop A crop plant that has been engineered by gene addition or gene subtraction.

Haplogroup One of the major sequence classes of mitochondrial DNA present in the human population.

Harvesting The removal of microorganisms from a culture, usually by centrifugation.

Helper phage A phage that is introduced into a host cell in conjunction with a related cloning vector, in order to provide enzymes required for replication of the cloning vector.

Heterologous probing The use of a labeled nucleic acid molecule to identify related molecules by hybridization probing.

Heteroplasmic Possessing two haplotypes of the mitochondrial genome.

Homologous recombination Recombination between two homologous double-stranded DNA molecules, i.e., ones which share extensive nucleotide sequence similarity.

Homology Refers to two genes from different organisms that have evolved from the same ancestral gene. Two homologous genes are usually sufficiently similar in sequence for one to be used as a hybridization probe for the other.

Homology search A technique in which genes with sequences similar to that of an unknown gene are sought, in order to confirm a gene identification or to understand the function of the unknown gene.

Homopolymer tailing The attachment of a sequence of identical nucleotides (e.g., AAAAA) to the end of a nucleic acid molecule, usually referring to the synthesis of single-stranded homopolymer extensions on the ends of a double-stranded DNA molecule.

Horseradish peroxidase An enzyme that can be complexed to DNA and which is used in a non-radioactive procedure for DNA labeling.

Host-controlled restriction A mechanism by which some bacteria prevent phage attack through the synthesis of a restriction endonuclease that cleaves the non-bacterial DNA.

Hybrid-arrest translation (HART) A method used to identify the polypeptide coded by a cloned gene.

Hybridization probe A labeled nucleic acid molecule that can be used to identify complementary or homologous molecules through the formation of stable base-paired hybrids.

Hybrid-release translation (HRT) A method used to identify the polypeptide coded by a cloned gene.

Hypervariable dispersed repetitive sequence The type of human repetitive DNA sequence used in genetic fingerprinting.

Immunological screening The use of an antibody to detect a polypeptide synthesized by a cloned gene.

In situ **hybridisation** A technique for gene mapping involving hybridization of a labeled sample of a cloned gene to a large DNA molecule, usually a chromosome.

In vitro **mutagenesis** Any one of several techniques used to produce a specified mutation at a predetermined position in a DNA molecule.

In vitro **packaging** Synthesis of infective λ particles from a preparation of λ capsid proteins and a catenane of DNA molecules separated by *cos* sites.

Inclusion body A crystalline or paracrystalline deposit within a cell, often containing substantial quantities of insoluble protein.

Incompatibility group Comprises a number of different types of plasmid, often related to each other, that are unable to coexist in the same cell.

Indel A position where a DNA sequence has been inserted into or deleted from a genome, so called because it is impossible from comparison of two sequences to determine which alternative has occurred: insertion into one genome or deletion from the other.

Induction (1) Of a gene: the switching on of the expression of a gene or group of genes in response to a chemical or other stimulus. (2) Of λ phage: the excision of the integrated form of λ and accompanying switch to the lytic mode of infection, in response to a chemical or other stimulus.

Insertion vector A λ vector constructed by deleting a segment of non-essential DNA.

Insertional inactivation A cloning strategy whereby insertion of a new piece of DNA into a vector inactivates a gene carried by the vector.

Interspersed repeat element PCR (IRE–PCR) A clone fingerprinting technique that uses PCR to detect the relative positions of repeated sequences in cloned DNA fragments.

Ion exchange chromatography A method for separating molecules according to how tightly they bind to electrically charged particles present in a chromatographic matrix.

Isoelectric focussing Separation of proteins in a gel that contains chemicals which establish a pH gradient when the electrical charge is applied.

Isoelectric point The position in a pH gradient where the net charge of a protein is zero.

Isotope coded affinity tag (ICAT) Markers, containing normal hydrogen and deuterium atoms, used to label individual proteomes.

Kinship analysis An examination of DNA profiles or other information to determine if two individuals are related.

Klenow fragment (of DNA polymerase I) A DNA polymerase enzyme, obtained by chemical modification of *E. coli* DNA polymerase I, used primarily in chain termination DNA sequencing.

Knockout mouse A mouse that has been engineered so that it carries an inactivated gene.

Labeling The incorporation of a marker nucleotide into a nucleic acid molecule. The marker is often, but not always, a radioactive or fluorescent label.

Lac selection A means of identifying recombinant bacteria containing vectors that carry the *lacZ'* gene. The bacteria are plated on a medium that contains an analog of lactose that gives a blue color in the presence of β-galactosidase activity.

Lambda (λ) A bacteriophage that infects *E. coli*, derivatives of which are used as cloning vectors.

Ligase (DNA ligase) An enzyme that, in the cell, repairs single-stranded discontinuities in double-stranded DNA molecules. Purified DNA ligase is used in gene cloning to join DNA molecules together.

Linkage analysis A technique for mapping the chromosomal position of a gene by comparing its inheritance pattern with that of genes and other loci whose map positions are already known.

Linker A synthetic, double-stranded oligonucleotide used to attach sticky ends to a blunt-ended molecule.

Liposome A lipid vesicle sometimes used to introduce DNA into an animal or plant cell.

Lysogen A bacterium that harbors a prophage.

Lysogenic infection cycle The pattern of phage infection that involves integration of the phage DNA into the host chromosome.

Lysozyme An enzyme that weakens the cell walls of certain types of bacteria.

Lytic infection cycle The pattern of infection displayed by a phage that replicates and lyses the host cell immediately after the initial infection. Integration of the phage DNA molecule into the bacterial chromosome does not occur.

M13 A bacteriophage that infects *E. coli*, derivatives of which are used as cloning vectors.

Mapping reagent A collection of DNA fragments spanning a chromosome or the entire genome and used in STS mapping.

Massively parallel A high throughput sequencing strategy in which many individual sequences are generated in parallel.

Mass spectrometry An analytical technique in which ions are separated according to their mass-to-charge ratios.

Matrix-assisted laser desorption ionization time-of-flight (MALDI-TOF) A type of mass spectrometry used in proteomics.

Melting temperature (T_m) The temperature at which a double-stranded DNA or DNA–RNA molecule denatures.

Messenger RNA (mRNA) The transcript of a protein-coding gene.

Microarray A set of genes or cDNAs immobilized on a glass slide and used in transcriptome studies.

Microinjection A method of introducing new DNA into a cell by injecting it directly into the nucleus.

Microsatellite A polymorphism comprising tandem copies of, usually, two-, three-, four- or five-nucleotide repeat units. Also called a short tandem repeat (STR).

Minimal medium A defined medium that provides only the minimum number of different nutrients needed for growth of a particular bacterium.

Mitochondrial DNA The DNA molecules present in the mitochondria of eukaryotes.

Mitochondrial Eve The woman who lived in Africa between 140,000 and 290,000 years ago and who carried the ancestral mitochondrial DNA that gave rise to all the mitochondrial DNAs in existence today.

Modification interference assay A technique that uses chemical modification to identify nucleotides involved in interactions with a DNA binding protein.

Molecular clock An analysis based on the inferred mutation rate that enables times to be assigned to the branch points in a gene tree.

Multicopy plasmid A plasmid with a high copy number.

Multigene family A number of identical or related genes present in the same organism, usually coding for a family of related polypeptides.

Multigene shuffling A directed evolution strategy that involves taking parts of each member of a multigene family and reassembling these parts to create new gene variants.

Multiplex PCR A PCR carried out with more than one pair of primers and hence targeting two or more sites in the DNA being studied.

Multiregional evolution A hypothesis that holds that modern humans in the Old World are descended from *Homo erectus* populations that left Africa over one million years ago.

Nick A single-strand break, involving the absence of one or more nucleotides, in a double-stranded DNA molecule.

Nick translation The repair of a nick with DNA polymerase I, usually to introduce labeled nucleotides into a DNA molecule.

Northern transfer A technique for transferring bands of RNA from an agarose gel to a nitrocellulose or nylon membrane.

Nuclear transfer A technique, used in the production of transgenic animals, that involves transfer of the nucleus of a somatic cell into an oocyte whose own nucleus has been removed.

Nucleic acid hybridisation Formation of a double-stranded molecule by base pairing between complementary or homologous polynucleotides.

Oligonucleotide A short, synthetic, single-stranded DNA molecule, such as one used as a primer in DNA sequencing or PCR.

Oligonucleotide-directed mutagenesis An *in vitro* mutagenesis technique that involves the use of a synthetic oligonucleotide to introduce the predetermined nucleotide alteration into the gene to be mutated.

Open-circular DNA (ocDNA) The non-supercoiled conformation taken up by a circular double-stranded DNA molecule when one or both polynucleotides carry nicks.

Open reading frame (ORF) A series of codons that is or could be a gene.

ORF scanning Examination of a DNA sequence for open reading frames in order to locate the genes.

Origin of replication The specific position on a DNA molecule where DNA replication begins.

Orphan An open reading frame thought to be a functional gene but to which no function has yet been assigned.

Orthogonal field alternation gel electrophoresis (OFAGE) A gel electrophoresis technique that employs a pulsed electric field to achieve separation of very large molecules of DNA.

Out of Africa hypothesis A hypothesis that holds that modern humans evolved in Africa, moving to the rest of the Old World between 100 000 and 50 000 years ago, displacing the descendants of *Homo erectus* that they encountered.

P element A transposon from *Drosophila melanogaster* used as the basis of a cloning vector for that organism.

P1 A bacteriophage that infects *E. coli*, derivatives of which are used as cloning vectors.

P1-derived artificial chromosome (PAC) A cloning vector based on the P1 bacteriophage, used for cloning relatively large fragments of DNA in *E. coli*.

Papillomaviruses A group of mammalian viruses, derivatives of which have been used as cloning vectors.

Partial digestion Treatment of a DNA molecule with a restriction endonuclease under such conditions that only a fraction of all the recognition sites are cleaved.

Pedigree analysis The use of a human family tree to analyze the inheritance of a genetic or DNA marker.

Peptide mass fingerprinting Identification of a protein by examination of the mass spectrometric properties of peptides generated by treatment with a sequence-specific protease.

Phage display A technique involving cloning in M13 that is used to identify proteins that interact with one another.

Phage display library A collection of M13 clones carrying different DNA fragments, used in phage display.

Phagemid A double-stranded plasmid vector that possesses an origin of replication from a filamentous phage and hence can be used to synthesize a single-stranded version of a cloned gene.

Pharming Genetic modification of a farm animal so that the animal synthesizes a recombinant pharmaceutical protein, often in its milk.

Physical map A genome map that has been obtained by direct examination of DNA molecules.

Pilus One of the structures present on the surface of a bacterium containing a conjugative plasmid, through which DNA is assumed to pass during conjugation.

Plaque A zone of clearing on a lawn of bacteria caused by lysis of the cells by infecting phage particles.

Plasmid A usually circular piece of DNA, primarily independent of the host chromosome, often found in bacteria and some other types of cells.

Plasmid amplification A method involving incubation with an inhibitor of protein synthesis aimed at increasing the copy number of certain types of plasmid in a bacterial culture.

Polyethylene glycol A polymeric compound used to precipitate macromolecules and molecular aggregates.

Polylinker A synthetic double-stranded oligonucleotide carrying a number of restriction sites.

Polymerase chain reaction (PCR) A technique that enables multiple copies of a DNA molecule to be generated by enzymatic amplification of a target DNA sequence.

Polymorphism Refers to a locus that is present as a number of different alleles or other variations in the population as a whole.

Positional cloning A procedure that uses information on the map position of a gene to obtain a clone of that gene.

Positional effect Refers to the variations in expression levels observed for genes inserted at different positions in a genome.

Post-genomics Studies aimed at identifying all the genes in a genome and determining their expression patterns and functions.

Primer A short single-stranded oligonucleotide which, when attached by base pairing to a single-stranded template molecule, acts as the start point for complementary strand synthesis directed by a DNA polymerase enzyme.

Primer extension A method of transcript analysis in which the 5′ end of an RNA is mapped by annealing and extending an oligonucleotide primer.

Principal component analysis A procedure that attempts to identify patterns in a large dataset of variable character states.

Processivity Refers to the amount of DNA synthesis that is carried out by a DNA polymerase before dissociation from the template.

Productive A virus infection cycle that is able to proceed to completion and result in synthesis and release of new virus particles.

Promoter The nucleotide sequence, upstream of a gene, that acts as a signal for RNA polymerase binding.

Prophage The integrated form of the DNA molecule of a lysogenic phage.

Protease An enzyme that degrades protein.

Protein A A protein from the bacterium *Staphylococcus aureus* that binds specifically to immunoglobulin G (i.e., antibody) molecules.

Protein electrophoresis Separation of proteins in an electrophoresis gel.

Protein engineering A collection of techniques, including but not exclusively gene mutagenesis, that result in directed alterations being made to protein molecules, often to improve the properties of enzymes used in industrial processes.

Protein profiling The methodology used to identify the proteins in a proteome.

Proteome The entire protein content of a cell or tissue.

Proteomics The collection of techniques used to study the proteome.

Protoplast A cell from which the cell wall has been completely removed.

Pyrosequencing A DNA sequencing method in which addition of a nucleotide to the end of a growing polynucleotide is detected directly by conversion of the released pyrophosphate into a flash of chemiluminescence.

Quantitative PCR A method for quantifying the amount of product synthesized during a test PCR by comparison with the amounts synthesized during PCRs with known amounts of starting DNA.

RACE (rapid amplification of cDNA ends) A PCR-based technique for mapping the end of an RNA molecule.

Radiation hybrid A collection of rodent cell lines that contain different fragments of the human genome, constructed by a technique involving irradiation and used as a mapping reagent in studies of the human genome.

Radioactive marker A radioactive atom used in the detection of a larger molecule into which it has been incorporated.

Random priming A method for DNA labeling that utilizes random DNA hexamers, which anneal to single-stranded DNA and act as primers for complementary strand synthesis by a suitable enzyme.

Reading frame One of the six overlapping sequences of triplet codons, three on each polynucleotide, contained in a segment of a DNA double helix.

Real-time PCR A modification of the standard PCR technique in which synthesis of the product is measured as the PCR proceeds through its series of cycles.

Recombinant A transformed cell that contains a recombinant DNA molecule.

Recombinant DNA molecule A DNA molecule created in the test tube by ligating together pieces of DNA that are not normally contiguous.

Recombinant DNA technology All of the techniques involved in the construction, study and use of recombinant DNA molecules.

Recombinant protein A polypeptide that is synthesized in a recombinant cell as the result of expression of a cloned gene.

Recombination The exchange of DNA sequences between different molecules, occurring either naturally or as a result of DNA manipulation.

Relaxed (1) Refers to a plasmid with a high copy number of perhaps 50 or more per cell. (2) The non-supercoiled conformation of open-circular DNA.

Repetitive DNA PCR A clone fingerprinting technique that uses PCR to detect the relative positions of repeated sequences in cloned DNA fragments.

Replacement vector A λ vector designed so that insertion of new DNA is by replacement of part of the non-essential region of the λ DNA molecule.

Replica plating A technique whereby the colonies on an agar plate are transferred *en masse* to a new plate, on which the colonies grow in the same relative positions as before.

Replicative form (RF) of M13 The double-stranded form of the M13 DNA molecule found within infected *E. coli* cells.

Reporter gene A gene whose phenotype can be assayed in a transformed organism, and which is used in, for example, deletion analyzes of regulatory regions.

Reporter probe A short oligonucleotide that gives a fluorescent signal when it hybridizes with a target DNA.

Repression The switching off of expression of a gene or a group of genes in response to a chemical or other stimulus.

Resin A chromatography matrix.

Restriction analysis Determination of the number and sizes of the DNA fragments produced when a particular DNA molecule is cut with a particular restriction endonuclease.

Restriction endonuclease An endonuclease that cuts DNA molecules only at a limited number of specific nucleotide sequences.

Restriction fragment length polymorphism (RFLP) A mutation that results in alteration of a restriction site and hence a change in the pattern of fragments obtained when a DNA molecule is cut with a restriction endonuclease.

Restriction map A map showing the positions of different restriction sites in a DNA molecule.

Retrovirus A virus with an RNA genome, able to insert into a host chromosome, derivatives of which have been used to clone genes in mammalian cells.

Reverse genetics The strategy by which the function of a gene is identified by mutating that gene and identifying the phenotypic change that results.

Reverse transcriptase An RNA-dependent DNA polymerase, able to synthesize a complementary DNA molecule on a template of single-stranded RNA.

Reverse transcription–PCR A PCR technique in which the starting material is RNA. The first step in the procedure is conversion of the RNA to cDNA with reverse transcriptase.

RFLP linkage analysis A technique that uses a closely linked RFLP as a marker for the presence of a particular allele in a DNA sample, often as a means of screening individuals for a defective gene responsible for a genetic disease.

Ribonuclease An enzyme that degrades RNA.

Ribosome binding site The short nucleotide sequence upstream of a gene, which after transcription forms the site on the mRNA molecule to which the ribosome binds.

Ri plasmid An *Agrobacterium rhizogenes* plasmid, similar to the Ti plasmid, used to clone genes in higher plants.

S1 nuclease mapping A method for RNA transcript mapping.

Selectable marker A gene carried by a vector and conferring a recognizable characteristic on a cell containing the vector or a recombinant DNA molecule derived from the vector.

Selection A means of obtaining a clone containing a desired recombinant DNA molecule.

Sequenase An enzyme used in chain termination DNA sequencing.

Sequence tagged site (STS) A DNA sequence whose position has been mapped in a genome.

Serial analysis of gene expression (SAGE) A method for studying the composition of a transcriptome.

Short tandem repeat (STR) A polymorphism comprising tandem copies of, usually, two-, three-, four- or five-nucleotide repeat units. Also called a microsatellite.

Shotgun approach A genome sequencing strategy in which the molecules to be sequenced are randomly broken into fragments which are then individually sequenced.

Shotgun cloning A cloning strategy that involves the insertion of random fragments of a large DNA molecule into a vector, resulting in a large number of different recombinant DNA molecules.

Shuttle vector A vector that can replicate in the cells of more than one organism (e.g., in *E. coli* and in yeast).

Simian virus 40 (SV40) A mammalian virus that has been used as the basis for a cloning vector.

Single nucleotide polymorphism (SNP) A point mutation that is carried by some individuals of a population.

Sonication A procedure that uses ultrasound to cause random breaks in DNA molecules.

Southern transfer A technique for transferring bands of DNA from an agarose gel to a nitrocellulose or nylon membrane.

Sphaeroplast A cell with a partially degraded cell wall.

Spin column A method for accelerating ion-exchange chromatography by centrifuging the chromatography column.

Stem–loop A hairpin structure, consisting of a base-paired stem and a non-base-paired loop, that may form in a polynucleotide.

Sticky end An end of a double-stranded DNA molecule where there is a single-stranded extension.

Stringent Refers to a plasmid with a low copy number of perhaps just one or two per cell.

Strong promoter An efficient promoter that can direct synthesis of RNA transcripts at a relatively fast rate.

Stuffer fragment The part of a λ replacement vector that is removed during insertion of new DNA.

Supercoiled The conformation of a covalently closed-circular DNA molecule, which is coiled by torsional strain into the shape taken by a wound-up elastic band.

Synteny Refers to a pair of genomes in which at least some of the genes are located at similar map positions.

Taq **DNA polymerase** The thermostable DNA polymerase that is used in PCR.

T-DNA The portion of the Ti plasmid transferred to the plant DNA.

Temperature-sensitive mutation A mutation that results in a gene product that is functional within a certain temperature range (e.g., at less than 30°C), but non-functional at different temperatures (e.g., above 30°C).

Template A single-stranded polynucleotide (or region of a polynucleotide) that directs synthesis of a complementary polynucleotide.

Terminator The short nucleotide sequence, downstream of a gene, that acts as a signal for termination of transcription.

Terminator technology A recombinant DNA process which results in synthesis of the ribosome inactivating protein in plant embryos, used to prevent GM crops from producing viable seeds.

Thermal cycle sequencing A DNA sequencing method that uses PCR to generate chain-terminated polynucleotides.

Ti plasmid The large plasmid found in those *Agrobacterium tumefaciens* cells able to direct crown gall formation on certain species of plants.

Topoisomerase An enzyme that introduces or removes turns from the double helix by breakage and reunion of one or both polynucleotides.

Total cell DNA Consists of all the DNA present in a single cell or group of cells.

Totipotent Refers to a cell that is not committed to a single developmental pathway and can hence give rise to all types of differentiated cell.

Transcript analysis A type of experiment aimed at determining which portions of a DNA molecule are transcribed into RNA.

Transcriptome The entire mRNA content of a cell or tissue.

Transfection The introduction of purified virus DNA molecules into any living cell.

Transformation The introduction of any DNA molecule into any living cell.

Transformation frequency A measure of the proportion of cells in a population that are transformed in a single experiment.

Transgenic Referring to an animal or plant, containing a cloned gene in all of its cells.

Transgenic animal An animal that possesses a cloned gene in all of its cells.

Transposon A DNA sequence that is able to move from place to place within a genome.

Two-dimensional gel electrophoresis A method for separation of proteins used especially in studies of the proteome.

Undefined medium A growth medium in which not all the components have been identified.

Universal primer A sequencing primer that is complementary to the part of the vector DNA immediately adjacent to the point into which new DNA is ligated.

UV absorbance spectrophotometry A method for measuring the concentration of a compound by determining the amount of ultraviolet radiation absorbed by a sample.

Vector A DNA molecule, capable of replication in a host organism, into which a gene is inserted to construct a recombinant DNA molecule.

Vehicle Sometimes used as a substitute for the word "vector", emphasizing that the vector transports the inserted gene through the cloning experiment.

Virus chromosome The DNA or RNA molecule(s) contained within a virus capsid and carrying the viral genes.

Virus-induced gene silencing (VIGS) A technique involving a geminivirus vector used to study the function of a plant gene.

Watson–Crick rules The base pairing rules that underlie gene structure and expression. A pairs with T, and G pairs with C.

Wave of advance model A hypothesis that holds that the spread of agriculture into Europe was accompanied by a large scale movement of human populations.

Weak promoter An inefficient promoter that directs synthesis of RNA transcripts at a relatively low rate.

Western transfer A technique for transferring bands of protein from an electrophoresis gel to a membrane support.

Yeast artificial chromosome (YAC) A cloning vector comprising the structural components of a yeast chromosome and able to clone very large pieces of DNA.

Yeast episomal plasmid (YEp) A yeast vector carrying the 2 μm circle origin of replication.

Yeast integrative plasmid (YIp) A yeast vector that relies on integration into the host chromosome for replication.

Yeast replicative plasmid (YRp) A yeast vector that carries a chromosomal origin of replication.

Yeast two hybrid system A technique involving cloning in *S. cerevisiae* that is used to identify proteins that interact with one another.

Zoo blot A nitrocellulose or nylon membrane carrying immobilized DNAs of several species, used to determine if a gene from one species has homologs in the other species.

Index